Birkhäuser

Applied and Numerical Harmonic Analysis

Series Editors

John Benedetto, University of Maryland, College Park, MD, USA
Wojciech Czaja, University of Maryland, College Park, MD, USA
Kasso Okoudjou, Tufts University, Medford, USA

Editorial Board Members

Akram Aldroubi, Vanderbilt University, Nashville, USA
Peter Casazza, University of Missouri, Columbia, USA
Douglas Cochran, Arizona State University, Phoenix, USA
Hans G. Feichtinger, University of Vienna, Vienna, Austria
Anna C. Gilbert, Yale University, New Haven, USA
Christopher Heil, Georgia Institute of Technology, Atlanta, USA
Stéphane Jaffard, Paris-East Créteil University, Paris, France
Gitta Kutyniok, Ludwig Maximilian University of Munich, München, Germany
Mauro Maggioni, Johns Hopkins University, Baltimore, USA
Ursula Molter, University of Buenos Aires, Buenos Aires, Argentina
Zuowei Shen, National University of Singapore, Singapore
Thomas Strohmer, University of California, Davis, USA
Michael Unser, École Polytechnique Fédérale de Lausanne, Lausanne, Switzerland
Yang Wang, Hong Kong University of Science & Technology, Kowloon, Hong Kong

Simon Foucart • Stephan Wojtowytsch
Editors

Explorations in the Mathematics of Data Science

The Inaugural Volume of the Center for Approximation and Mathematical Data Analytics

TEXAS A&M UNIVERSITY
Center for Approximation
& Mathematical Data Analytics

Birkhäuser

Editors
Simon Foucart
Department of Mathematics
Texas A&M University
College Station, TX, USA

Stephan Wojtowytsch
Department of Mathematics
University of Pittsburgh
Pittsburgh, PA, USA

ISSN 2296-5009 ISSN 2296-5017 (electronic)
Applied and Numerical Harmonic Analysis
ISBN 978-3-031-66496-0 ISBN 978-3-031-66497-7 (eBook)
https://doi.org/10.1007/978-3-031-66497-7

This book is published under the imprint Birkhäuser, www.birkhauser-science.com by the registered company Springer Nature Switzerland AG
The registered company address is: Gewerbestrasse 11, 6330 Cham, Switzerland

ANHA Series Preface

The *Applied and Numerical Harmonic Analysis* (ANHA) book series aims to provide the engineering, mathematical, and scientific communities with significant developments in harmonic analysis, ranging from abstract harmonic analysis to basic applications. The title of the series reflects the importance of applications and numerical implementation, but richness and relevance of applications and implementation depend fundamentally on the structure and depth of theoretical underpinnings. Thus, from our point of view, the interleaving of theory and applications and their creative symbiotic evolution is axiomatic.

Harmonic analysis is a wellspring of ideas and applicability that has flourished, developed, and deepened over time within many disciplines and by means of creative cross-fertilization with diverse areas. The intricate and fundamental relationship between harmonic analysis and fields such as signal processing, partial differential equations (PDEs), and image processing is reflected in our state-of-the-art ANHA series.

Our vision of modern harmonic analysis includes a broad array of mathematical areas, e.g., wavelet theory, Banach algebras, classical Fourier analysis, time-frequency analysis, deep learning, and fractal geometry, as well as the diverse topics that impinge on them.

For example, wavelet theory can be considered an appropriate tool to deal with some basic problems in digital signal processing, speech and image processing, geophysics, pattern recognition, biomedical engineering, and turbulence. These areas implement the latest technology from sampling methods on surfaces to fast algorithms and computer vision methods. The underlying mathematics of wavelet theory depends not only on classical Fourier analysis but also on ideas from abstract harmonic analysis, including von Neumann algebras and the affine group. This leads to a study of the Heisenberg group and its relationship to Gabor systems, and of the metaplectic group for a meaningful interaction of signal decomposition methods.

The unifying influence of wavelet theory in the aforementioned topics illustrates the justification for providing a means for centralizing and disseminating information from the broader, but still focused, area of harmonic analysis. This will be a key role of ANHA. We intend to publish with the scope and interaction that such a host of issues demands.

Along with our commitment to publish mathematically significant works at the frontiers of harmonic analysis, we have a comparably strong commitment to publish major advances in the following applicable topics in which harmonic analysis plays a substantial role:

*Analytic Number theory * Antenna Theory * Artificial Intelligence * Biomedical Signal Processing * Classical Fourier Analysis * Coding Theory * Communications Theory * Compressed Sensing * Crystallography and Quasi-Crystals * Data Mining * Data Science * Deep Learning * Digital Signal Processing * Dimension Reduction and Classification * Fast Algorithms * Frame Theory and Applications * Gabor Theory and Applications * Geophysics * Image Processing * Machine Learning * Manifold Learning * Numerical Partial Differential Equations * Neural Networks * Phaseless Reconstruction * Prediction Theory * Quantum Information Theory * Radar Applications * Sampling Theory (Uniform and Non-uniform) and Applications * Spectral Estimation * Speech Processing * Statistical Signal Processing * Super-resolution * Time Series * Time-Frequency and Time-Scale Analysis * Tomography * Turbulence * Uncertainty Principles *Waveform Design * Wavelet Theory and Applications

The above point of view for the ANHA book series is inspired by the history of Fourier analysis itself, whose tentacles reach into so many fields.

In the last two centuries Fourier analysis has had a major impact on the development of mathematics, on the understanding of many engineering and scientific phenomena, and on the solution of some of the most important problems in mathematics and the sciences. Historically, Fourier series were developed in the analysis of some of the classical PDEs of mathematical physics; these series were used to solve such equations. In order to understand Fourier series and the kinds of solutions they could represent, some of the most basic notions of analysis were defined, e.g., the concept of "function." Since the coefficients of Fourier series are integrals, it is no surprise that Riemann integrals were conceived to deal with uniqueness properties of trigonometric series. Cantor's set theory was also developed because of such uniqueness questions.

A basic problem in Fourier analysis is to show how complicated phenomena, such as sound waves, can be described in terms of elementary harmonics. There are two aspects of this problem: first, to find, or even define properly, the harmonics or spectrum of a given phenomenon, for example, the spectroscopy problem in optics; second, to determine which phenomena can be constructed from given classes of harmonics, as done, for example, by the mechanical synthesizers in tidal analysis.

Fourier analysis is also the natural setting for many other problems in engineering, mathematics, and the sciences. For example, Wiener's Tauberian theorem in Fourier analysis not only characterizes the behavior of the prime numbers but is a fundamental tool for analyzing the ideal structures of Banach algebras. It also provides the proper notion of spectrum for phenomena such as white light. This latter process leads to the Fourier analysis associated with correlation functions in filtering and prediction problems. These problems, in turn, deal naturally with Hardy spaces in complex analysis, as well as inspiring Wiener to consider communications engineering in terms of feedback and stability, his cybernetics. This latter theory develops concepts to understand complex systems such as learning and cognition and neural networks; and it is arguably a precursor of deep learning and its spectacular interactions with data science and AI.

Nowadays, some of the theory of PDEs has given way to the study of Fourier integral operators. Problems in antenna theory are studied in terms of unimodular trigonometric polynomials. Applications of Fourier analysis abound in signal processing, whether with the fast Fourier transform (FFT), or filter design, or the adaptive modeling inherent in time-frequency-scale methods such as wavelet theory.

The coherent states of mathematical physics are translated and modulated Fourier transforms, and these are used, in conjunction with the uncertainty principle, for dealing with signal reconstruction in communications theory. We are back to the raison d'etre of the ANHA series!

College Park, MD, USA — John Benedetto
College Park, MD, USA — Wojciech Czaja
Boston, MA, USA — Kasso Okoudjou

Preface

Thank you, dear reader, for looking into the inaugural volume of the Center for Approximation and Mathematical Data Analytics (CAMDA). Although you can certainly jump right away to the interesting part, i.e., the mathematical writings found after this preface, we feel compelled to start by sharing a few particulars putting into context this book entitled "Explorations in the Mathematics of Data Science."

Here is the historical background first: CAMDA is a continuation of the Center for Approximation Theory (CAT), which was founded at Texas A&M University in 1981 by Charles K. Chui, Larry L. Schumaker, and Joseph D. Ward. The center quickly became a preeminent research hub, hosting hundreds of visitors over the years. Its domain of expertise included linear and nonlinear approximations by polynomials, rational functions, splines, wavelets, radial and spherical basis functions. A proud milestone was the election of Ronald A. DeVore to the National Academy of Sciences in 2017.

In 2021, the center changed its name and Simon Foucart became the first director of CAMDA, succeeding Francis J. Narcowich, who was the last director of CAT. The name change was dictated in response to a growing interest in the theoretical aspects of Data Science, where many concepts—e.g., neural networks—are rooted in Approximation Theory. CAMDA's scientific aspiration is to promote a sound comprehension of the success and limitations of methods substantiated empirically in other parts of Data Science. This is captured in its mission statement, which reads:

- Enrich the mathematical foundations of Data Science via precise formulation of data-inspired problems and rigorous evaluation of algorithmic solutions;
- Foster cooperation between Mathematics and data-rich disciplines at Texas A&M;
- Disseminate emerging techniques from Mathematical Data Science through courses, seminars, and collaborations.

In 2023, CAMDA marked its inception with an inaugural conference held from May 22 to 25 on Texas A&M University's main campus. It took place fully in-person—remember that the COVID pandemic constrained many gatherings to occur virtually until then, so a sense of relief accompanied this return to normal.[1] The event was quite successful, with about one hundred participants making the trip to College Station. The line-up of plenary speakers showcased the interdisciplinary nature of Data Science. It consisted of

- Yuejie Chi (Electrical & Computer Engineering, Carnegie Mellon University)
- Roman Vershynin (Mathematics, University of California, Irvine)
- Martin Wainwright (Electrical Engineering & Computer Science, Mathematics, Massachusetts Institute of Technology)
- Stephen Wright (Computer Science, University of Wisconsin-Madison)

Complementing the plenary talks, five special sessions were organized to align with the five pilot projects selected to set CAMDA's research activities in motion. These projects were:

1. Contemporary Optimal Recovery
2. Neural Network Approximation
3. Learning in Data-Poor Conditions
4. Deep Learning Methods for Parametric PDEs
5. Approximation Theory for Gaussian Process Models

The first three of these projects are represented in this volume (article 1 for project 1, articles 2 and 3 for project 2, and article 4 for project 3). The conference also comprised 18 contributed talks, two of which made their ways to this volume (articles 5 and 6). Another contribution directly tied to the conference is the survey by plenary speaker Yuejie Chi and coauthors (article 7).

This volume, however, is not only a conference proceedings, as it does include one work (article 8) presented at the CAMDA seminar. The latter took place within Texas A&M's Department of Mathematics roughly every other week since Fall 2021. Moreover, the volume also features satellite works (articles 9 and 10) performed by CAMDA members. In fact, several authors are or have been associated with CAMDA: Chunyang Liao and Srinivas Subramanian as PhD students; Josiah Park and Ming Zhong as postdocs; Simon Foucart, Guergana Petrova, Jonathan Siegel, and Stephan Wojtowytsch as faculty members. The PhD students and postdocs have now gone to greener pastures, as well as Stephan Wojtowytsch, who co-organized the inaugural CAMDA conference.

We now hope, dear reader, that you will appreciate the articles you are about to peruse. In anticipation, we wish to show our appreciation to all the contributors for turning this inaugural CAMDA volume into a reality. We are indebted to Texas A&M's Department of Mathematics, (late) College of Science, and Institute of Data Science for financially supporting CAMDA's beginnings. We also thank Cara Barton and Reagan Scott for their tireless efforts to ensure a smooth inaugural conference, as well as all the speakers and participants whose enthusiasm made

[1] Actually, virtual meetings may be the normal to our readers from the future "thanks" to alternative realities developed alongside advances in . . . Data Science.

it a success. Finally, we express our gratitude to the National Science Foundation for the generous support (grant DMS-2329268), which enabled the engagement of numerous young and underrepresented researchers.

College Station, TX, USA
Pittsburgh, PA, USA
February 2024

Simon Foucart
Stephan Wojtowytsch

Contents

S-procedure Relaxation: A Case of Exactness Involving Chebyshev Centers

Simon Foucart and Chunyang Liao

Abstract Optimal recovery is a mathematical framework for learning functions from observational data by adopting a worst-case perspective tied to model assumptions on the functions to be learned. Working in a finite-dimensional Hilbert space, we consider model assumptions based on approximability and observation inaccuracies modeled as additive errors bounded in ℓ_2. We focus on the local recovery problem, which amounts to the determination of Chebyshev centers. Earlier work by Beck and Eldar presented a semidefinite recipe for the determination of Chebyshev centers. The result was valid in the complex setting only, but not necessarily in the real setting, since it relied on the S-procedure with two quadratic constraints, which offers a tight relaxation only in the complex setting. Our contribution consists in proving that this semidefinite recipe is exact in the real setting, too, at least in the particular instance where the quadratic constraints involve orthogonal projectors. Our argument exploits a previous work of ours, where exact Chebyshev centers were obtained in a different way. We conclude by stating some open questions and by commenting on other recent results in optimal recovery.

1 Rundown on Optimal Recovery

The field of optimal recovery, arguably shaped by the influence of Kolmogorov [1] and deeply rooted in approximation theory [2], is experiencing a revival, thanks to newly available optimization tools. Chosen as one of the core topics first investigated at CAMDA—the Center for Approximation and Mathematical Data Analytics—optimal recovery can be understood as a nonstatistical learning theory. Indeed,

S. Foucart (✉)
Department of Mathematics, Texas A&M University, College Station, TX, USA
e-mail: foucart@tamu.edu

C. Liao
University of California, Los Angeles, CA, USA
e-mail: liaochunyang@math.ucla.edu

S. Foucart, S. Wojtowytsch (eds.), *Explorations in the Mathematics of Data Science*, Applied and Numerical Harmonic Analysis,
https://doi.org/10.1007/978-3-031-66497-7_1

a function f acquired through point evaluations $y_i = f(x^{(i)})$, $i = 1, \dots, m$, needs to be learned—or recovered, in the parlance preferred here. But one does not abide by the postulate behind statistical learning theory, stipulating that the $x^{(i)}$'s are independent realizations of a random variable. Thus, the performance of a learning/recovery procedure cannot be assessed in an average case, and one opts for an assessment focusing on the worst case, relative to a model expressing some prior scientific knowledge about f.

To be more precise, the task at hand consists in recovering an element f from a Banach space F. It is typically thought of as a space of functions, although it does not have to be. Rather than recovering f in full, it can be more relevant to recover a quantity of interest $Q(f)$, where $Q : F \to Z$ represents a linear map in this article. The element f is only available through partial information, specifically:

- some a priori information conveyed by a modeling assumption taking the form

$$f \in \mathcal{K},$$

  where the so-called model set $\mathcal{K} \subseteq F$ reflects an educated guess about realistic objects to be recovered;
- some a posteriori information obtained through observational data of the form

$$y_i = \lambda_i(f), \qquad i = 1, \dots, m,$$

  for some linear functionals $\lambda_1, \dots, \lambda_m : F \to \mathbb{R}$. More concisely, one writes $y = \Lambda f \in \mathbb{R}^m$, where the so-called observation map $\Lambda : F \to \mathbb{R}^m$ is a linear map.

A recovery procedure is a process, perhaps partially cognizant of the model set $\mathcal{K}$, which takes in the observational data in $\mathbb{R}^m$ and returns an estimation to $Q(f)$ in Z. In other words, it is nothing but a map $\Delta : \mathbb{R}^m \to Z$. Its recovery performance can evidently be quantified by $\|Q(f) - \Delta(y)\|_Z = \|Q(f) - \Delta(\Lambda f)\|_Z$ for a fixed $f \in \mathcal{K}$ satisfying $\Lambda f = y$. However, f being unknown, one takes a worst-case perspective over all consistent f's, leading to a recovery performance quantified by

- the local worst-case error, at a fixed $y \in \mathbb{R}^m$, defined as

$$\mathrm{lwce}(\Delta, y) = \sup_{\substack{f \in \mathcal{K} \\ \Lambda f = y}} \|Q(f) - \Delta(y)\|_Z; \tag{1}$$

- the global worst-case error defined as

$$\mathrm{gwce}(\Delta) = \sup_{f \in \mathcal{K}} \|Q(f) - \Delta(\Lambda f)\|_Z. \tag{2}$$

A locally, respectively globally, optimal recovery map $\Delta^{\rm opt} : \mathbb{R}^m \to Z$ is a map $\Delta : \mathbb{R}^m \to Z$ that minimizes $\mathrm{lwce}(\Delta, y)$ at every $y \in \mathbb{R}^m$, respectively $\mathrm{gwce}(\Delta)$. Noticing that $\mathrm{gwce}(\Delta) = \sup\{\mathrm{lwce}(\Delta, y), y \in \Lambda(\mathcal{K})\}$, one realizes that a locally optimal recovery map is automatically globally optimal. This somewhat makes the global setting "easier" than the local setting, conceivably explaining the prevalence of the latter in the standard theory of optimal recovery. There, a coveted result often consists of the assertion that there exists an optimal recovery map which is linear—of course, one strives to construct it! In the global setting, this existence result classically holds when the model set $\mathcal{K}$ is symmetric and convex and Q is a linear functional (see [3, Theorem 4.7] or [4, Theorem 9.3]) and when the model set is a centered hyperellipsoid in a Hilbert space (see [3, Theorem 4.11] or [4, Theorem 9.4]). In the latter situation, the optimal recovery map, dubbed spline algorithm, is also locally optimal. There are other situations where global optimality via linear maps stands, for instance, in the space $C(\mathcal{X})$ of continuous functions on a compact space $\mathcal{X}$ relative to the model set:

$$\mathcal{K}_{\mathcal{V}} = \{f \in C(\mathcal{X}) : \mathrm{dist}_{C(\mathcal{X})}(f, \mathcal{V}) \leq \varepsilon\} \tag{3}$$

subordinate to a linear subspace $\mathcal{V}$ of $C(\mathcal{X})$ and an approximability parameter $\varepsilon \geq 0$. More details will be given in Sect. 6.

2 Our Contribution: Local Optimality from Inaccurate Data

From now on, we leave the global setting behind and tackle the harder local setting, starting by highlighting its geometric interpretation. Namely, considering a locally optimal recovery map $\Delta^{\rm opt} : \mathbb{R}^m \to Z$ and a fixed $y \in \mathbb{R}^m$, since $\Delta^{\rm opt}(y) \in Z$ minimizes $\sup\{\|Q(f) - z\|_Z : f \in \mathcal{K}, \Lambda f = y\}$, one can write, almost tautologically, that

$$\Delta^{\rm opt}(y) \in \operatorname*{argmin}_{z \in Z, r \geq 0} \; r \qquad \text{s.to} \quad \|Q(f) - z\|_Z \leq r \text{ whenever } f \in \mathcal{K} \text{ and } \Lambda f = y.$$

This shows that $\Delta^{\rm opt}(y)$ is a center of a smallest-radius ball containing $Q(\mathcal{K}_y)$, where $\mathcal{K}_y := \{f \in F : f \in \mathcal{K} \text{ and } \Lambda f = y\}$. It is said that $\Delta^{\rm opt}(y)$ is a Chebyshev center for the set $Q(\mathcal{K}_y)$, often eluding to mention the norm on Z.

But the above context is not quite where our current investigations take place. Indeed, the discussion so far assumed that the observational data were accurate. In realistic situations, they are contaminated by additive noise, so that $y_i = \lambda_i(f) + e_i$, $i = 1, \ldots, m$. In short, we write $y = \Lambda f + e$ for some error vector $e \in \mathbb{R}^m$. We shall model this vector deterministically through $e \in \mathcal{E}$ for a so-called uncertainty set $\mathcal{E} \subseteq \mathbb{R}^m$. Thus, the *a posteriori* information now takes the form:

$$y - \Lambda f \in \mathcal{E}.$$

This leads to an updated notion of local worst-case error, as defined by

$$\mathrm{lwce}(\Delta, y) = \sup_{\substack{f \in \mathcal{K} \\ y - \Lambda f \in \mathcal{E}}} \|Q(f) - \Delta(y)\|_Z.$$

Our objective of finding an optimal recovery map $\Delta^{\mathrm{opt}} : \mathbb{R}^m \to Z$ now becomes the determination, for each $y \in \mathbb{R}^m$, of a solution $\Delta^{\mathrm{opt}}(y)$ to the optimization program:

$$\underset{z \in Z}{\mathrm{minimize}} \quad \sup_{\substack{f \in \mathcal{K} \\ y - \Lambda f \in \mathcal{E}}} \|Q(f) - z\|_Z.$$

As before, one interprets geometrically $\Delta^{\mathrm{opt}}(y)$ as a Chebyshev center for the set $Q(\mathcal{K}_{y,\mathcal{E}})$, where $\mathcal{K}_{y,\mathcal{E}} = \{f \in F : f \in \mathcal{K} \text{ and } y - \Lambda f \in \mathcal{E}\}$, i.e., as a center for a smallest-radius ball containing $Q(\mathcal{K}_{y,\mathcal{E}})$.

To achieve our objective—at least partially—we place ourselves in a Hilbert framework, i.e., we assume from now on that F is a finite-dimensional Hilbert space; hence, it is denoted by H instead of a generic F. Let us state our first contribution, before placing it in the context of the current knowledge. Note that we can safely talk about *the* Chebyshev center in this statement, as it was known as early as [5] that a bounded set in a uniformly convex Banach space possesses a unique Chebyshev center.

Theorem 1 *In a finite-dimensional Hilbert space H, consider a model set $\mathcal{K}$ and an uncertainty set $\mathcal{E}$ given by*

$$\mathcal{K} = \{f \in H : \|Pf\|_H \le \varepsilon\} \qquad \text{and} \qquad \mathcal{E} = \{e \in \mathbb{R}^m : \|e\|_2 \le \eta\}.$$

If P is an orthogonal projection, if $\Lambda\Lambda^* = \mathrm{Id}_m$, and if $\ker(P) \cap \ker(\Lambda) = \{0\}$, then, for any $y \in \mathbb{R}^m$, the Chebyshev radius of $\mathcal{K}_{y,\mathcal{E}} = \{f \in H : f \in \mathcal{K} \text{ and } y - \Lambda f \in \mathcal{E}\}$ is equal to the optimal value of the semidefinite program:

$$\underset{\substack{c,d \ge 0 \\ t \in \mathbb{R}}}{\mathrm{minimize}} \; c\varepsilon^2 + d\eta^2 - d\|y\|_2^2 + t \qquad \text{s.to} \quad cP + d\Lambda^*\Lambda \succeq \mathrm{Id}_H, \tag{4}$$

$$\text{and} \quad \left[\begin{array}{c|c} cP + d\Lambda^*\Lambda & d\Lambda^* y \\ \hline dy^*\Lambda & t \end{array}\right] \succeq 0.$$

Moreover, the Chebyshev center is the solution to a regularization program with specified parameters, namely, it is given by

$$f_{\widetilde{c},\widetilde{d}} := \underset{f \in H}{\mathrm{argmin}} \; \widetilde{c}\,\|Pf\|_H^2 + \widetilde{d}\,\|y - \Lambda f\|_2^2, \tag{5}$$

where $\widetilde{c}, \widetilde{d} \ge 0$ solve the semidefinite program (4).

For the sake of the following discussion, we select the model and uncertainty sets as arbitrary hyperellipsoids. Specifically, with R, S representing linear maps from H into other Hilbert spaces (all norms now being written as $\|\cdot\|$ for ease of notation) and with ε, η representing positive parameters, we consider:

$$\mathcal{K} = \{f \in H : \|Rf\| \leq \varepsilon\} \subseteq H, \tag{6}$$

$$\mathcal{E} = \{e \in \mathbb{R}^m : \|Se\| \leq \eta\} \subseteq \mathbb{R}^m. \tag{7}$$

In this case, Micchelli and Melkman [6, 7] already observed that some regularization (with unspecified parameters) provides linear recovery maps that are optimal, albeit globally. Locally, a similar conclusion was derived by Beck and Eldar [8] for the full estimation problem, i.e., when $Q = \mathrm{Id}_H$. In other words, they—almost—established Theorem 1 in the more general situation of the model and uncertainty sets (6)–(7). There is a subtlety, and an important one: technically, the Chebyshev center was determined in the complex setting only, but not necessarily in the real setting. This incongruity occurs because the main tool used in the argument, i.e., the S-procedure with two constraints, is only exact in the complex setting; see next section for details. In the real setting, the S-procedure is only a relaxation, merely leading to an overestimation of the Chebyshev radius rather than to the genuine Chebyshev radius. Our main contribution in this article therefore consists in showing that the supposed overestimation actually agrees with the Chebyshev radius (i.e., the minimal local worst-case error), at least in a particular instance.

The basis of the argument is the fact that the exact Chebyshev radius and center have already been obtained, in a different form, in case P is an orthogonal projection and $\Lambda\Lambda^* = \mathrm{Id}_m$. Indeed, as an extension to the local optimal recovery problem in Hilbert spaces with approximability model (3) and accurate data (i.e., with $P = P_{\mathcal{V}^\perp}$ and $\eta = 0$), which was settled in [9], we established in [10, Theorem 8] that the Chebyshev center arises from a regularization program with explicitly described parameter $\tau_\sharp \in (0, 1)$. Precisely, for $\tau \in (0, 1)$, let us define:

$$f_\tau := \underset{f \in H}{\operatorname{argmin}}\ (1-\tau)\|Pf\|^2 + \tau\|y - \Lambda f\|^2 \tag{8}$$

as the solution to a regularization program akin to (5). The Chebyshev center is $f_{\tau_\sharp}$, where the desired parameter $\tau_\sharp$ is the unique τ between $1/2$ and $\varepsilon/(\varepsilon+\eta)$ satisfying the implicit equation:

$$\lambda_{\min}((1-\tau)P + \tau\Lambda^*\Lambda) = \frac{(1-\tau)^2\varepsilon^2 - \tau^2\eta^2}{(1-\tau)\varepsilon^2 - \tau\eta^2 + (1-\tau)\tau(1-2\tau)\delta^2}, \tag{9}$$

in which δ is precomputed as $\delta = \min\{\|Pf\| : \Lambda f = y\} = \min\{\|\Lambda f - y\| : Pf = 0\}$. As explained in [10, Appendix], the above equation can be solved efficiently via the Newton method. According to the yet-to-be-established Theorem 1, the Chebyshev center can alternatively be determined by solving the semidefinite program (4). We have not seriously compared these two options, but we would instinctively favor solving (9) to bypass the black-box nature of semidefinite solvers.

3 Overestimate of the Chebyshev Radius via the S-procedure

Our goal in this section is to extend a result of [8] to a quantity of interest Q which is an arbitrary linear map between two Hilbert spaces, instead of just $Q = \mathrm{Id}_H$. This extension is not really difficult, but our arguments differ slightly from the ones of [8]. The result itself, which provides an upper bound for the Chebyshev radius of $Q(\mathcal{K}_{y,\mathcal{E}})$, as well a candidate for its Chebyshev center, is based on the S-procedure relaxation. When this relaxation is exact, the upper bound agrees with the Chebyshev radius, and the candidate is the genuine Chebyshev center. It is in the next section that we establish the exactness of the S-procedure relaxation in our particular instance. Here, we simply explain where the upper bound is coming from.

To this end, we start by recalling the gist of the S-procedure and point to the survey [11] for more details. Given $K+1$ quadratic functions defined on H, say

$$q_k(h) = \langle A_k h, h\rangle + 2\langle a_k, h\rangle + \alpha_k, \qquad k = 0, 1, \dots, K,$$

where the A_k's are self-adjoint operators, the a_k's are vectors, and the α_k's are scalars, we consider the two assertions:

$$q_0(h) \le 0 \text{ whenever } q_1(h) \le 0, \dots, q_K(h) \le 0, \tag{10}$$

$$\text{there exist } c_1, \dots, c_K \ge 0 \colon q_0(h) \le c_1 q_1(h) + \cdots + c_K q_K(h) \text{ for all } h \in H. \tag{11}$$

Obviously, if assertion (11) holds, then assertion (10) holds as well. This, in essence, is the S-procedure. The question of its exactness is whether (10) and (11) are in reality equivalent. It is the case for $K = 1$, this is Yakubovich S-lemma [12]. We are actually interested in $K = 2$ here. In this situation, it was shown in [13] that the S-procedure is exact when the scalar field is $\mathbb{C}$, but not necessarily when the scalar field is $\mathbb{R}$, which is our primary concern. Still, under mild assumptions, exactness holds for $\mathbb{R}$ and $K = 2$ in the absence of linear terms, i.e., when $a_0 = a_1 = a_2 = 0$. The latter result, established by Polyak in [14], was the key for us to settle the global optimality problem in [10]. But in general, we make do with the mere relaxation: this leads to the overestimation derived below (which turns into an exact evaluation if the scalar field is $\mathbb{C}$).

Theorem 2 *Let $Q : H \to Z$ be a linear map between finite-dimensional Hilbert spaces. Given Hilbert-space-valued linear maps R and S defined on H and $\mathbb{R}^m$, respectively, and satisfying* $\ker(R) \cap \ker(S\Lambda) = \{0\}$, *consider the model set* (6) *and uncertainty set* (7), *i.e.,*

$$\mathcal{K} = \{f \in H : \|Rf\| \le \varepsilon\} \subseteq H \qquad \textit{and} \qquad \mathcal{E} = \{e \in \mathbb{R}^m : \|Se\| \le \eta\} \subseteq \mathbb{R}^m.$$

For $y \in \mathbb{R}^m$, consider $\widetilde{\gamma}$ and $\widetilde{c}, \widetilde{d}, \widetilde{t}$ to be the minimal value and minimizers of the semidefinite program:

$$\underset{\substack{c,d \geq 0 \\ t \in \mathbb{R}}}{\text{minimize}} \; c\varepsilon^2 + d\eta^2 - d\|Sy\|_2^2 + t \quad \textit{s.to} \;\; cR^*R + d\Lambda^*S^*S\Lambda \succeq Q^*Q, \tag{12}$$

$$\textit{and} \quad \left[\begin{array}{c|c} cR^*R + d\Lambda^*S^*S\Lambda & d\Lambda^*S^*Sy \\ \hline dy^*S^*S\Lambda & t \end{array}\right] \succeq 0.$$

Then, the Chebyshev radius of $Q(\mathcal{K}_{y,\mathcal{E}})$, $\mathcal{K}_{y,\mathcal{E}} := \{f \in H : f \in \mathcal{K}$ and $y - \Lambda f \in \mathcal{E}\}$, is upper bounded as

$$\inf_{z \in Z} \sup_{\substack{f \in \mathcal{K} \\ y - \Lambda f \in \mathcal{E}}} \|Q(f) - z\| \leq \sup_{\substack{f \in \mathcal{K} \\ y - \Lambda f \in \mathcal{E}}} \|Q(f) - Q(f_{\widetilde{c},\widetilde{d}})\| = \widetilde{\gamma}^{1/2}, \tag{13}$$

where $f_{\widetilde{c},\widetilde{d}}$ is the solution to a regularization program with parameters $\widetilde{c}$, $\widetilde{d}$, namely:

$$f_{\widetilde{c},\widetilde{d}} := \underset{f \in H}{\text{argmin}} \; \widetilde{c}\,\|Rf\|^2 + \widetilde{d}\,\|S(y - \Lambda f)\|^2. \tag{14}$$

Proof We drop the dependence on y throughout the argument below. For $c, d \geq 0$, we also use the notation:

$$f_{c,d} := \underset{f \in H}{\text{argmin}} \; c\,\|Rf\|^2 + d\,\|S(y - \Lambda f)\|^2.$$

It is essential to keep in mind that $f_{c,d}$ is characterized by

$$cR^*Rf_{c,d} + d\Lambda^*S^*S(\Lambda f_{c,d} - y) = 0. \tag{15}$$

As $cR^*R + d\Lambda^*S^*S\Lambda$ is invertible, thanks to the assumption $\ker(R) \cap \ker(S\Lambda) = \{0\}$, this characterization can be rewritten as

$$f_{c,d} = [cR^*R + d\Lambda^*S^*S\Lambda]^{-1}(d\Lambda^*S^*Sy). \tag{16}$$

Moreover, working preferentially with squared norms, we introduce the quantities (sv is for squared value, ub is for upper bound, and lub is for least upper bound):

$$\mathsf{sv}(z) = \sup_{f \in H} \|Q(f) - z\|^2 \quad \text{s.to} \quad f \in \mathcal{K} \text{ and } y - \Lambda f \in \mathcal{E},$$

$$\mathsf{ub}(z) = \inf_{\gamma, c, d \geq 0} \gamma \quad \text{s.to} \quad \text{the constraint elucidated in (17) below},$$

$$\mathsf{lub} = \inf_{c,d \geq 0} \mathsf{Obj} \quad \text{s.to} \quad \text{the constraint elucidated in (18) below},$$

where the objective function is $\mathsf{Obj} = c(\varepsilon^2 - \|Rf_{c,d}\|^2) + d(\eta^2 - \|S(y - \Lambda f_{c,d})\|^2)$. As for the constraints, they read:

$$\gamma - c(\varepsilon^2 - \|Rf_{c,d}\|^2) - d(\eta^2 - \|S(y - \Lambda f_{c,d})\|^2) + c\|Rh\|^2 + d\|S\Lambda h\|^2 - \|Qh\|^2 \\ \geq \|Q(f_{c,d}) - z\|^2 + 2\langle Q(f_{c,d}) - z, Qh\rangle \qquad \text{for all } h \in H, \tag{17}$$

and

$$c\|Rh\|^2 + d\|S\Lambda h\|^2 - \|Qh\|^2 \geq 0 \qquad \text{for all } h \in H. \tag{18}$$

We now divide the argument into the proofs of several facts, namely:

(i) for all $z \in Z$, $\mathsf{sv}(z) \leq \mathsf{ub}(z)$;
(ii) $\inf_{z\in Z} \mathsf{ub}(z) = \mathsf{lub} = \mathsf{ub}(Q(f_{\widetilde{c},\widetilde{d}}))$;
(iii) the program defining lub is equivalent to the program (12).

Once all these facts are justified, we are able to conclude via

$$\inf_{z\in Z} \mathsf{sv}(z) \leq \mathsf{sv}(Q(f_{\widetilde{c},\widetilde{d}})) \underset{(i)}{\leq} \mathsf{ub}(Q(f_{\widetilde{c},\widetilde{d}})) \underset{(ii)}{=} \mathsf{lub} \underset{(iii)}{=} \widetilde{\gamma}$$

which, up to taking square roots, is the result announced in (13).

Justification of (i). For $z \in Z$, the definition of $\mathsf{sv}(z)$ specified to our situation yields:

$$\begin{aligned} \mathsf{sv}(z) &= \sup_{f\in H} \{\|Q(f) - z\|^2 : \|Rf\|^2 \leq \varepsilon^2, \|S(y - \Lambda f)\|^2 \leq \eta^2\} \\ &= \inf_{\gamma\geq 0} \{\gamma : \|Q(f) - z\|^2 \leq \gamma \text{ whenever } \|Rf\|^2 \leq \varepsilon^2 \text{ and } \|S(y - \Lambda f)\|^2 \leq \eta^2\}. \end{aligned}$$

In the spirit of the S-procedure, the latter constraint is satisfied if, for some $c, d \geq 0$:

$$\|Q(f) - z\|^2 - \gamma \leq c(\|Rf\|^2 - \varepsilon^2) + d(\|S(y - \Lambda f)\|^2 - \eta^2) \quad \text{for all } f \in H. \tag{19}$$

Thus, fixing such $c, d \geq 0$, we obtain:

$$\mathsf{sv}(z) \leq \inf_{\gamma} \gamma \quad \text{s.to the constraint (19).}$$

Reparametrizing $f \in H$ as $f = f_{c,d} + h$ with variable $h \in H$, (19) becomes:

$$\begin{aligned} \|Q(f_{c,d}) - z\|^2 - \gamma + 2\langle Q(f_{c,d}) - z, Qh\rangle + \|Qh\|^2 & \\ \leq c(\|Rf_{c,d}\|^2 - \varepsilon^2) + d(\|S(y - \Lambda f_{c,d})\|^2 - \eta^2) & \\ + 2(c\langle Rf_{c,d}, Rh\rangle + d\langle S(\Lambda f_{c,d} - y), S\Lambda h\rangle) & \\ + c\|Rh\|^2 + d\|S\Lambda h\|^2 & \qquad \text{for all } h \in H. \end{aligned}$$

The linear term in the right-hand side is, according to (15),

$$2\big\langle cR^*Rf_{c,d} + d\Lambda^* S^* S(\Lambda f_{c,d} - y),\, h\big\rangle = 0.$$

Therefore, for any $c, d \geq 0$, we arrive at

$$\begin{aligned} \mathsf{sv}(z) \leq \inf_{\gamma \geq 0} \quad & \gamma \\ \text{s.to} \quad & \|Q(f_{c,d}) - z\|^2 + 2\langle Q(f_{c,d}) - z, Qh\rangle \\ & \leq \gamma - c(\varepsilon^2 - \|Rf_{c,d}\|^2) - d(\eta^2 - \|S(y - \Lambda f_{c,d})\|^2) \\ & + c\|Rh\|^2 + d\|S\Lambda h\|^2 - \|Qh\|^2 \qquad \text{for all } h \in H. \end{aligned}$$

Taking the infimum over $c, d \geq 0$, we recognize the desired inequality $\mathsf{sv}(z) \leq \mathsf{ub}(z)$. Note that this inequality turns into an equality in case where the S-procedure is exact—in particular, if the scalar field is $\mathbb{C}$. The subsequent steps (ii)-(iii) would then establish that $Q(f_{\widetilde{c},\widetilde{d}})$ is the Chebyshev center.

Justification of (ii): Part 1. Here, we prove that $\inf_{z \in Z} \mathsf{ub}(z) \geq \mathsf{lub}$. Towards this end, for $z \in Z$, we consider the constraint (17) in the defining expression of $\mathsf{ub}(z)$, which we write succinctly as

$$\mathrm{LHS}(h) \geq \|Q(f_{c,d}) - z\|^2 + 2\langle Q(f_{c,d}) - z, Qh\rangle \qquad \text{for all } h \in H.$$

Since $\mathrm{LHS}(-h) = \mathrm{LHS}(h)$, averaging the above inequality for h and $-h$ leads to $\mathrm{LHS}(h) \geq \|Q(f_{c,d}) - z\|^2$ for all $h \in H$, and hence to $\mathrm{LHS}(h) \geq 0$ for all $h \in H$. Having loosened the constraint, we deduce that

$$\mathsf{ub}(z) \geq \inf_{\gamma, c, d \geq 0} \gamma \qquad \text{s.to} \quad \mathrm{LHS}(h) \geq 0 \quad \text{for all } h \in H.$$

Taking the explicit form of $\mathrm{LHS}(h)$ into account, we see that the above constraint decouples as

$$\gamma - c(\varepsilon^2 - \|Rf_{c,d}\|^2) - d(\eta^2 - \|S(y - \Lambda f_{c,d})\|^2) \geq 0$$

and

$$c\|Rh\|^2 + d\|S\Lambda h\|^2 - \|Qh\|^2 \geq 0 \qquad \text{for all } h \in H.$$

The former reads $\gamma \geq \mathsf{Obj}$ and the latter is the constraint (18). Thus, we arrive at

$$\mathsf{ub}(z) \geq \inf_{\gamma, c, d \geq 0} \big\{\gamma \quad \text{s.to} \quad \gamma \geq \mathsf{Obj} \text{ and } (18)\big\} = \inf_{c, d \geq 0} \big\{\mathsf{Obj} \quad \text{s.to} \quad (18)\big\}.$$

This is the desired inequality $\mathsf{ub}(z) \geq \mathsf{lub}$, valid for any $z \in Z$.

Justification of (iii). In view of

$$c\|Rh\|^2 + d\|S\Lambda h\|^2 - \|Qh\|^2 = \big\langle (cR^*R + d\Lambda^*S^*S\Lambda - Q^*Q)h,\, h \big\rangle,$$

we instantly see that the constraint (18) is equivalent to $cR^*R{+}d\Lambda^*S^*S\Lambda{-}Q^*Q{\succeq}0$. Therefore, the program defining lub is equivalent to

$$\underset{c,d\geq 0}{\text{minimize}}\ \mathsf{Obj} \qquad \text{s.to} \quad cR^*R + d\Lambda^*S^*S\Lambda \succeq Q^*Q. \tag{20}$$

We now transform Obj by observing that

$$\begin{aligned}
c\varepsilon^2{+}d\eta^2 - \mathsf{Obj} &= c\|Rf_{c,d}\|^2 + d\|S(\Lambda f_{c,d} - y)\|^2 \\
&= c\langle R^*Rf_{c,d}, f_{c,d}\rangle + d\langle S^*S(\Lambda f_{c,d} - y), \Lambda f_{c,d} - y\rangle \\
&= \langle cR^*Rf_{c,d} + d\Lambda^*S^*S(\Lambda f_{c,d} - y), f_{c,d}\rangle - d\langle S^*S\Lambda f_{c,d}, y\rangle + d\langle S^*Sy, y\rangle \\
&= -d\langle S^*S\Lambda f_{c,d}, y\rangle + d\langle S^*Sy, y\rangle,
\end{aligned}$$

where the last step made use of the characterization (15). It follows that

$$\begin{aligned}
\mathsf{Obj} &= c\varepsilon^2 + d\eta^2 - d\|Sy\|^2 + d\langle S^*S\Lambda f_{c,d}, y\rangle \\
&= \inf_t\ c\varepsilon^2 + d\eta^2 - d\|Sy\|^2 + t \quad \text{s.to} \quad t \geq d\langle S^*S\Lambda f_{c,d}, y\rangle.
\end{aligned} \tag{21}$$

According to (16), the latter inequality can be written as

$$t \geq dy^*S^*S\Lambda f_{c,d} = (dy^*S^*S\Lambda)[cR^*R + d\Lambda^*S^*S\Lambda]^{-1}(d\Lambda^*S^*Sy),$$

or equivalently as the positive semidefiniteness of a Schur complement, namely, as

$$\left[\begin{array}{c|c} cR^*R + d\Lambda^*S^*S\Lambda & d\Lambda^*S^*Sy \\ \hline dy^*S^*S\Lambda & t \end{array}\right] \succeq 0. \tag{22}$$

Substituting (21) into (20) while imposing the additional constraint (22) shows that the program defining lub is indeed equivalent to (12).

Justification of (ii): Part 2. It now remains to prove that $\mathsf{ub}(Q(f_{\widetilde{c},\widetilde{d}})) \leq \mathsf{lub}$, where we recall that $\widetilde{c}, \widetilde{d}, \widetilde{t}$ represent minimizers of (12). By (iii), this also means that $\widetilde{c}, \widetilde{d}$ are minimizers of the problem defining lub. Thus, the feasibility constraint (18) is met for $c = \widetilde{c}$ and $d = \widetilde{d}$, so choosing $\gamma = \widetilde{c}(\varepsilon^2 - \|Rf_{\widetilde{c},\widetilde{d}}\|^2) + \widetilde{d}(\eta^2 - \|S(y - \Lambda f_{\widetilde{c},\widetilde{d}})\|^2)$, we see that the constraint (17) associated with $\mathsf{ub}(z)$ is met with $z = Q(f_{\widetilde{c},\widetilde{d}})$. We deduce that $\mathsf{ub}(Q(f_{\widetilde{c},\widetilde{d}})) \leq \gamma = \widetilde{c}(\varepsilon^2 - \|Rf_{\widetilde{c},\widetilde{d}}\|^2) + \widetilde{d}(\eta^2 -$

$\|S(y - \Lambda f_{\widetilde{c},\widetilde{d}})\|^2)$, which is the minimal value of Obj under the constraint (17). In other words, we have shown that $\mathsf{ub}(Q(f_{\widetilde{c},\widetilde{d}})) \leq \mathsf{lub}$, as desired. □

4 Exactness of the S-procedure: Proof of Theorem 1

Our goal in this section is to show that the overestimation of Theorem 2 becomes an exact evaluation if $Q = \mathrm{Id}_H$, P is an orthogonal projection, $S = \mathrm{Id}_m$, and $\Lambda\Lambda^* = \mathrm{Id}_m$, thus proving Theorem 1. We rely on duality in semidefinite programming. Retaining full generality for the moment, our primal program is the rewriting of (12) in the form:

$$\mathsf{lub} = \min_{\substack{c,d\geq 0\\ t\in\mathbb{R}}} c\varepsilon^2 + d(\eta^2 - \|Sy\|^2) + t \tag{23}$$

$$\text{s.to } M_{c,d,t} := \left[\begin{array}{c|c|c} cR^*R + d\Lambda^*S^*S\Lambda - Q^*Q & 0 & 0 \\ \hline 0 & cR^*R + d\Lambda^*S^*S\Lambda & d\Lambda^*S^*Sy \\ \hline 0 & dy^*S^*S\Lambda & t \end{array}\right] \succeq 0.$$

According to, e.g., [15, Example 5.11], its dual program reads:

$$\mathsf{lub}' = \max_{X\succeq 0} \operatorname{tr}\left(\left[\begin{array}{c|c|c} Q^*Q & 0 & 0 \\ \hline 0 & 0 & 0 \\ \hline 0 & 0 & 0 \end{array}\right] X\right) \quad \text{s.to} \quad \operatorname{tr}\left(\left[\begin{array}{c|c|c} R^*R & 0 & 0 \\ \hline 0 & R^*R & 0 \\ \hline 0 & 0 & 0 \end{array}\right] X\right) = \varepsilon^2, \tag{24}$$

$$\operatorname{tr}\left(\left[\begin{array}{c|c|c} \Lambda^*S^*S\Lambda & 0 & 0 \\ \hline 0 & \Lambda^*S^*S\Lambda & \Lambda^*S^*Sy \\ \hline 0 & y^*S^*S\Lambda & 0 \end{array}\right] X\right) = \eta^2 - \|Sy\|^2, \ \operatorname{tr}\left(\left[\begin{array}{c|c|c} 0 & 0 & 0 \\ \hline 0 & 0 & 0 \\ \hline 0 & 0 & 1 \end{array}\right] X\right) = 1.$$

It is well known (and easy to verify) that the inequality $\mathsf{lub} \geq \mathsf{lub}'$ always holds. Furthermore, if $\widehat{c}, \widehat{d}, \widehat{t}$ are feasible for (23) and if $\widehat{X}$ is feasible for (24), while $\operatorname{tr}(M_{\widehat{c},\widehat{d},\widehat{t}}\widehat{X}) = 0$, then the equality $\mathsf{lub} = \mathsf{lub}'$ actually holds, and in fact

$$\mathsf{lub} = \widehat{c}\varepsilon^2 + \widehat{d}(\eta^2 - \|Sy\|^2) + \widehat{t} = \operatorname{tr}\left(\left[\begin{array}{c|c|c} Q^*Q & 0 & 0 \\ \hline 0 & 0 & 0 \\ \hline 0 & 0 & 0 \end{array}\right] \widehat{X}\right) = \mathsf{lub}'.$$

This simply is a consequence of equalities throughout the chain of inequalities:

$$
\begin{aligned}
0 \le \mathsf{lub} - \mathsf{lub}' &\le \widehat{c}\varepsilon^2 + \widehat{d}(\eta^2 - \|Sy\|^2) + t - \operatorname{tr}\left(\left[\begin{array}{c|c|c} Q^*Q & 0 & 0 \\ \hline 0 & 0 & 0 \\ \hline 0 & 0 & 0 \end{array}\right]\widehat{X}\right) \\
&= \widehat{c}\operatorname{tr}\left(\left[\begin{array}{c|c|c} R^*R & 0 & 0 \\ \hline 0 & R^*R & 0 \\ \hline 0 & 0 & 0 \end{array}\right]\widehat{X}\right) + \widehat{d}\operatorname{tr}\left(\left[\begin{array}{c|c|c} \Lambda^*S^*S\Lambda & 0 & 0 \\ \hline 0 & \Lambda^*S^*S\Lambda & \Lambda^*S^*Sy \\ \hline 0 & y^*S^*S\Lambda & 0 \end{array}\right]\widehat{X}\right) \\
&\quad + \widehat{t}\operatorname{tr}\left(\left[\begin{array}{c|c|c} 0 & 0 & 0 \\ \hline 0 & 0 & 0 \\ \hline 0 & 0 & 1 \end{array}\right]\widehat{X}\right) \\
&= \operatorname{tr}(M_{\widehat{c},\widehat{d},\widehat{t}}\widehat{X}) = 0.
\end{aligned}
$$

With the aim of choosing such suitable $\widehat{c}, \widehat{d}, \widehat{t}$ and $\widehat{X}$, it is now time to lose generality and consider our specific situation where $Q = \mathrm{Id}_H$, P is an orthogonal projection, $S = \mathrm{Id}_m$, and $\Lambda\Lambda^* = \mathrm{Id}_m$. As pointed out in Sect. 2, this situation was settled in [10]. There, we showed that the set $\mathcal{K}_{y,\mathcal{E}} = \{f \in H : \|Pf\| \le \varepsilon \text{ and } \|\Lambda f - y\| \le \eta\}$ admits $f_\sharp \in H$ as its Chebyshev center as soon as one can find $c_\sharp, d_\sharp \ge 0$ and $h_\sharp \in H$ such that:

(a) $c_\sharp P + d_\sharp \Lambda^*\Lambda \succeq \mathrm{Id}$,
(b) $c_\sharp P f_\sharp + d_\sharp \Lambda^*(\Lambda f_\sharp - y) + (c_\sharp P + d_\sharp \Lambda^*\Lambda)h_\sharp = h_\sharp$,
(c) $\langle P f_\sharp, h_\sharp\rangle = 0, \qquad \langle \Lambda^*(\Lambda f_\sharp - y), h_\sharp\rangle = 0$,
(d) $\|Pf_\sharp + Ph_\sharp\|^2 = \varepsilon^2, \qquad \|\Lambda f_\sharp - y + \Lambda h_\sharp\|^2 = \eta^2$.

These four sufficient conditions were verified for $f_\sharp = f_{\tau_\sharp}$, where $\tau_\sharp \in (0, 1)$ was selected as the solution to the implicit equation (9) and where $f_{\tau_\sharp}$ was selected as in (8) with $\tau = \tau_\sharp$. We also made the choices $c_\sharp = (1 - \tau_\sharp)/\lambda_{\min}$ and $d_\sharp = \tau_\sharp/\lambda_{\min}$, where $\lambda_{\min}$ was the smallest eigenvalue of $(1 - \tau_\sharp)P + \tau_\sharp\Lambda^*\Lambda$. Finally, we took $h_\sharp$ as an associated eigenvector, so that $(c_\sharp P + d_\sharp\Lambda^*\Lambda)h_\sharp = h_\sharp$. It was normalized to satisfy (d)—note that it is the specific choice of $\tau_\sharp$ that made it possible to satisfy both equalities in (d). Keeping these recollections in mind, we now set:

$$
\widehat{c} := c_\sharp, \quad \widehat{d} := d_\sharp, \quad \widehat{t} = d_\sharp\langle\Lambda f_\sharp, y\rangle, \qquad \text{and} \quad \widehat{X} = \left[\begin{array}{c|c|c} h_\sharp h_\sharp^* & 0 & 0 \\ \hline 0 & f_\sharp f_\sharp^* & -f_\sharp \\ \hline 0 & -f_\sharp^* & 1 \end{array}\right] \succeq 0.
$$

The feasibility of $\widehat{c}, \widehat{d}, \widehat{t}$ for (23) follows from (a), combined with the inequality $\widehat{t} \ge d_\sharp\langle\Lambda f_\sharp, y\rangle$ reformulated via the Schur complement as in the justification of (iii)—note that $f_\sharp = f_{c_\sharp, d_\sharp}$. The feasibility of $\widehat{X}$ for (24) is a consequence of (c) and (d): while the third part of the constraint is obvious, the first two parts require some work. For the first part, we observe that

$$\operatorname{tr}\left(\left[\begin{array}{c|c|c} P & 0 & 0 \\ \hline 0 & P & 0 \\ \hline 0 & 0 & 0 \end{array}\right] \widehat{X}\right) = \operatorname{tr}(Ph_\sharp h_\sharp^*) + \operatorname{tr}(Pf_\sharp f_\sharp^*) = \|Ph_\sharp\|^2 + \|Pf_\sharp\|^2$$

$$\underset{(c)}{=} \|Ph_\sharp + Pf_\sharp\|^2 \underset{(d)}{=} \varepsilon^2.$$

For the second part, we observe that

$$\operatorname{tr}\left(\left[\begin{array}{c|c|c} \Lambda^*\Lambda & 0 & 0 \\ \hline 0 & \Lambda^*\Lambda & \Lambda^* y \\ \hline 0 & y^*\Lambda & 0 \end{array}\right] \widehat{X}\right) = \operatorname{tr}(\Lambda^*\Lambda h_\sharp h_\sharp^*) + \operatorname{tr}(\Lambda^*\Lambda f_\sharp f_\sharp^*) - 2\langle \Lambda^* y, f_\sharp\rangle$$

$$= \|\Lambda h_\sharp\|^2 + \|\Lambda f_\sharp\|^2 - 2\langle y, \Lambda f_\sharp\rangle = \|\Lambda h_\sharp\|^2 + \|\Lambda f_\sharp - y\|^2 - \|y\|^2$$

$$\underset{(c)}{=} \|\Lambda f_\sharp - y + \Lambda h_\sharp\|^2 - \|y\|^2 \underset{(d)}{=} \eta^2 - \|y\|^2.$$

Having verified the feasibility conditions, we turn to the complementary slackness condition. Using the eigenvalue property of $h_\sharp$ and the characterization of $f_\sharp$ (which together accounted for (b)), we obtain:

$$\begin{aligned}
&\operatorname{tr}(M_{\widehat{c},\widehat{d},\widehat{t}}\widehat{X}) \\
&\quad = \operatorname{tr}((\widehat{c}P + \widehat{d}\Lambda^*\Lambda - \mathrm{Id})h_\sharp h_\sharp^*) + \operatorname{tr}((\widehat{c}P + \widehat{d}\Lambda^*\Lambda) f_\sharp f_\sharp^*) - 2\widehat{d}\langle \Lambda^* y, f_\sharp\rangle + \widehat{t} \\
&\quad = \langle (c_\sharp P + d_\sharp \Lambda^*\Lambda - \mathrm{Id})h_\sharp, h_\sharp\rangle + \langle (c_\sharp P + d_\sharp \Lambda^*(\Lambda f_\sharp - y), f_\sharp\rangle \\
&\quad = 0,
\end{aligned}$$

as expected. At this point, we are guaranteed that

$$\mathsf{lub} = \mathsf{lub}' = \operatorname{tr}\left(\left[\begin{array}{c|c|c} \mathrm{Id}_H & 0 & 0 \\ \hline 0 & 0 & 0 \\ \hline 0 & 0 & 0 \end{array}\right] \widehat{X}\right) = \operatorname{tr}(h_\sharp h_\sharp^*) = \|h_\sharp\|^2 = \|f_\sharp + h_\sharp - f_\sharp\|^2$$

$$\underset{(d)}{\leq} \sup_{f\in H} \|f - f_\sharp\|^2 \quad \text{s.to} \quad \|Pf\|^2 \leq \varepsilon^2 \text{ and } \|\Lambda f - y\|^2 \leq \eta^2$$

$$= \mathsf{sv}(f_\sharp) = \inf_{g\in H} \mathsf{sv}(g),$$

where the last equality expresses the result of [10] that $f_\sharp$ is the Chebyshev. Now, with $f_{\widetilde{c},\widetilde{d}}$ denoting the candidate Chebyshev center from Theorem 2, we recall that $\mathsf{sv}(f_{\widetilde{c},\widetilde{d}}) \leq \mathsf{ub}(f_{\widetilde{c},\widetilde{d}}) = \mathsf{lub}$. Thus, we finally derive that

$$\mathsf{sv}(f_{\widetilde{c},\widetilde{d}}) \leq \inf_{g\in H} \mathsf{sv}(g),$$

i.e., that $f_{\widetilde{c},\widetilde{d}}$ is a Chebyshev center of $\{f \in H : \|Pf\| \le \varepsilon \text{ and } \|y - \Lambda f\| \le \eta\}$—actually, by uniqueness, it is the Chebyshev center. This concludes the full proof of Theorem 1.

5 Some Open Questions

This work, as well as the previous work [10], considered the model set $\mathcal{K}$ and the uncertainty set $\mathcal{E}$ given by

$$\mathcal{K} = \{f \in H : \|Pf\|_H \le \varepsilon\} \qquad \text{and} \qquad \mathcal{E} = \{e \in \mathbb{R}^m : \|e\|_2 \le \varepsilon\}. \tag{25}$$

For the quantity of interest $Q = \mathrm{Id}_H$, they settled the Chebyshev center issue under the specific conditions that P is an orthogonal projector and that $\Lambda\Lambda^* = \mathrm{Id}_m$. Without these conditions, however, there are several issues that remain unsettled. Here is a list of questions, all concerning the model set and the uncertainty set from (25), to which we do not know the answers—we cannot even foretell if any of them are likely to be easily solved:

1. *Does the Chebyshev center always belong to* $\mathcal{V} + \mathrm{ran}(\Lambda^*)$?
   Under the specific conditions that P is an orthogonal projector and that $\Lambda\Lambda^* = \mathrm{Id}_m$, we showed in [10] that the solution f_τ to (8) depends affinely on $\tau \in [0, 1]$. Since $f_0 \in \mathcal{V} + \mathrm{ran}(\Lambda^*)$ and $f_1 \in \mathcal{V}$, any f_τ belongs to $\mathcal{V} + \mathrm{ran}(\Lambda^*)$ and in particular the Chebyshev center does. Can this be proved directly without the specific conditions, even if the Hilbert space H is not finite-dimensional? An affirmative answer would allow one to deal with the infinite-dimensional setting by restricting the problem to the finite-dimensional space $\mathcal{V} + \mathrm{ran}(\Lambda^*)$.
2. *Is the Chebyshev center always obtained via regularization?*
   From [10], we know that the answer is affirmative under the specific conditions mentioned above, since the Chebyshev center equals the regularizer $f_{\tau_\sharp}$ with parameter $\tau_\sharp \in (0, 1)$ satisfying the implicit equation (9). Without the specific conditions, we also know from [8] that the answer is affirmative if one considered the complex setting rather than the real setting.
3. *What is the Chebyshev center relative to a quantity of interest* $Q \neq \mathrm{Id}_H$?
   Here, we do not have any partial answers in the real setting, even under the specific conditions above. Indeed, the result of [10] only considered the case $Q = \mathrm{Id}_H$, and initial attempts to extend it to the case $Q \neq \mathrm{Id}_H$ were not conclusive. An obvious guess (valid in the complex setting) is that the Chebyshev center equals $Q(f_{\widetilde{c},\widetilde{d}})$, where $f_{\widetilde{c},\widetilde{d}}$ is the regularizer obtained from the relaxation of Theorem 2.
4. *Does relaxation always lead to near-optimality?*
   Should one fail to find the Chebyshev center, one may be content with the statement that there exists a parameter $\tau_\natural \in (0, 1)$ such that the regularizer $f_{\tau_\natural}$ admits a near-minimal local worst-case error, in the sense that

$$\sup_{\substack{f \in \mathcal{K} \\ y - \Lambda f \in \mathcal{E}}} \|Q(f) - Q(f_{\tau_\sharp})\| \leq C \times \inf_{z \in Z} \sup_{\substack{f \in \mathcal{K} \\ y - \Lambda f \in \mathcal{E}}} \|Q(f) - z\|$$

for some constant $C \geq 1$. This was proved in [16] under no specific conditions. We wonder if the same holds for $f_{\widetilde{c},\widetilde{d}}$ instead of $f_{\tau_\sharp}$.

6 Some Other Optimal Recovery Results Obtained at CAMDA

As alluded to in the introduction, investigations in the field of optimal recovery constituted one of the pilot research projects during the academic year 2022–2023, i.e., the initial year of the Center for Approximation and Mathematical Data Analytics, better known as CAMDA. Below is a synopsis of results generated during this time:

- The article [16] considered a scenario of optimal recovery from inaccurate data very close to the present one, the only difference being that it targeted a global worst-case error akin to (2) rather than a local worst-case error akin to (1). In spite of a focus on graph signals (i.e., functions defined on the set of vertices of a graph), it settled the global optimality question in the general setting of the model and uncertainty sets from (25) without any specific conditions on P and Λ and for an arbitrary linear map $Q : H \to Z$ as quantity of interest. It established that the a globally optimal recovery map is given by the linear map $Q \circ \Delta_{\tau_\sharp} : \mathbb{R}^m \to Z$, where

$$\Delta_{\tau_\sharp} : y \in \mathbb{R}^m \mapsto \left[\operatorname*{argmin}_{f \in H} \; (1 - \tau_\sharp)\|Pf\|^2 + \tau_\sharp \|y - \Lambda f\|^2 \right] \in H$$

  and where the parameter $\tau_\sharp \in (0, 1)$ is the ratio $\tau_\sharp = d_\sharp/(c_\sharp + d_\sharp)$ featuring the solutions $c_\sharp, d_\sharp \geq 0$ to the semidefinite program

$$\operatorname*{minimize}_{c,d \geq 0} \; c\varepsilon^2 + d\eta^2 \qquad \text{s.to } cP^*P + d\Lambda^*\Lambda \succeq Q^*Q.$$

  Precisely, this means that

$$\sup_{\substack{f \in \mathcal{K} \\ e \in \mathcal{E}}} \|Q(f) - (Q \circ \Delta_{\tau_\sharp})(\Lambda f + e)\| = \inf_{\Delta : \mathbb{R}^m \to Z} \sup_{\substack{f \in \mathcal{K} \\ e \in \mathcal{E}}} \|Q(f) - \Delta(\Lambda f + e)\|.$$

- The article [17] viewed the observation errors differently, since $e \in \mathbb{R}^m$ was considered to be a random vector there. It advocated for the global error:

$$\mathrm{ge}_p^{\mathrm{or}}(\Delta) := \left(\mathbb{E}\Big[\sup_{f\in\mathcal{K}} \|Q(f) - \Delta(\Lambda f + e)\|_Z^p\Big]\right)^{1/p}, \qquad p \geq 1,$$

to be used in optimal recovery, as opposed to another notion appearing more often in statistical estimation, namely:

$$\mathrm{ge}_p^{\mathrm{se}}(\Delta) := \left(\sup_{f\in\mathcal{K}} \mathbb{E}\Big[\|Q(f) - \Delta(\Lambda f + e)\|_Z^p\Big]\right)^{1/p}, \qquad p \geq 1.$$

In an arbitrary Banach space F, if $Q \in F^*$ is a linear functional, if the model set $\mathcal{K}$ is symmetric and convex, and if the random vector $e \in \mathbb{R}^m$ is merely log-concave, it was proved that linear maps are near-optimal relative to $\mathrm{ge}^{\mathrm{or}}$, in the sense that

$$\inf_{\Delta:\mathbb{R}^m\to\mathbb{R}\text{ linear}} \mathrm{ge}_p^{\mathrm{or}}(\Delta^{\mathrm{lin}}) \leq C_p \times \inf_{\Delta:\mathbb{R}^m\to\mathbb{R}} \mathrm{ge}_p^{\mathrm{or}}(\Delta) \tag{26}$$

for some constant C_p depending on p. This extends the seminal result of [18], which dealt with $\mathrm{ge}^{\mathrm{se}}$ instead of $\mathrm{ge}^{\mathrm{or}}$ and considered only Gaussian vectors $e \in \mathbb{R}^m$.

- The article [20] did not take place in a Hilbert space H but in the Banach space of continuous functions on a compact set $\mathcal{X}$, i.e., $F = C(\mathcal{X})$. It targeted, again from a global optimality perspective, the full recovery problem ($Q = \mathrm{Id}_{C(\mathcal{X})}$) relative to a model set based on approximation capabilities relative to some linear subspace $\mathcal{V} \subseteq C(\mathcal{X})$ and to a parameter $\varepsilon > 0$, namely:

$$\mathcal{K}_{\mathcal{V}} = \{f \in C(\mathcal{X}) : \mathrm{dist}_{C(\mathcal{X})}(f, \mathcal{V}) \leq \varepsilon\}.$$

  The observational data were assumed to be acquired as accurate evaluations at points $x^{(1)}, \ldots, x^{(m)} \in \mathcal{X}$. In this situation, it was known (see [19]) how to optimally estimate point evaluations $\delta_x \in C(\mathcal{X})^*$ at each $x \in \mathcal{X}$, and how to subsequently produce an optimal recovery map for $Q = \mathrm{Id}_{C(\mathcal{X})}$—a linear one, to boot. However, the construction was not practical. A practical construction, which amounts to solving about m standard-form linear programs via the simplex method, was conceived in [20] under the proviso that $\mathcal{V}$ is a Chebyshev space. Although this is a strong proviso—by Mairhuber—Curtis theorem, it essentially excludes multivariate functions—it brings hope that genuinely optimal recovery maps in $C(\mathcal{X})$ can be computationally constructed in specific instances.

- The article [21] studied the numerical solutions to partial differential equations. With Ω being a bounded Lipschitz domain in $\mathbb{R}^d$ with $d = 2$ or $d = 3$, it considered the elliptic problem:

$$-\Delta u = f \quad \text{in } \Omega, \qquad\qquad u = g \quad \text{on } \partial\Omega,$$

where $f \in H^{-1}(\Omega)$ is known but $g \in H^{1/2}(\partial\Omega)$ is not known and when linear observations $y = \Lambda u$ made on the solution $u \in H^1(\Omega)$ are available. Finding an approximant $\widehat{u}$ is viewed as an optimal recovery problem relative to the model set

$$\mathcal{K}_s = \{u \in H^1(\Omega) : -\Delta u = f \text{ and } u_{|\partial\Omega} = g \text{ for some } g \in B_{H^s(\partial\Omega)}\}, \quad s > 1/2.$$

Based on minimum-norm interpolation (aka spline algorithm), an implementable procedure was conceived to generate an output to $\widehat{u}$ which is locally near-optimal, in the sense that $\sup\{\|u - \widehat{u}\|_{H^1(\Omega)} : u \in \mathcal{K}_s, \Lambda u = y\}$ is a most a multiplicative constant away from the Chebyshev radius of the set $\{u \in \mathcal{K}_s : \Lambda u = y\}$.

Acknowledgments S. F. is partially supported by grants from the NSF (DMS-2053172) and from the ONR (N00014-20-1-2787).

References

1. Tikhomirov, V. M.: A. N. Kolmogorov and Approximation Theory. Russian Mathematical Surveys, **44**, 101 (1989).
2. Micchelli, C. A., Rivlin, T. J.: A survey of optimal recovery. In: Optimal estimation in approximation theory. pp.1–54. (1977)
3. Novak, E., Wozniakowski, H. : Tractability of Multivariate Problems, Volume I. European Mathematical Society, Zürich (2010)
4. Foucart, S.: Mathematical Pictures at a Data Science Exhibition. Cambridge University Press (2022).
5. Garkavi., A. L. : On the optimal net and best cross-section of a set in a normed space. Izvestiya Rossiiskoi Akademii Nauk. Seriya Matematicheskaya **26**, 87–106 (1962)
6. Melkman, A. A., Micchelli, C. A. : Optimal estimation of linear operators in Hilbert spaces from inaccurate data. SIAM Journal on Numerical Analysis **16**, 87–105 (1979)
7. Micchelli, C. A. : Optimal estimation of linear operators from inaccurate data: a second look. Numerical Algorithms **5**, 375–390 (1993)
8. Beck, A., Eldar, Y.C.: Regularization in regression with bounded noise: a Chebyshev center approach. SIAM Journal on Matrix Analysis and Applications **29**, 606–625 (2007)
9. Binev, P., Cohen, A., Dahmen, W., DeVore, R., Petrova, G., Wojtaszczyk, P.: Data assimilation in reduced modeling. SIAM/ASA Journal on Uncertainty Quantification **5**, 1–29 (2017)
10. Foucart, S., Chunyang, L. : Optimal recovery from inaccurate data in Hilbert spaces: regularize, but what of the parameter? Constructive Approximation **57**, 489–520 (2023)
11. Pólik. I., Terlaky, T.: A survey of the S-lemma. SIAM Review **49**, 371–418 (2007)
12. Yakubovich, V. A. : S-procedure in nonlinear control theory. Vestnik Leningrad. Univ. **1**, 62–77 (1971)
13. Beck, A., Eldar, Y. C.: Strong duality in nonconvex quadratic optimization with two quadratic constraints. SIAM J. Optim. **17**, 844–860 (2006)
14. Polyak, B. T.: Convexity of quadratic transformations and its use in control and optimization. J. Optim. Theory Appl. **99**, 553–583 (1998)
15. Boyd, S. P., Vandenberghe, L. : Convex Optimization. Cambridge University Press (2004)
16. Foucart, S., Liao, C., and Veldt, N.: On the optimal recovery of graph signals. In: International Conference on Sampling Theory and Applications (SampTA), 2023.

17. Foucart, S. and Paouris, G.: Near-optimal estimation of linear functionals with log-concave observation errors. Information and Inference, **12/4**, 2546–2561 (2023)
18. Donoho, D. L.: Statistical estimation and optimal recovery. The Annals of Statistics **22**, 238–270 (1994).
19. DeVore, R., Foucart, S., Petrova, G., Wojtaszczyk, P.: Computing a quantity of interest from observational data. Constructive Approximation **49**, 461–508 (2019)
20. Foucart, S.: Full recovery from point values: an optimal algorithm for Chebyshev approximability prior. Advances in Computational Mathematics **49**, 57 (2023)
21. Binev, P., Bonito, A., Cohen, A., Dahmen, W., DeVore, R., Petrova, G.: Solving PDEs with incomplete information. Numer. Anal., **62**(3), 1278–1312 (2024)

Neural Networks: Deep, Shallow, or in Between?

Guergana Petrova and Przemysław Wojtaszczyk

Abstract We give estimates from below for the error of approximation of a compact subset from a Banach space by the outputs of feed-forward neural networks with width W, depth ℓ, and Lipschitz activation functions. We show that modulo logarithmic factors, rates better than entropy numbers' rates are possibly attainable only for neural networks for which the depth $\ell \to \infty$, and that there is no gain if we fix the depth and let the width $W \to \infty$.

1 Introduction

The fascinating new developments in the area of artificial intelligence (AI) and other important applications of neural networks prompt the need for a theoretical mathematical study of their potential to reliably approximate complicated objects. Various network architectures have been used in different applications with substantial success rates without significant theoretical backing of the choices made. Thus, a natural question to ask is whether and how the architecture chosen affects the approximation power of the outputs of the resulting neural network.

In this paper, we attempt to clarify how the width and the depth of a feed-forward neural network affect its worst performance. More precisely, we provide estimates from below for the error of approximation of a compact subset $\mathcal{K} \subset X$ of a Banach space X by the outputs of feed-forward neural networks (NNs) with width W, depth ℓ, bound $w(W, \ell)$ on their parameters, and Lipschitz activation functions. Note that the ReLU function is included in our investigation since it is a Lipschitz function with a Lipschitz constant $L = 1$.

G. Petrova (✉)
Department of Mathematics, Texas A&M University, College Station, TX, USA
e-mail: gpetrova@tamu.edu

P. Wojtaszczyk
Institut of Mathematics, Polish Academy of Sciences, Warszawa, Poland
e-mail: wojtaszczyk@impan.pl

S. Foucart, S. Wojtowytsch (eds.), *Explorations in the Mathematics of Data Science*, Applied and Numerical Harmonic Analysis,
https://doi.org/10.1007/978-3-031-66497-7_2

To prove our results, we assume that we know lower bounds on the entropy numbers of the compact sets $\mathcal{K}$ that we approximate by the outputs of feed-forward NNs. Such bounds are known for a wide range of classical and novel classes $\mathcal{K}$ and Banach spaces X, and are usually of the form $n^{-\alpha}[\log n]^{\beta}$, $\alpha > 0$, $\beta \in \mathbb{R}$. We refer the reader to [10, Chapters 3,4], [12, Chapter 15], [5, Section 5], [20, Theorem 9], or [6, 11], where such examples are provided.

It is a well-known fact that the number n of parameters of a feed-forward NN with width W and depth ℓ is

$$n \asymp \begin{cases} W^2\ell, & \text{when} \quad \ell > 1, \\ W, & \text{when} \quad \ell = 1. \end{cases} \tag{1}$$

Let us denote by $\Sigma(W, \ell, \sigma; w)$ the set of functions that are outputs of such a NN with bounds $w = w(W, \ell)$ on its parameters and with Lipschitz activation function σ. We prove estimates from below for the error $E(\mathcal{K}, \Sigma(W, \ell, \sigma; w))_X$ of approximation of a class $\mathcal{K}$ by the functions from $\Sigma(W, \ell, \sigma; w)$; see Theorem 3. Our conclusion is that under a moderate growth of the bound $w \asymp n^{\delta}$, $\delta \geq 0$, one can possibly obtain rates of approximation that are better than the corresponding entropy numbers' rates only when the depth of the NN is let to grow. If the rate of approximation of $\mathcal{K}$ by outputs of feed-forward NNs is better than the decay rate of its entropy numbers, then we say that we have super convergence. In fact, since we only obtain estimates from below, we claim that super convergence is possibly attainable in such cases. If the depth ℓ is fixed, then the rates of decay of $E(\mathcal{K}, \Sigma(W, \ell, \sigma; w))_X$ cannot be better (modulo logarithmic factors) than the rates of the entropy numbers of $\mathcal{K}$. If both the width W and depth ℓ are allowed to grow, then an improvement of the rates of decay of $E(\mathcal{K}, \Sigma(W, \ell, \sigma; w))_X$ in comparison with the entropy numbers' decay is possible. Of course, the bound w on the NN's parameters also has an effect, and a fast growing bound, for example, $w \asymp 2^n$, could lead to improved convergence in all cases. However, one needs to be aware of the fact that NNs with such bounds are computationally infeasible.

We show that the mapping assigning to each choice of neural network parameters the function that is an output of a feed-forward NN with these parameters is a Lipschitz mapping; see Theorem 2. This allows us to study the approximation properties of such NNs via the recently introduced Lipschitz widths; see [16, 17]. We have utilized this approach in [17] to discuss deep ($W = W_0$ is fixed and $\ell \to \infty$) and shallow ($W \to \infty$ and $\ell = 1$) NNs with bounded Lipschitz or ReLU activation functions and their limitations in approximating compact sets $\mathcal{K}$. Here, we implement the developed technique to treat NNs for which both $W, \ell \to \infty$. Results in this direction are available for shallow and deep NNs, and we refer the reader to the series of works [1, 2, 8, 14, 15, 18, 21–24], where various estimates from below are given for the error of approximation for particular classes $\mathcal{K}$ and Banach spaces X.

The paper is organized as follows. In Sect. 2, we introduce our notation; recall the definitions of NNs, entropy numbers, and Lipschitz widths; and state some known results about them. We show in Sect. 3 that feed-forward NNs are Lipschitz mappings. Finally, in Sect. 4, we use results for Lipschitz widths to derive estimates from below for the error of neural network approximation for a compact class $\mathcal{K}$.

2 Preliminaries

In this section, we introduce our notation and recall some known facts about NNs, Lipschitz widths, and entropy numbers. In what follows, we will denote by $A \gtrsim B$ the fact that there is an absolute constant $c > 0$ such that $A \geq cB$, where A, B are expressions that depend on some variable which tends to infinity. Note that the value of c may change from line to line, but is always independent on that variable. Similarly, we use the notation $A \lesssim B$ (defined in an analogues way) and $A \asymp B$ if $A \gtrsim B$ and $A \lesssim B$.

We also write $A = A(B)$ to stress the fact that the quantity A depends on B. For example, if C is a constant, the expression $C = C(d, \sigma)$ means that C depends on d and σ.

2.1 *Entropy Numbers*

We recall (see, e.g., [3, 4, 12]) that the *entropy numbers* $\epsilon_n(\mathcal{K})_X, n \geq 0$, of a compact set $\mathcal{K} \subset X$ are defined as the infimum of all $\epsilon > 0$ for which 2^n balls with centers from X and radius ϵ cover $\mathcal{K}$. Formally, we write:

$$\epsilon_n(\mathcal{K})_X = \inf\{\epsilon > 0 \ : \ \mathcal{K} \subset \bigcup_{j=1}^{2^n} B(g_j, \epsilon), \ g_j \in X, \ j = 1, \dots, 2^n\}.$$

2.2 *Lipschitz Widths*

A classical notion of the quality of approximation of a compact subset $K \subset X$ of a Banach space X by linear spaces of dimension n is the Kolmogorov width $d_n(K)_X$, defined as

$$d_0(K)_X := \sup_{f \in K} \|f\|_X, \quad d_n(K)_X = \inf_{V:\dim V = n} \ \sup_{f \in K} \ \inf_{g \in V} \|f - g\|_X, \quad n \geq 1,$$

where the first infimum is taken over all n-dimensional subspaces V of X. Modern applications however, require new ways of approximating the elements of K, where the approximants do not come from linear spaces but rather from nonlinear manifolds. Thus, a need for different types of widths arises, where these new widths would give a theoretical benchmark of the performance of the most recent types of approximation. Examples of such widths include the manifold widths (see [9]), and stable manifold widths (see [7]). Recently, a new notion of width, the so-called Lipschitz width, was proposed in [16]. It measures how well one can approximate the set K by images of Lipschitz mappings. In order to have a meaningful definition that would exclude space-filling manifolds, the domain of these mappings is restricted to the ball in $\mathbb{R}^n$ with radius r_n with respect to a norm $\|\cdot\|_{Y_n}$. In the original definition of Lipschitz widths (see [16]), we required that $r_n = 1$ for all n. However, there is an interplay between the Lipschitz constant of the mapping and the radius of the ball on which it is defined. To address this issue, we have decided to state an equivalent definition that takes this dependence into account; see [17]. The latter definition is more suitable for the applications of the theory of Lipschitz widths to NNs, since it could be directly applied to general networks and not to only those whose parameters are bounded by 1.

We denote by $(\mathbb{R}^n, \|.\|_{Y_n})$, $n \in \mathbb{N}$, the n-dimensional Banach space with a fixed norm $\|\cdot\|_{Y_n}$, by

$$B_{Y_n}(r) := \{y \in \mathbb{R}^n : \ \|y\|_{Y_n} \leq r\},$$

its ball with radius r, and by

$$\|y\|_{\ell_\infty^n} := \max_{j=1,\dots,n} |y_j|,$$

the ℓ_∞ norm of $y = (y_1, \dots, y_n) \in \mathbb{R}^n$. For any fixed n, the Lipschitz width $d_n^{\gamma_n}(\mathcal{K})_X$ of the compact set $\mathcal{K}$ with respect to the norm $\|\cdot\|_X$ is defined as

$$d_n^{\gamma_n}(\mathcal{K})_X := \inf_{\mathcal{L}_n,\, r_n>0,\, \|\cdot\|_{Y_n}} \ \sup_{f\in\mathcal{K}} \ \inf_{y\in B_{Y_n}(r_n)} \|f - \mathcal{L}_n(y)\|_X, \tag{2}$$

where the infimum is taken over all γ_n/r_n-Lipschitz maps $\mathcal{L}_n : (B_{Y_n}(r_n), \|\cdot\|_{Y_n}) \to X$, all $r_n > 0$, and all norms $\|\cdot\|_{Y_n}$ in $\mathbb{R}^n$. We have proven (see Theorem 9 in [17]) the following result which relates the behavior of the entropy numbers of $\mathcal{K}$ and its Lipschitz widths with a Lipschitz constant $\gamma_n = 2^{\varphi(n)}$.

Theorem 1 *For any compact set $\mathcal{K} \subset X$, we consider the Lipschitz width $d_n^{\gamma_n}(\mathcal{K})_X$ with Lipschitz constant $\gamma_n = 2^{\varphi(n)}$, where $\varphi(n) \geq c\log_2 n$ for some fixed constant $c > 0$. Let $\alpha > 0$ and $\beta \in \mathbb{R}$. Then, the following holds:*

$$\text{(i) } \epsilon_n(\mathcal{K})_X \gtrsim \frac{(\log_2 n)^\beta}{n^\alpha}, \quad n \in \mathbb{N} \quad \Rightarrow \quad d_n^{\gamma_n}(\mathcal{K})_X \gtrsim \frac{[\log_2(n\varphi(n))]^\beta}{[n\varphi(n)]^\alpha}, \quad n \in \mathbb{N}; \tag{3}$$

$$\text{(ii) } \epsilon_n(\mathcal{K})_X \gtrsim [\log_2 n]^{-\alpha}, \quad n \in \mathbb{N} \Rightarrow \quad d_n^{\gamma_n}(\mathcal{K})_X \gtrsim [\log_2(n\varphi(n))]^{-\alpha}, \quad n \in \mathbb{N}. \tag{4}$$

2.3 Neural Networks

Let us denote by $C(\Omega)$ the set of continuous functions defined on the compact set $\Omega \subset \mathbb{R}^d$, equipped with the uniform norm.

A feed-forward NN with activation function $\sigma : \mathbb{R} \to \mathbb{R}$, width W, depth ℓ, and bound $w = w(W, \ell)$ on its parameters generates a family $\Sigma(W, \ell, \sigma; w)$ of continuous functions:

$$\Sigma(W, \ell, \sigma; w) := \{\Phi_\sigma^{W,\ell}(y) : \ y \in \mathbb{R}^n, \ \|y\|_{\ell_\infty^n} \le w\} \subset C(\Omega), \quad \Omega \subset \mathbb{R}^d,$$

where the number n of its parameters satisfies (1). Each $y \in \mathbb{R}^n$, $\|y\|_{\ell_\infty^n} \le w$, determines a continuous function $\Phi_\sigma^{W,\ell}(y) \in \Sigma(W, \ell, \sigma; w)$, defined on Ω, of the form:

$$\Phi_\sigma^{W,\ell}(y) := A^{(\ell)} \circ \bar{\sigma} \circ A^{(\ell-1)} \circ \ldots \circ \bar{\sigma} \circ A^{(0)}, \tag{5}$$

where $\bar{\sigma} : \mathbb{R}^W \to \mathbb{R}^W$ is given by

$$\bar{\sigma}(z_1, \ldots, z_W) = (\sigma(z_1), \ldots, \sigma(z_W)), \tag{6}$$

and $A^{(0)} : \mathbb{R}^d \to \mathbb{R}^W$, $A^{(j)} : \mathbb{R}^W \to \mathbb{R}^W$, $j = 1, \ldots, \ell - 1$, and $A^{(\ell)} : \mathbb{R}^W \to \mathbb{R}$ are affine mappings. The coordinates of $y \in \mathbb{R}^n$ are the entries of the matrices and offset vectors (biases) of the affine mappings $A^{(j)}$, $j = 0, \ldots, \ell$, taken in a pre-assigned order. The entries of $A^{(j)}$ appear before those of $A^{(j+1)}$ and the ordering for each $A^{(j)}$ is done in the same way. We refer the reader to [8] and the references therein for detailed study of such NNs with fixed width $W = W_0$ and depth $\ell \to \infty$.

We view a feed-forward NN as a mapping that to each vector of parameters $y \in \mathbb{R}^n$ assigns the output $\Phi_\sigma^{W,\ell}(y) \in \Sigma(W, \ell, \sigma; w)$ of this network:

$$y \to \Phi_\sigma^{W,\ell}(y), \tag{7}$$

where all parameters (entries of the matrices and biases) are bounded by $w(W, \ell)$, namely:

$$\Sigma(W, \ell, \sigma; w) = \Phi_\sigma^{W,\ell}(B_{\ell_\infty^n}(w(W, \ell))),$$

with $\Phi_\sigma^{W,\ell}$ being defined in (5).

Lower bounds for the error of approximation of a class $\mathcal{K} \subset X$ by the outputs of DNNs (when $W = W_0$ for a fixed W_0 and $\ell \to \infty$, in which $n \asymp \ell$) and SNNs (when $\ell = 1$ and $W \to \infty$, in which $n \asymp W$) have been discussed in [17] in the case of bounded Lipschitz or ReLU activation functions. In this paper, we state similar results for any feed-forward NN with general Lipschitz activation function. We use the approach from [17] and first investigate the mapping (7) that assigns to each vector of parameters $y \in B_{\ell_\infty^n}(w(W, \ell))$ the continuous function $\Phi_\sigma^{W,\ell}(y)$.

3 Feed-Forward NNs Are Lipschitz Mappings

In this section, we prove that the mapping

$$\Phi_\sigma^{W,\ell} : (B_{\ell_\infty^n}(w(W,\ell)), \|\cdot\|_{\ell_\infty^n}) \to X, \quad B_{\ell_\infty^n}(w(W,\ell)) \subset \mathbb{R}^n,$$

defined in (5) is a Lipschitz mapping with a Lipschitz constant L_n, provided the activation function σ is Lipschitz. Note that $\Phi_\sigma^{W,\ell}$ depends on n parameters (the weights and biases of the NN), which are expressed in terms of the width W and depth ℓ via formula (1). We also determine its Lipschitz constant L_n. More precisely, if

$$L := \max\{L', |\sigma(0)|\}, \tag{8}$$

where L' is the Lipschitz constant of σ, then the following theorem holds.

Theorem 2 *Let X be a Banach space such that $C([0,1]^d) \subset X$ is continuously embedded in X. Then, the mapping $\Phi_\sigma^{W,\ell}$, defined in (5) with a Lipschitz function σ, is an L_n-Lipschitz mapping, that is,*

$$\|\Phi_\sigma^{W,\ell}(y) - \Phi_\sigma^{W,\ell}(y')\|_X \le L_n \|y - y'\|_{\ell_\infty^n}, \quad y, y' \in B_{\ell_\infty^n}(w(W,\ell)).$$

Moreover, there are constants $c_1, c_2 > 0$ such that

$$2^{c_1 \ell \log_2(W(w+1)))} < L_n < 2^{c_2 \ell \log_2(W(w+1)))}, \quad w = w(W,\ell),$$

provided $LW \ge 2$.

Proof Let us first set up the notation by denoting $\|g\| := \max_{1\le i\le W} \|g_i\|_{C(\Omega)}$, where g is the vector function $g = (g_1, \dots, g_W)^T$ whose coordinates $g_i \in C(\Omega)$. We also will use

$$w := w(W,\ell), \quad \text{and} \quad \tilde{w} := w + 1.$$

Let y, y' be the two parameters from $B_{\ell_\infty^n}(w(W,\ell))$ that determine the continuous functions $\Phi_\sigma^{W,\ell}(y), \Phi_\sigma^{W,\ell}(y') \in \Sigma(W,\ell,\sigma; w)$. We fix $x \in \Omega$ and denote by

$$\eta^{(0)}(x) := \overline{\sigma}(A_0 x + b^{(0)}), \quad \eta'^{(0)}(x) := \overline{\sigma}(A'_0 x + b'^{(0)}),$$

$$\eta^{(j)} := \overline{\sigma}(A_j \eta^{(j-1)} + b^{(j)}), \quad \eta'^{(j)} := \overline{\sigma}(A'_j \eta'^{(j-1)} + b'^{(j)}), \quad j = 1, \dots, \ell - 1,$$

$$\eta^{(\ell)} := A_\ell \eta^{(\ell-1)} + b^{(\ell)}, \quad \eta'^{(\ell)} := A'_\ell \eta'^{(\ell-1)} + b'^{(\ell)}.$$

Note that $A_0, A'_0 \in \mathbb{R}^{W\times d}$, $A_j, A'_j \in \mathbb{R}^{W\times W}$, $b^{(j)}, b'^{(j)} \in \mathbb{R}^W$, for $j = 0, \dots, \ell-1$, while $A_\ell, A'_\ell \in \mathbb{R}^{1\times W}$, and $b^{(\ell)}, b'^{(\ell)} \in \mathbb{R}$. Each of the $\eta^{(j)}, \eta'^{(j)}$, $j = 0, \dots, \ell-1$, is a continuous vector function with W coordinates, while $\eta^{(\ell)}, \eta'^{(\ell)}$ are the outputs of the NN with activation function σ and parameters y, y', respectively.

Since (see (8))

$$|\sigma(t)| \le |\sigma(t)-\sigma(0)|+|\sigma(0)| \le L(|t|+1), \quad |\sigma(t_1)-\sigma(t_2)| \le L|t_1-t_2|, \quad t_1, t_2 \in \mathbb{R},$$

it follows that for any m, vectors $\bar{y}, \hat{y}, \eta \in \mathbb{R}^m$ and numbers $y_0, \hat{y}_0 \in \mathbb{R}$, where $\bar{y}, y_0$ and $\hat{y}, \hat{y}_0$ are subsets of the coordinates of $y, y' \in \mathbb{R}^n$, respectively, we have:

$$\begin{aligned} |\sigma(\bar{y}\cdot\eta + y_0)| &\le L(|\bar{y}\cdot\eta + y_0| + 1) \le L(m\|\eta\|_{\ell_\infty^m} + 1)\|y\|_{\ell_\infty^n} + L \\ &\le L(m\|\eta\|_{\ell_\infty^m} + 1)w + L < L\tilde{w}m\|\eta\|_{\ell_\infty^m} + L\tilde{w} \end{aligned} \tag{9}$$

and

$$|\sigma(\bar{y}\cdot\eta + y_0) - \sigma(\hat{y}\cdot\eta + \hat{y}_0)| \le L(m\|\eta\|_{\ell_\infty^m} + 1)\|y - y'\|_{\ell_\infty^n}. \tag{10}$$

Then, we have $\|\eta'^{(0)}\| < L\tilde{w}d + L\tilde{w}$ (when $m = d$ and $\eta = x$) and

$$\|\eta'^{(j)}\| < LW\tilde{w}\|\eta'^{(j-1)}\| + L\tilde{w}, \quad j = 1, \dots, \ell,$$

(when $m = W$ and $\eta = \eta'^{(j-1)}$). One can show by induction that for $j = 1, \dots, \ell$,

$$\|\eta'^{(j)}\| \le dW^j[L\tilde{w}]^{j+1} + L\tilde{w}\sum_{i=0}^{j}[LW\tilde{w}]^i.$$

Therefore, we have that

$$\|\eta'^{(j)}\| \le dW^j[L\tilde{w}]^{j+1} + 2L\tilde{w}[LW\tilde{w}]^j = (d+2)L\tilde{w}[LW\tilde{w}]^j, \tag{11}$$

since $LW\tilde{w} > LW \ge 2$. The above inequality also holds for $j = 0$.

Clearly, we have:

$$\|\eta^{(0)} - \eta'^{(0)}\| \le L(d+1)\|y - y'\|_{\ell_\infty^n} =: C_0\|y - y'\|_{\ell_\infty^n}.$$

Suppose we have proved the inequality:

$$\|\eta^{(j-1)} - \eta'^{(j-1)}\| \le C_{j-1}\|y - y'\|_{\ell_\infty^n},$$

for some constant C_{j-1}. Then we derive that

$$\begin{aligned}
&\|\eta^{(j)} - \eta'^{(j)}\| \\
&\le L\|A_j\eta^{(j-1)} + b^{(j)} - A'_j\eta'^{(j-1)} - b'^{(j)}\| \\
&\le L\|A_j(\eta^{(j-1)} - \eta'^{(j-1)})\| + L\|(A_j - A'_j)\eta'^{(j-1)}\| + L\|b^{(j)} - b'^{(j)}\| \\
&\le LW\|y\|_{\ell_\infty^n}\|\eta^{(j-1)} - \eta'^{(j-1)}\| + LW\|y - y'\|_{\ell_\infty^n}\|\eta'^{(j-1)}\| + L\|y - y'\|_{\ell_\infty^n} \\
&\le (LW\tilde{w}C_{j-1} + LW(d+2)L\tilde{w}[LW\tilde{w}]^{j-1} + L)\|y - y'\|_{\ell_\infty^n} \\
&= L(W\tilde{w}C_{j-1} + (d+2)[LW\tilde{w}]^j + 1)\|y - y'\|_{\ell_\infty^n} \\
&=: C_j\|y - y'\|_{\ell_\infty^n},
\end{aligned}$$

where we have used that $\|y\|_{\ell_\infty^{\tilde{n}}} \le w$, the bound (11), and the induction hypothesis. The relation between C_j and C_{j-1} can be written as

$$C_0 = L(d+1), \quad C_j = L(W\tilde{w}C_{j-1} + (d+2)[LW\tilde{w}]^j + 1), \quad j = 1, \dots, \ell.$$

Clearly,

$$C_1 = L((d+1)LW\tilde{w} + (d+2)LW\tilde{w} + 1) < (d+2)L(2LW\tilde{w} + 1),$$

and we obtain by induction that

$$C_\ell < (d+2)L\left(\ell[LW\tilde{w}]^\ell + \sum_{i=0}^{\ell}[LW\tilde{w}]^i\right).$$

If we use the fact $2 \le LW < LW\tilde{w}$, we derive the inequality:

$$C_\ell < (d+2)L(\ell+2)[LW\tilde{w}]^\ell.$$

Finally, we have:

$$\begin{aligned}
\|\Phi_\sigma^{W,\ell}(y) - \Phi_\sigma^{W,\ell}(y')\|_{C(\Omega)} &= \|\eta^{(\ell)} - \eta'^{(\ell)}\| \le C_\ell\|y - y'\|_{\ell_\infty^n} \\
&< (d+2)L(\ell+2)[LW\tilde{w}]^\ell\|y - y'\|_{\ell_\infty^n},
\end{aligned}$$

and therefore

$$\|\Phi_\sigma^{W,\ell}(y) - \Phi_\sigma^{W,\ell}(y')\|_X \le c_0\|\Phi_\sigma^{W,\ell}(y) - \Phi_\sigma^{W,\ell}(y')\|_{C(\Omega)} \le \tilde{C}\ell[LW\tilde{w}]^\ell\|y - y'\|_{\ell_\infty^n},$$

where $\tilde{C} = \tilde{C}(d, \sigma)$. Clearly, the Lipschitz constant $L_n := \tilde{C}\ell[LW\tilde{w}]^\ell$ is such that $2^{c_1\ell\log_2(W(w+1))} < L_n < 2^{c_2\ell\log_2(W(w+1))}$ for some $c_1, c_2 > 0$, and the proof is completed. □

Remark 1 Theorem 2 is a generalization of Theorems 3 and 5 from [17] to the case of any feed-forward NN. Note that its proof holds also in the case when every coordinate of $\bar{\sigma}$ (see (6)) is chosen to be a different Lipschitz function σ as long as $LW \geq 2$, where L is defined via (8).

4 Estimates from Below for Neural Network Approximation

In this section, we consider Banach spaces X such that $C([0,1]^d)$ is continuously embedded in X. Let us denote by

$$E(f, \Sigma(W, \ell, \sigma; w))_X := \inf_{y \in B^n_{\ell_\infty}(w)} \|f - \Phi^{W,\ell}_\sigma(y)\|_X,$$

the error of approximation in the norm $\|\cdot\|_X$ of the element $f \in \mathcal{K}$ by the set of outputs $\Sigma(W, \ell, \sigma; w)$ of a feed-forward NN with width W, depth ℓ, activation function σ, and a bound w on its parameters y, that is, $\|y\|_{\ell^n_\infty} \leq w$. We also denote by

$$E(\mathcal{K}, \Sigma(W, \ell, \sigma; w))_X := \sup_{f \in \mathcal{K}} E(f, \Sigma(W, \ell, \sigma; w))_X,$$

the error for the class $\mathcal{K} \subset X$. It follows from Theorem 2 that

$$E(\mathcal{K}, \Sigma(W, \ell, \sigma; w))_X \geq d_n^{\gamma_n}(\mathcal{K})_X, \quad \text{with} \quad \gamma_n = 2^{c\ell \log_2(W(w+1))} =: 2^{\varphi(n)}, \tag{12}$$

for some $c > 0$. Therefore, see (1),

$$n\varphi(n) = \begin{cases} cn\ell \log_2(W(w+1)), & n \asymp W^2\ell, \quad \ell > 1, \\ cn \log_2(n(w+1)), & n \asymp W, \quad \ell = 1, \end{cases}$$

and we can state the following corollary of (12) and Theorem 1.

Theorem 3 *Let $\Sigma(W, \ell, \sigma; w)$ be the set of outputs of an n parameter NN with width W, depth ℓ, Lipschitz activation function σ, and weights and biases bounded by w, where $LW \geq 2$. Then, the error of approximation $E(\mathcal{K}, \Sigma(W, \ell, \sigma; w))_X$ of a compact subset $\mathcal{K}$ of a Banach space X by $\Sigma(W, \ell, \sigma; w)$ satisfies the following estimates from below, provided we know the following information about the entropy numbers $\epsilon_n(\mathcal{K})_X$ of $\mathcal{K}$:*

- *if for $\alpha > 0$ and $\beta \in \mathbb{R}$ we have*

$$\epsilon_n(\mathcal{K})_X \gtrsim \frac{[\log_2 n]^\beta}{n^\alpha}, \ n \in \mathbb{N},$$

then

$$E(\mathcal{K}, \Sigma(W, \ell, \sigma; w))_X \gtrsim \begin{cases} \frac{1}{n^\alpha \ell^\alpha} \cdot \frac{[\log_2(n\ell \log_2(W(w+1)))]^\beta}{[\log_2(W(w+1))]^\alpha}, & n \asymp W^2\ell, \quad \ell > 1, \\ \frac{1}{n^\alpha} \cdot \frac{[\log_2(n \log_2(nw))]^\beta}{[\log_2(n(w+1))]^\alpha}, & n \asymp W, \quad \ell = 1. \end{cases}$$

- *if for $\alpha > 0$ we have*

$$\epsilon_n(\mathcal{K})_X \gtrsim [\log_2 n]^{-\alpha}, \; n \in \mathbb{N},$$

then

$$E(\mathcal{K}, \Sigma(W, \ell, \sigma; w))_X \gtrsim \begin{cases} [\log_2(n\ell \log_2(W(w+1)))]^{-\alpha}, & n \asymp W^2\ell, \; \ell > 1, \\ [\log_2(n \log_2(n(w+1)))]^{-\alpha}, & n \asymp W, \; \ell = 1. \end{cases}$$

Proof The proof follows directly from (12) and Theorem 1. □

Remark 2 Theorem 3 gives various estimates from below depending on the behavior of the bound $w = w(W, \ell)$ on the absolute values of the parameters of the NN. Here we state only one particular case. Under the conditions of Theorem 3 with $w = w(W, \ell) = \text{const}$, we have:

- if for $\alpha > 0$ and $\beta \in \mathbb{R}$ we have

$$\epsilon_n(\mathcal{K})_X \gtrsim \frac{[\log_2 n]^\beta}{n^\alpha}, \; n \in \mathbb{N},$$

then

$$E(\mathcal{K}, \Sigma(W, \ell, \sigma; w))_X \gtrsim \begin{cases} \frac{1}{n^\alpha \ell^\alpha} \cdot \frac{[\log_2(n\ell \log_2 W)]^\beta}{[\log_2 W]^\alpha}, & n \asymp W^2\ell, \quad \ell > 1, \\ \frac{1}{n^\alpha} \cdot [\log_2 n]^{\beta-\alpha}, & n \asymp W, \quad \ell = 1. \end{cases}$$

- if for $\alpha > 0$ we have

$$\epsilon_n(\mathcal{K})_X \gtrsim [\log_2 n]^{-\alpha}, \; n \in \mathbb{N},$$

then

$$E(\mathcal{K}, \Sigma(W, \ell, \sigma; w))_X \gtrsim \begin{cases} [\log_2(n\ell \log_2 W)]^{-\alpha}, & n \asymp W^2\ell, \quad \ell > 1, \\ [\log_2 n]^{-\alpha}, & n \asymp W, \quad \ell = 1. \end{cases}$$

Acknowledgments G.P. was supported by the NSF Grant DMS 2134077 and ONR Contract N00014-20-1-278.

References

1. Achour E-M., Foucault A., Gerchinovitz S., Malgouyres F. (2022). A general approximation lower bound in L_p norm, with applications to feed-forward neural networks. arXiv:2206.04360.
2. Bartlett P., Harvey N., Liaw C., Mehrabian A. (2019). Nearly-tight vc-dimension and pseudo dimension bounds for piecewise linear neural networks. *The Journal of Machine Learning Research*, 20(1), 2285–2301.
3. Carl B. (1981). Entropy numbers, s-numbers, and eigenvalue problems. *J. Funct. Anal.*, 41, 290–306.
4. Carl B., Stephani I. (1990). *Entropy, compactness and the approximation of operators.* Cambridge University Press.
5. Cobos F., Dominguez O., Kuhn T. (2018). Approximation and entropy numbers of embeddings between approximation spaces. *Constructive Approximation*, 47, 453–486.
6. Cobos F., Kuhn T. (2009). Approximation and entropy numbers in Besov spaces of generalized smoothness. *J. Approx. Theory*, 160, 56–70.
7. Cohen A., DeVore, R., Petrova, G., Wojtaszczyk, P. (2022). Optimal Stable Nonlinear Approximation, *J. FOCM*, 22, 607–647.
8. DeVore R., Hanin B., Petrova G. (2021). Neural Network Approximation. *Acta Numerica*, 30, 327–444.
9. DeVore, R., Howard, R., Micchelli, C. (1989). Optimal nonlinear approximation, *Manuscripta Mathematica*, 63(4), 469–478.
10. Edmunds D., Triebel H. (1996). *Function spaces, Entropy numbers and differential operators.* Cambridge Tracts in Mathematics 120.
11. Gao F. (2008). Entropy estimate for k-monotone functions via small ball probability of integrated Brownian motion, *Elect. Comm. in Probab.*, 13, 121–130.
12. Lorentz G., Golitschek M., Makovoz Y. (1996). *Constructive Approximation.* Springer Verlag.
13. Lu J., Shen Z., Yang H., Zhang S. (2020). Deep network approximation for smooth functions, *SIAM Journal on Mathematical Analysis* 53(5), 5465–5506.
14. Maiorov V. (1999). On best approximation by ridge functions. *J. Approx. Theory*, 99(1), 68–94.
15. Maiorov V., Meir R., Ratsaby J. (1999). On the approximation of functional classes equipped with a uniform measure using ridge functions, *J. Approx. Theory*, 99, 95–111.
16. Petrova G., Wojtaszczyk P. (2023). Lipschitz widths, *Constructive Approximation*, 7, 759–805.
17. Petrova G., Wojtaszczyk P. (2023). Limitations on approximation by deep and shallow neural networks, *JMLR*, 24(353), 1–38.
18. Shen Z., Yang H., Zhang S. (2022). Optimal approximation rate of ReLU networks in terms of width and depth, *Journal de Mathematiques Pures et Appliquees*, 9(157), 136–144.
19. Siegel J. (2023). Optimal Approximation Rates for Deep ReLU Neural Networks on Sobolev and Besov Spaces, *JMLR*, 24, 1–52.
20. Siegel J., Xu J. (2022). Sharp bounds on the approximation rates, metric entropy and n widths of shallow neural networks, *Journal of FOCM*. arXiv:2101.12365v9.
21. Yang Y., Barron A. (1999). Information-theoretic determination of minimax rates of convergence, *The Annals of Statistics*, 27(5), 1564–1599.
22. Yarotsky D. (2017). Error bounds for approximations with deep ReLU networks, *Neural networks*, 97, 103–114.
23. Yarotsky D. (2018). Optimal approximation of continuous functions by very deep ReLU networks, *Proceedings of the 31st Conference On Learning Theory*, PMLR, 75, 639–649.
24. Yarotsky D., Zhevnerchuk A. (2020). The phase diagram of approximation rates for deep neural networks., *Advances in neural information processing systems*, 33, 13005–13015.

Qualitative Neural Network Approximation over $\mathbb{R}$ and $\mathbb{C}$: Elementary Proofs for Analytic and Polynomial Activation

Josiah Park and Stephan Wojtowytsch

Abstract In this article, we prove approximation theorems in classes of deep and shallow neural networks with analytic activation functions by elementary arguments. We prove for both real and complex networks with non-polynomial activation that the closure of the class of neural networks coincides with the closure of the space of polynomials. The closure can further be characterized by the Stone-Weierstrass theorem (in the real case) and Mergelyan's theorem (in the complex case). In the real case, we further prove approximation results for networks with higher-dimensional harmonic activation and orthogonally projected linear maps.

We further show that fully connected and residual networks of large depth with polynomial activation functions can approximate any polynomial under certain width requirements. All proofs are entirely elementary.

1 Introduction

Neural networks are becoming increasingly popular tools in fields outside of the classical domain of data science. Chief among these applications are experiments in scientific computing, which often involve the solution of potentially high-dimensional partial differential equations. In many important problems—e.g., computations involving quantum systems with many particles—the output of the neural network may be complex valued. For such problems, the approximation power of complex neural networks with activation functions $\sigma : \mathbb{C} \to \mathbb{C}$ has recently come under investigation [7, 78, 80, 81], and sufficient conditions were established for σ under which (shallow, deep) neural networks can approximate *any* continuous function on a compact subset of $\mathbb{C}^d$ [82].

It is a well-known fact that the field of complex numbers and the algebra of complex-dif- ferentiable functions behave in surprising ways when compared to

J. Park · S. Wojtowytsch (✉)
Department of Mathematics, University of Pittsburgh, Pittsburgh, PA, USA
e-mail: s.woj@pitt.edu

S. Foucart, S. Wojtowytsch (eds.), *Explorations in the Mathematics of Data Science*, Applied and Numerical Harmonic Analysis,
https://doi.org/10.1007/978-3-031-66497-7_3

the analogous objects of real analysis. For instance, if $D \subseteq \mathbb{C}$ is a bounded open set, the space of functions $f : \overline{D} \to \mathbb{C}$ which are complex differentiable in D and continuous on $\overline{D}$ form a proper closed subspace of the space of continuous functions on $\overline{D}$. In particular there exists a continuous function $f : \overline{D} \to \mathbb{C}$ such that $\|f - p\|_{C^0(\overline{D})} \geq 1$ for all polynomials p. This contrasts drastically with the situation in real analysis, where polynomials are dense in the space of continuous functions on a compact set as established by the Stone-Weierstrass theorem.

A similar observation can be made in classes of neural networks. For a function $\sigma : \mathbb{C} \to \mathbb{C}$ and a set of parameters

$$\Theta := \{(a_k, w_k, b_k) \in \mathbb{C} \times \mathbb{C}^d \times \mathbb{C} : k = 1, \dots, n\},$$

we define:

$$f_\Theta(z) = \sum_{k=1}^{n} a_k \, \sigma\big(\langle w_k, z\rangle + b_k\big) \tag{1}$$

where $z = (z_1, \dots, z_d) \in \mathbb{C}^d$ is a complex vector and $\langle w, z\rangle = \sum_{i=1}^{d} \overline{w_i} \cdot z_i$. We denote the set of all such functions with a fixed number n of "neurons" by $\mathcal{M}_{\sigma,n}$ and $\mathcal{M}_\sigma = \bigcup_{n=1}^{\infty} \mathcal{M}_{\sigma,n}$. The following observations are immediate:

1. If σ is a polynomial of degree m in one complex variable, then f_Θ is a polynomial of degree m in d complex variables.
2. If σ can be represented by a convergent power series on $\mathbb{C}$, then f_Θ can be represented by a convergent power series in d complex variables.

In particular, if $D \subseteq \mathbb{C}$ is open and σ is an entire function, there exists a continuous function $f : \overline{D} \to \mathbb{C}$ which *cannot* be approximated by functions of the form (1). Depending on the application, this "rigidity" may be an asset or an obstacle. Encoding a priori information about the solution of a problem (often referred to as "domain knowledge") in the neural network architecture has proved invaluable in many tasks, for example, by designing convolutional neural networks to approximately respect translation invariance [63], using periodic activation functions in signal processing [1, 76, 85, 86], designing specialized neural networks for data in hyperbolic spaces [9, 31, 53, 66], or directly enforcing physical symmetries in computational chemistry [12, 89]. By analogy, if we can show that the solution to a problem in scientific computing is given by a holomorphic function or operator, it serves us well to encode this into the design of our neural network.

The question remains: When trying to approximate a holomorphic function $f : D \subseteq \mathbb{C}^d \to \mathbb{C}$, which activation functions can be used? We present an approximation theorem which treats complex and real shallow neural networks in a unified fashion.

Theorem 1 *Let* $\mathbb{K} \in \{\mathbb{R}, \mathbb{C}\}$ *and let*

$$\sigma : \mathbb{K} \to \mathbb{K}, \qquad \sigma(z) = \sum_{k=0}^{\infty} \alpha_k z^k$$

be an analytic function defined by a power series with infinite radius of convergence. Consider the class of shallow neural networks of finite width n with activation σ*:*

$$\mathcal{M}_{\sigma,n} = \left\{ \sum_{k=1}^{n} a_k\, \sigma\big(\langle w_k, z\rangle + b_k\big) : (a_k, w_k, b_k) \in \mathbb{K} \times \mathbb{K}^d \times \mathbb{K} \right\}.$$

and of arbitrary finite width $\mathcal{M}_\sigma = \bigcup_{n=1}^{\infty} \mathcal{M}_{\sigma,n}$*. Let* $D \subseteq \mathbb{K}^d$ *be an open bounded subset and denote by* $C^0(\overline{D})$ *the space of continuous functions from* $\overline{D}$ *to* $\mathbb{K}$*:*

1. *If* σ *is a polynomial of degree m in* $z \in \mathbb{K}$*, then* $\mathcal{M}_\sigma$ *is the space* $\mathcal{P}_m$ *of polynomials of degree m in* $z \in \mathbb{K}^d$*.*
2. *If* σ *is a not a polynomial, then the closures of* $\mathcal{M}_\sigma$ *and* $\mathcal{P} := \bigcup_{m=0}^{\infty} \mathcal{P}_m$ *in* $C^0(\overline{D})$ *coincide.*

If $\mathbb{K} = \mathbb{R}$, we can now recover the classical universal approximation theorem for networks with analytic activation by appealing the Stone-Weierstrass theorem [50, Section 15.7]:

Theorem (Stone-Weierstrass Theorem) Let $K \subset \mathbb{R}^d$ be compact and $\mathcal{P}$ the vector space of polynomials. Then, $\mathcal{P}$ is dense in $C^0(K)$. □

While the result is restrictive due to the strong assumptions on σ, our proof is entirely elementary and does not require advanced techniques beyond an introductory class in (real, complex) analysis.

If $\mathbb{K} = \mathbb{C}$, we emphasize that σ is analytic in z, not $(z, \bar{z})$ or (x, y), and that the elements of $\mathcal{P}_m$ are equally polynomials in the complex variable z. The closure of $\mathcal{P}$ depends on the topology of the set D. If $D = \{z \in \mathbb{C} : r < |z| < R\}$ is an annular domain and $k \in \mathbb{N}$, then the function $f(z) = z^{-k}$ is holomorphic on U, but cannot be approximated by polynomials. This follows from Cauchy's integral formula [30, Theorem II.3.2] as

$$\int_\gamma z^{k-1} f(z) \cdot \mathrm{d}z = 2\pi i, \qquad \int_\gamma z^{k-1}\, p(z) \cdot \mathrm{d}z = 0$$

for all polynomials p and all curves γ in D which loop around the origin. If there were polynomials which could approximate f uniformly, also the integrals would have to converge.

On the positive side, in 1951 Mergelyan [57] showed the following [70, Theorem 20.5].

Theorem (Mergelyan's Theorem) Let $K \subseteq \mathbb{C}$ be compact such that $\mathbb{C} \setminus K$ is connected and $f : K \to \mathbb{C}$ a continuous function which is holomorphic in the interior K° of K. Then, for every $\varepsilon > 0$, there exists a complex polynomial P such that $\sup_{z\in K} |f(z) - P(z)| < \varepsilon$. □

The class of "good" domains in particular includes (the closure of) any simply connected bounded open set with Lipschitz boundary. The situation in many complex variables is more complicated and not entirely understood. While multivariate Mergelyan-type theorems [14, 28, 35] and related results like the Oka-Weil theorem [61, 84] have been obtained, there are obstructions to proving the statement in full generality. Notably, Diederich and Fornaess [19] constructed an example of a pseudoconvex domain $D \subseteq \mathbb{C}^2$ with smooth boundary and a continuous function $f : \overline{D} \to \mathbb{C}$ such that f is holomorphic in D, but cannot be approximated uniformly by polynomials in $\overline{D}$. Thus, even for functions of multiple complex variables which can be shown to be holomorphic on a "good" domain, there may be deep obstructions to approximation by both polynomials and holomorphic neural networks. A recent survey of holomorphic approximation can be found, e.g., in [29].

While holomorphic functions can be thought of simultaneously as a generalization of differentiable functions and infinitely differentiable functions, the perhaps closest analogue in real analysis is the class of *harmonic* functions. In two dimensions, a correspondence between harmonic and holomorphic functions on $\mathbb{C} = \mathbb{R}^2$ can be constructed by taking the real part of a function, which is one-to-one up to an affine shift in the imaginary part. Like holomorphic functions, harmonic functions in any dimension form a closed proper subspace of $C^0(\overline{D})$ for open bounded $D \subseteq \mathbb{R}^d$. For a deeper understanding of complex-analytic rigidity, we study real neural networks with harmonic activation functions.

Since harmonic functions in one real dimension are just affine linear, the interesting case concerns activation functions σ of two or more real variables. Furthermore, we restrict the linear representation of data to be angle-preserving projections. Namely, if $\sigma : \mathbb{R}^k \to \mathbb{R}$ is harmonic, we consider a class of functions on $\mathbb{R}^d$ for $d \geq k$ given by $\mathcal{M}_\sigma^{\text{harm}} = \bigcup_{n=1}^\infty \mathcal{M}_{\sigma,n}^{\text{harm}}$ where

$$\mathcal{M}_{\sigma,n}^{\text{harm}} = \left\{ \sum_{i=1}^n a_i \sigma\big(\rho_i\, P_i x + b_i\big) : \begin{array}{c} a_i, \rho_i \in \mathbb{R},\ b_i \in \mathbb{R}^k, \\ P_i \in \mathbb{R}^{k\times d} \text{ s.t. } P_i P_i^T = I_{k\times k} \end{array} \right\}. \tag{2}$$

Geometrically, the linear maps P_i are orthogonal projections from $\mathbb{R}^d$ to $\mathbb{R}^k$ for all i. It is easy to see that $f \in \mathcal{M}_\sigma^{\text{harm}}$ is harmonic and thus that $\mathcal{M}_\sigma^{\text{harm}}$ is at most dense in the space of harmonic functions. We show the following.

Theorem 2 *Let $\sigma : \mathbb{R}^k \to \mathbb{R}$ be a harmonic function, $d \geq k$ and $D \subseteq \mathbb{R}^d$ open and bounded:*

1. *If σ is a harmonic polynomial of degree m, then $\mathcal{M}_\sigma^{harm}$ is the class of harmonic polynomials of degree m:*

$$\mathcal{HP}_m(\mathbb{R}^d) = \{p \in \mathcal{P}_m(\mathbb{R}^d) : \Delta p = 0\}.$$

2. *If σ is not a polynomial, then the closures of $\mathcal{M}_\sigma^{harm}$ and $\mathcal{HP} = \bigcup_{m=0}^\infty \mathcal{HP}_m$ in $C^0(\overline{D})$ coincide.*

So far, we only considered *shallow* neural networks with two layers (i.e., one hidden layer). While this case is historically well-studied, modern neural networks are "deep," i.e., they have many layers. To keep things simple, we focus on functions $f : \mathbb{K}^d \to \mathbb{K}$ which are represented by residual neural networks (ResNets), but comparable results can be obtained for classical fully connected feedforward networks and DenseNets (see Appendix "Appendix: Classical Multilayer Perceptra and DenseNets"). Residual neural networks form a function class which is comparable to feedforward networks in terms of approximation power, but with a parametrization that facilitates gradient-based optimization to find appropriate network parameters. The incremental nature of the change to the internal state in every layer alleviates the *vanishing and exploding gradient phenomenon*, which is the main motivation of the ResNets in [38]. For this reason, truly "deep" networks typically have a form of residual structure. Continuum limits for infinitely deep neural networks have been studied in [20, 52] and later in [11] as "neural ODEs."

A ResNet can be understood as follows:

- Let $d_0 \in \mathbb{N}$. For a given input $z \in \mathbb{K}^d$ and parameters $A^0 \in \mathbb{K}^{d_0 \times d}$, $b^0 \in \mathbb{K}^{d_0}$, designate $z^0 = A^0 z + b^0 \in \mathbb{K}^{d_0}$.
- For $\ell \in \{1, \ldots, L-1\}$ and parameters $A^\ell \in \mathbb{K}^{d_0 \times D_\ell}$, $W^\ell \in \mathbb{K}^{D_\ell \times d_0}$, $b^\ell \in \mathbb{K}^{D_\ell}$ for some $D_\ell \in \mathbb{N}$, set:

$$z^\ell = z^{\ell-1} + A^\ell \sigma(W^\ell z^{\ell-1} + b^\ell) \tag{3}$$

  where σ is applied coordinate-wise.
- For parameters $A^L \in \mathbb{K}^{1 \times d_0}$, set $z^L = A^L z^{L-1}$.

As previously, we collect the weights in a single vector:

$$\Theta = \left(A^0, b^0, A^1, b^1, W^1, \ldots, A^{L-1}, b^{L-1}, W^{L-1}, A^L\right)$$

and denote $f_\Theta(z) = z^L$, where we suppressed the dependence of z^L on the input z and weights Θ for the sake of compact notation. It should be noted that sometimes in ResNets, the architecture is specified further by taking $D_\ell \equiv d_0$ and A^ℓ as the unit matrix.

Deeper neural networks have multiple parameters that govern their complexity: the depth L and the vector of widths $d^0, D^1, \ldots, D^{L-1}$. The widths of the input layer d and the output layer $d_L = 1$ are given by the problem statement. The notation therefore becomes somewhat less compact compared to shallow networks. We denote the classes of ResNets with a fixed architecture by $\mathcal{R}_{L,\sigma}(d_0, D_1, \ldots, D_{L-1})$. We can now present our third main result.

Theorem 3 *Let* $\mathbb{K} \in \{\mathbb{R}, \mathbb{C}\}$ *and let*

$$\sigma : \mathbb{K} \to \mathbb{K}, \qquad \sigma(z) = \sum_{n=0}^{\infty} \alpha_n z^n$$

be an analytic function defined by a power series with infinite radius of convergence. Let $U \subseteq \mathbb{K}^d$ be an open bounded subset:

1. *Let $L \geq 2$ be fixed, $d_0 \geq d+1$ and $n := D_1 + \cdots + D_{L-1}$. Then $\mathcal{M}_{\sigma,n} \subseteq \mathcal{R}_{L,\sigma}(d_0, D_1, \ldots, D_{L-1})$. In particular, if σ is not a polynomial, then the closures of*

$$\mathcal{N}_1 := \bigcup_{D_1=1}^{\infty} \mathcal{R}_{L,\sigma}(d+1, D_1, \underbrace{1, \ldots, 1}_{L-2 \text{ times}}), \quad \mathcal{N}_2 := \bigcup_{L=1}^{\infty} \mathcal{R}_{L,\sigma}(d+1, \underbrace{m, m, \ldots, m}_{L-1 \text{ times}})$$

and the closure of the space of polynomials $\mathcal{P}$ in $C^0(\overline{D})$ coincide for all $m \geq 1$.

2. *If σ is not a linear function, then the closure of*

$$\widetilde{\mathcal{N}_2} := \bigcup_{L=1}^{\infty} \mathcal{R}_{L,\sigma}(d+2, m, \ldots, m)$$

and the closure of the space of polynomials $\mathcal{P}$ in $C^0(\overline{D})$ coincide for all $m \geq 4$.

Roughly speaking, we only need to have the width of one layer go to infinity, while all others remain bounded, or the depth go to infinity if $d_0 \geq d+1$. This is not surprising: In the first case, we approximate the target function with the first hidden layer and the remaining layers are the identity map, while in the second case, we "turn the shallow neural network on its side." This idea is classical and will be explained in greater detail below for the reader's convenience. The restriction that $d_0 \geq d+1$ is equally classical if the input dimension is $d \geq 2$, as there are obstructions to universal approximation for thinner networks [45].

The most interesting of the results given in Theorem 3 is the fact that polynomial activation functions are admissible if the neural network is slightly wider than before, and we vary the depth rather than the width. This is not entirely surprising since elements of $\mathcal{R}_{L,\sigma}(d_0, D_1, \ldots, D_{L-1})$ are polynomials of degree at most m^L if σ is a polynomial of degree m. As $L \to \infty$, the upper bound becomes less and less restrictive. Nevertheless, it is not immediately obvious that we can in fact approximate *all* polynomials. In the real case, this has been made rigorous by Kidger and Lyons [47].

The approximation properties of neural networks have been studied in great detail by many authors over the course of at least three decades in [3, 6, 10, 15–18, 22, 23, 26, 34, 40, 48, 48, 51, 54, 56, 58, 64, 65, 67, 68, 71, 73–75, 87, 88] to name only a few. The main goal of this article is to provide a self-contained and elementary introduction to qualitative versions of universal and qualified approximation theorems (approximation theorems in smaller function classes like holomorphic functions), not to improve upon the state of the art in specific classes of functions. As such, the proofs in the remainder of the article are elementary and require little or no knowledge beyond undergraduate real and complex analysis. Nevertheless, we maintain that several results, including the unified treatment of real and complex networks, as well as the treatment of polynomial activation functions, are at most folklore to the best of our knowledge. The main novel contribution of these notes, the proof of Theorem 2 in Sect. 3, is the only place where deeper results are used.

2 Proof of Theorem 1: Shallow Networks

In the real case, a version of this proof goes back to [59].

Proof **Step 1**. Assume that $\sigma(z) = \sum_{j=0}^{\infty} \alpha_j z^j$. In this step, we show that for any $m \in \mathbb{N}_0$, the function $z \mapsto \alpha_m z^m$ can be approximated by a shallow neural network $f_m \in \mathcal{M}_{\sigma,m+1}$ of the form (1) in one dimension. This holds trivially in the case $m = 0$, since $\sigma(0) = \alpha_0 z^0$ and $z \mapsto \sigma(0) = 1 \cdot \sigma(\langle 0, z\rangle + 0) \in \mathcal{M}_{\sigma,1}$. For $m \geq 1$, note that

$$\frac{d^m}{dh^m}\bigg|_{h=0} \sigma(hz) = \frac{d^m}{dh^m}\bigg|_{h=0} \sum_{j=0}^{\infty} \alpha_j (hz)^j = m!\,\alpha_m z^m$$

since power series and their derivatives converge locally uniformly, so summation and differentiation commute [50, Sections 6.4 and 15.2]. The m-th derivative of σ is the limit of iterated difference quotients:

$$\frac{d^m}{dh^m}\sigma(hz) = \lim_{\gamma \to 0} \frac{\sum_{\ell=0}^{m}(-1)^{\ell}\binom{m}{\ell}\sigma\big((h+\ell\gamma)z\big)}{\gamma^m}$$

thus

$$\lim_{\gamma \to 0} \frac{\sum_{\ell=0}^{m}(-1)^{\ell}\binom{m}{\ell}\,\sigma\big(\ell\gamma z\big)}{\gamma^m} = m!\,\alpha_m z^m,$$

where the limit holds uniformly in the set $|z| \leq R$ for any $R > 0$. While known to the experts, we have been unable to find a reference for this fact. A proof can be found in Appendix "Appendix: On the Uniform Convergence of Difference Quotients". Since

$$z \mapsto \frac{1}{\gamma^m} \sum_{\ell=0}^{m} (-1)^{\ell} \binom{m}{\ell} \sigma\big(\ell\gamma z\big) \in \mathcal{M}_{\sigma,m+1}$$

by definition, the result is proved.

Step 2. In this step, we show that if $a_M \neq 0$, then $z \mapsto z^m$ can be approximated uniformly on any compact set by elements of $\mathcal{M}_{\sigma}$ for any $m \leq M$. The result is trivial for $m = M$, since we can divide by a_M. For $m < M$, the result follows as previously by noting that

$$\begin{aligned} z^m &= \frac{m!}{M!}\frac{d^{M-m}}{dz^{M-m}}z^M = \frac{m!}{M!}\lim_{\beta\to 0}\frac{\sum_{n=0}^{M-m}(-1)^n\binom{M-m}{n}(z+n\beta)^M}{\beta^{M-m}} \\ &= \frac{m!}{a_M\cdot M!}\lim_{\beta\to 0}\lim_{\gamma\to 0}\frac{\sum_{n=0}^{M-m}\sum_{\ell=0}^{M}(-1)^{n+\ell}\binom{M}{\ell}\binom{M-m}{n}\sigma(\ell\gamma(z+n\beta))}{\beta^{M-m}\gamma^M}, \end{aligned}$$

and noting that the approximating functions are elements of $\mathcal{M}_{\sigma,2M-m+2}$. Uniform convergence on compact sets can be established analogously to Appendix "Appendix: On the Uniform Convergence of Difference Quotients".

Step 3. From now on, we will consider the general case $d \geq 1$. Assume that σ is a polynomial of degree m. Then clearly $\mathcal{M}_\sigma \subseteq \mathcal{P}_m$ since

$$\sum_{k=1}^{n} a_k\,\sigma(\langle w_k, z\rangle + b_k) = \sum_{k=1}^{n} a_k \sum_{j=0}^{m}\alpha_j\big(\langle w_k, z\rangle + b_k\big)^j$$

is a polynomial of degree at most m in d variables. On the other hand, we claim that any polynomial p of degree $\leq m$ can be approximated uniformly by elements of $\mathcal{M}_\sigma$. We already know that any *ridge polynomial*, i.e., every function of the form $P(x) = p(\langle x, v\rangle)$ with $p \in \mathcal{P}_m$x, is an element of $\overline{\mathcal{M}_\sigma}$ by applying step 2. Namely, if $\sum_{k=1}^{M} a_k\sigma(w_k z + b_k)$ approximates p, then $\sum_{k=1}^{M} a_k\sigma(\langle w_k v, z\rangle + b_k)$ approximates P. It remains to show that the class of ridge polynomials of degree at most m spans the space of all polynomials of degree $\leq m$. It suffices to approximate monomials, which form a basis of the space of polynomials in any dimension:

$$p(z) = z_1^{m_1}\ldots z_d^{m_d}.$$

We set $\overline{m} := m_1 + \cdots + m_d \leq m$ and note that

$$p(z) = \frac{1}{\overline{m}!}\frac{\partial^{\overline{m}}}{\partial_{h_1}^{m_1}\ldots\partial_{h_d}^{m_d}}\big(h_1 z_1 + \cdots + h_d z_d\big)^{\overline{m}}$$

at any point $(h_1, \ldots, h_d) \in \mathbb{K}^d$. In fact, as the expression on the right is polynomial in h, the derivative is achieved exactly by a difference quotient even for h not close to zero since difference quotients of polynomials annihilate all lower-order polynomials. It follows that all monomials on $\mathbb{K}^d$ can be expressed as a sum of ridge polynomials, and therefore can be approximated by elements of $\mathcal{M}_\sigma$. Since monomials can be approximated and $\mathcal{M}_\sigma$ is a linear class, we find that $\mathcal{P}_m \subseteq \overline{\mathcal{M}_\sigma}$.

Recall that $\mathcal{M}_\sigma$ is a vector space. Since $\mathcal{M}_\sigma \subseteq \mathcal{P}_m \subseteq \overline{\mathcal{M}_\sigma}$, we conclude that $\mathcal{M}_\sigma$ is finite-dimensional and thus $\mathcal{M}_\sigma = \overline{\mathcal{M}_\sigma}$. Hence $\mathcal{P}_m = \mathcal{M}_\sigma$.

Step 4. If $d \geq 1$ and σ is not a polynomial, then for every $m \in \mathbb{N}_0$, there exists $M \geq m$ such that $a_M \neq 0$. As in step 3, we can show that every polynomial of degree at most m can be approximated uniformly by elements $f_m \in \mathcal{M}_\sigma$. Taking the union over $m \in \mathbb{N}$, we find that any polynomial can be approximated uniformly by $f \in \mathcal{M}_\sigma$. Hence $\overline{\bigcup_{m=1}^{\infty} \mathcal{P}_m} \subseteq \overline{\overline{\mathcal{M}_\sigma}} = \overline{\mathcal{M}_\sigma}$ since the closure of a closed set is the set itself.

On the other hand, let $f \in \overline{\mathcal{M}_\sigma}$. Then, for every $\varepsilon > 0$, there exist $n \in \mathbb{N}$ and $f_n \in \mathcal{M}_{\sigma,n}$ such that $\|f_n - f\|_{C^0(\overline{D})} < \varepsilon/2$. Since

$$f_n(z) = \sum_{k=1}^{n} a_k\, \sigma(\langle w_k, z\rangle + b_k) = \lim_{m\to\infty} \sum_{j=1}^{m}\sum_{i=1}^{n} \alpha_j a_k \left(\langle w_k, z\rangle + b_k\right)^m$$

uniformly in the set $|z| \leq R$ for any given $R > 0$, we can truncate the series for f_n at an index m such that

$$\|f_{n,m} - f\|_{C^0(\overline{D})} \leq \|f_n - f\|_{C^0(\overline{D})} + \|f_n - f_{n,m}\|_{C^0(\overline{D})} < \varepsilon.$$

As this can be done for any $\varepsilon > 0$, we have $f \in \overline{\bigcup_{m=0}^{\infty} \mathcal{P}_m}$. □

Since power series converge to their limit in C^k for any $k \in \mathbb{N}$,[1] we find that $\overline{\mathcal{M}_\sigma} \subseteq \overline{\mathcal{P}}$ also if the closure is taken with respect to the C^k-topology. Conversely, since the m-th difference quotient of a C^{m+k}-function also converges in C^k, and since $\sigma \in C^\infty$, we can use the proof of Theorem 1 also to see that $\overline{P} \subseteq \overline{\mathcal{M}_\sigma}$ in the C^k-topology.

Corollary 1 *Let* $\mathbb{K} \in \{\mathbb{R}, \mathbb{C}\}$ *and let*

$$\sigma : \mathbb{K} \to \mathbb{K}, \qquad \sigma(z) = \sum_{n=0}^{\infty} \alpha_n z^n$$

be an analytic function defined by a power series with infinite radius of convergence. Consider the class $\mathcal{M}_\sigma$ *of shallow neural networks of arbitrary finite width n and activation* σ *as before. Let* $D \subseteq \mathbb{K}^d$ *be an open bounded subset,* $k \geq 1$*, and denote by* $C^k(\overline{D})$ *the space of k times differentiable functions from D to* $\mathbb{K}$ *such that the derivatives of all orders extend continuously to the closure* $\overline{D}$ *equipped with the norm:*

$$\|f\|_{C^k(\overline{D})} = \max_{m} \max_{z\in\overline{D}} \left| \frac{\partial^{\overline{m}}}{\partial_{z_1}^{m_1}\dots\partial_{z_d}^{m_d}} f(z)\right|$$

where the maximum is taken over $m \in \mathbb{N}_0^d$ *such that* $\overline{m} = \sum_{i=1}^{d} m_i \leq k$*:*

1. *If* σ *is a polynomial of degree m, then* $\mathcal{M}_\sigma$ *is the space* $\mathcal{P}_m$ *of polynomials of degree m.*

[1] Since their derivatives, which are also power series, converge uniformly.

2. *If σ is a not a polynomial, then the closures of $\mathcal{M}_\sigma$ and $\mathcal{P} := \bigcup_{m=0}^{\infty} \mathcal{P}_m$ in $C^k(\overline{D})$ coincide.*

In Corollary 1, we replaced the C^0-topology by the stronger C^k-topology. Similarly, we could pass from a stronger topology like C^0 or C^k to a weaker one, like L^p or $W^{k,p}$, and conclude that the closures of $\mathcal{P}$ and $\mathcal{M}_\sigma$ coincide.

There exist a finite number of neurons n that a shallow neural network of the form (1) requires to approximate the function $z \mapsto z^m$ to *arbitrary* accuracy. Reaching higher precision requires increasing the magnitude of weights, but not the number of neurons. In particular, n depends only on m and on which coefficients of σ in a power series expansion are nonzero. It seems advantageous to choose activation functions in which all coefficients are nonzero (such as exp), or at least such that there are no long gaps in the set of nonzero coefficients (such as sin, cos, sinh, cosh). If *all* coefficients are nonzero, then the proof of Theorem 1 also illustrates that the bias term b_i in $\langle w_i, z\rangle + b_i$ is not needed to prove approximation results. This explains the density of Fourier series in the space of continuous functions, which are formally neural networks with a single hidden layer and the activation function $\sigma(z) = \exp(2\pi i z)$. In a Fourier series, all biases are set to zero.

The conditions on σ in Theorem 1 can be weakened somewhat. To approximate z^m by elements of $\mathcal{M}_\sigma$, we “zoomed in” suitably at the origin to utilize the power series expansion. In particular, if f can be represented by a convergent non-polynomial power series in a neighborhood of the origin on the real line, then every polynomial can be approximated arbitrarily well by elements of $\mathcal{M}_\sigma$. This applies in particular to real analytic activation functions like $\tanh(x) = \frac{e^x - e^{-x}}{e^x + e^{-x}}$ or the sigmoid function $\sigma(x) = \frac{1}{1+e^{-x}}$, whose power series representation does not converge globally. Since polynomials are dense in the space of continuous functions due to the Stone-Weierstrass theorem [50, Section 15.7], it follows that $\overline{\mathcal{M}_\sigma} = \overline{\mathcal{P}} = C^0(\overline{D})$, if the closure is taken in the uniform topology.

In the complex plane, the assumption that the radius of convergence of σ is ∞ is used to prove that elements of $\mathcal{M}_\sigma$ can be approximated by polynomials. It is implied by the assumption that σ is holomorphic on the entire plane $\mathbb{C}$, since the radius of convergence can be characterized as the distance to the closest singularity (which, in this case, is infinity) [70, Theorem 16.2]. If $\mathbb{K} = \mathbb{C}$ and σ is a rational function, then the closure of $\mathcal{M}_\sigma$ may be strictly larger than that of the space of polynomials (e.g., if U is an annular domain and the weights of the network are chosen such that a pole of σ is inside the hole in U). In this situation “neural networks” $f : \mathbb{C} \to \mathbb{C}$ with activation $\sigma(z) = z^{-1}$ can approximate any meromorphic function $f : \mathbb{C} \to \overline{\mathbb{C}}$ in $C^0(K)$, if f does not have a singularity in K. The key observation is that any type of pole $z \mapsto z^{-m}$ can be generated by a superposition of derivatives of σ. Difference quotients approximate these derivatives uniformly away from the singularity.

Of course, if σ is a rational function (and not a polynomial), then there exists $z^* \in \mathbb{C}$ such that $\lim_{z\to z^*} |f(z)| = \infty$. When approximating a target function which is bounded on $\overline{D}$, the weights should be chosen such that these poles lie outside of $\overline{D}$. However, especially in the initial phase of training, the infinite gradients of f

may lead to greater numerical instability, and there is no guarantee that the domain U is captured accurately by finite amounts of data. The greater expressivity therefore comes with a not-so-hidden cost.

In the context of machine learning, obstructions to polynomial approximation (in the complex case) would not be immediately visible: If $\{z_1, \ldots, z_N\} \subset \mathbb{C}$ and $\{y_1, \ldots, y_N\} \subset \mathbb{C}$ are finite data sets, then there always exists a unique polynomial P_N of degree N such that $P_N(z_i) = y_i$ (assuming that all z_i are different). In Lagrange representation, we can write:

$$P_N(z) = \sum_{i=1}^{N} y_i \prod_{j \neq i} \frac{z - z_j}{z_i - z_j}.$$

An exact interpolant can also be found in $\mathcal{M}_{\sigma,n}$ for sufficiently large n, which may scale linearly with N (depending on which power series coefficients of σ vanish). Even for functions which are holomorphic on the data domain D, it is therefore imperative to understand whether f can be approximated by polynomials, as this cannot be determined from a finite data set. At most we may notice that P_N does not approximate the function we expected at previously unseen data points (the test set).

As another consequence of Cauchy's integral formula, we recall the following Liouville theorem.

Theorem (Liouville's Theorem) If

$$\limsup_{r\to\infty} \frac{\max\{|\sigma(z)| : z \in B_r(0)\}}{r^\alpha} < \infty$$

for some $\alpha > 0$, then σ is a polynomial of degree at most $\lfloor \alpha \rfloor$ since the $\lfloor \alpha \rfloor + 1$-th derivative of σ vanishes. □

This version of Liouville's Theorem can be proved by appealing to Cauchy's integral formula in a fashion virtually identical to the classical case $\alpha = 0$, which states that every bounded holomorphic function is constant. Surprisingly, it is skipped in many standard texts on complex analysis. For a reference, see, e.g., [41, Exercise 7.11].

Consequently, any holomorphic activation function which generates universal approximators (in the class of holomorphic functions) fails to be Lipschitz continuous. This lack of quantitative global continuity has undesirable consequences from the perspective of statistical learning and gradient-based optimization.

In analogy to [82], rather than the closure in the $C^0(K)$-topology for a fixed compact set K, we can consider the closure of $\mathcal{M}_\sigma$ in the topology of *locally uniform convergence* (or compact-open topology) on $\mathbb{K}^d$, i.e., the set of functions $f : \mathbb{K}^d \to \mathbb{K}$ such that there exists a sequence $f_n \in \mathcal{M}_\sigma$ such that $f_n \to f$ uniformly on every compact set $K \subset \mathbb{K}^d$. The relationship between the closures is more subtle in our case and discussed in Appendix "Appendix: Locally Uniform Approximation and Approximation on Compact Sets".

3 Proof of Theorem 2: Harmonic Shallow Networks

3.1 Approximating Harmonic Functions

The application of neural networks in applications not traditionally associated with data science, such as scientific computing, has received intense attention in recent years—see, e.g., [37, 69], and many others. Most commonly, such applications involve finding an approximate solution of a partial differential equation (PDE) in a class of neural networks. In the spirit of concepts like input-convex neural networks [2, 55], we sketch here how compliance with a PDE can be encoded as structural principle in the design of the neural network class.

The possibly most common partial differential equation is the Poisson equation:

$$\begin{cases} -\Delta u = f \text{ in } D \\ \quad u = g \text{ on } \partial D \end{cases} \quad \text{or} \quad \begin{cases} -\Delta u = f \text{ in } D \\ \partial_\nu u = g \text{ on } \partial D \end{cases}$$

where $D \subseteq \mathbb{R}^d$ is a bounded domain and $\partial_\nu u$ denotes the derivative of u in the direction orthogonal to ∂D. The Poisson equation links stochastic calculus to partial differential equations [62, Chapter 9] and characterizes minimizers of the Dirichlet energy:

$$E(u) = \int_D \frac{1}{2} |\nabla u|^2 - f u \, \mathrm{d}x,$$

which serves as a simplified model for an elastic membrane with clamped boundary condition g and applied force f on the membrane. The eigenvalues of the Laplacian operator Δ govern both heat diffusion and vibrational frequencies of a domain [46].

It is not difficult to find a solution of the equation $-\Delta u = f$, which can be obtained by convolving with the Newton potential:

$$u_{part}(x) = \int_{\mathbb{R}^d} f(y) \, \Phi_d(x - y) \, \mathrm{d}y, \qquad \text{where} \quad \Phi_d(y) = c_d \begin{cases} \log(|y|) & \text{if } d = 2 \\ |y|^{2-d} & \text{if } d > 2 \end{cases}$$

and c_d is a dimension-dependent constant [27, Section 2.2.1] or [33, Chapter 4]. If f is only known at a finite number of points $x_1, \ldots, x_n$, then u_{part} is given by a finite sum of (potentially smoothed out) Newton potentials. The general solution to the equation $-\Delta u = f$ is given by $u_{part} + h$, where h is a harmonic function (i.e., $\Delta h = 0$) to be determined by the boundary condition g. An *input-harmonic* neural network is designed to output a function h such that $-\Delta h \equiv 0$.

For harmonic functions $h : \mathbb{R}^2 \to \mathbb{R}$, an easy strategy is to identify $\mathbb{R}^2$ with $\mathbb{C}$, compute a holomorphic neural network function with suitable point values, and then take the real part for the harmonic function. For deeper networks, this strategy builds on the fact that compositions of holomorphic functions are holomorphic. In dimensions $d > 2$, we only discuss *shallow* input-harmonic neural networks.

3.2 A Primer on Harmonic Polynomials

Before we come to the main proof, we review some properties of harmonic functions and harmonic polynomials.

Definition 1

1. A C^2-function $f : \mathbb{R}^d \to \mathbb{R}$ is called *harmonic* if its Laplacian vanishes: $\Delta f = (\partial_1^2 + \cdots + \partial_d^2) f = 0$.
2. A function $f : \mathbb{R}^d \to \mathbb{R}$ is called *homogeneous of degree* j if $f(\lambda x) = \lambda^j f(x)$ for all $\lambda \in \mathbb{R}$ and $x \in \mathbb{R}^d$.

It is an easy exercise to see that if $p : \mathbb{R}^d \to \mathbb{R}$ is harmonic, then so is $x \mapsto p(Ox)$ for any orthogonal matrix O; see also (1). For polynomials, being homogeneous means that there are no lower-order terms, and the degree of the polynomial and degree of homogeneity coincide. Every (harmonic) polynomial can be decomposed uniquely as $p = \sum_{j=0}^{\deg p} p_j$, where p_j is a homogeneous (harmonic) polynomial of degree j. The fact that the j-homogeneous part p_j of a harmonic polynomial p is harmonic follows from the fact that Δp_j is $j-2$-homogeneous and that the terms of different homogeneity must vanish separately. We denote:

$$\mathcal{HP}_j^d = \{p : \mathbb{R}^d \to \mathbb{R} : p \text{ harmonic homogeneous polynomial of degree } j\}.$$

Before we come to the proof of Theorem 2, we require the following auxiliary result about harmonic homogeneous polynomials and coordinate rotations.

Lemma 1 *Let $p \in \mathcal{HP}_j^d$ for some d, n and denote*

$$V_p = \text{span}\{p(Ox) : O \in O(d)\} \subseteq \mathcal{HP}_j^d.$$

If $p \not\equiv 0$, then $V_p = \mathcal{HP}_j^d$.

In particular, Lemma 1 illustrates that there is no substantial difference between harmonic polynomials in many variables and few variables, since, e.g., $x_1^2 - x_2^2$ and its rotations can be used to generate *any* homogeneous harmonic polynomial of degree 2 on a high-dimensional space.

We prove Lemma 1 in Appendix "Appendix: Proof of Lemma 1".

3.3 Approximation by Harmonic Neural Networks

Let $P_i \in O(d,k) = \{P \in \mathbb{R}^{k\times d} : PP^T = I_{k\times k}\}$. If $\sigma : \mathbb{R}^k \to \mathbb{R}$ is harmonic, then any $f \in \mathcal{M}_\sigma^{\text{harm}}$ as defined in (2) is harmonic for any $d \geq k$ since the Laplace operator Δ is linear and

$$
\begin{aligned}
\Delta a_i\sigma\big(\rho_i\, P_i x + b_i\big) &= a_i \sum_{j=1}^{d} \partial_{x_j}\partial_{x_j}\sigma\big(\rho_i\, P_i x + b_i\big) \\
&= a_i\rho_i \sum_{j=1}^{d} \partial_{x_j}\left(\sum_{\ell=1}^{k} P_{i,\ell j}(\partial_{y_\ell}\sigma)\big(\rho_i\, P_i x + b_i\big)\right) \\
&= a_i\rho_i^2 \sum_{j=1}^{d}\sum_{\ell,m=1}^{k} P_{i,\ell j}P_{i,mj}(\partial_{y_\ell}\partial_{y_m}\sigma)\big(\rho_i\, P_i x + b_i\big) \\
&= a_i\rho_i^2 \sum_{\ell,m=1}^{k}\left(\sum_{j=1}^{d} P_{i,\ell j}P_{i,mj}\right)(\partial_{y_\ell}\partial_{y_m}\sigma)\big(\rho_i\, P_i x + b_i\big) \\
&= a_i\rho_i^2 \sum_{\ell,m=1}^{k} (P_i P_i^T)_{\ell m}(\partial_{y_\ell}\partial_{y_m}\sigma)\big(\rho_i\, P_i x + b_i\big) \\
&= a_i\rho_i^2 \sum_{\ell,m=1}^{k} \delta_{\ell m}(\partial_{y_\ell}\partial_{y_m}\sigma)\big(\rho_i\, P_i x + b_i\big) \\
&= a_i\rho_i^2(\Delta\sigma)\big(\rho_i P_i x + b_i\big) = 0.
\end{aligned}
\tag{1}
$$

This implies that elements of $\mathcal{M}_\sigma^{\text{harm}}$ cannot approximate any function which is not harmonic due to [33, Theorem 2.8] in close analogy to the observations for holomorphic functions.

Proof (Proof of Theorem 2) **Step 1**. Let $q \in \mathcal{HP}_m^d$ and $\sigma : \mathbb{R}^k \to \mathbb{R}$ harmonic such that σ is not a polynomial of degree at most $m-1$. In this step, we show that $q \in \overline{\mathcal{M}_\sigma^{\text{harm}}}$.

Note that for any $t \in \mathbb{R}$, the function $x \in \mathbb{R}^k \mapsto \sigma(tx)$ is harmonic by the same argument as (1). Since sums of harmonic functions are harmonic, this means that

$$x \mapsto \frac{\sigma(tx) - \sigma(0)}{t}$$

is harmonic for every $t \neq 0$, and since the uniform limit of harmonic functions is harmonic, also $x \mapsto \frac{d}{dt}\big|_{t=0}\sigma(tx)$ is harmonic. The same is true for iterated difference quotients, and therefore higher-order derivatives. Since σ is harmonic, it is analytic and can be written as

$$\sigma(x) = \sum_{j=0}^{\infty} p_j(x)$$

in a neighborhood of the origin [27, Section 2.2, Theorem 10], where p_j is a homogeneous polynomial of degree j. As for holomorphic functions, we find that

$$\left.\frac{d^\ell}{dt^\ell}\right|_{t=0} \sigma(tx) = \sum_{j=0}^{\infty} \left.\frac{d^\ell}{dt^\ell}\right|_{t=0} p_j(tx) = \sum_{j=0}^{\infty} \left.\frac{d^\ell}{dt^\ell}\right|_{t=0} t^j\, p_j(x) = \ell! \cdot p_\ell(x).$$

In particular, we see that p_j is harmonic for all $j \in \mathbb{N}$. Since σ is not a polynomial of degree at most $m-1$, there exists some $j \geq m$ such that $p_j \not\equiv 0$. By the same analysis as in Theorem 1 and the definition of $\mathcal{M}_\sigma^{\mathrm{harm}}$, we find that

$$x \mapsto p_j(Px) \in \overline{\mathcal{M}_\sigma^{\mathrm{harm}}}$$

for any $P \in O(d,k)$. Thus, fixing $P : \mathbb{R}^d \to \mathbb{R}^k$ as $x \mapsto (x_1, \ldots, x_k)$ and setting $\tilde{p}_j(x) = p_j(Px)$, we find that $x \mapsto \tilde{p}_j(Ox) \in \overline{\mathcal{M}_\sigma^{\mathrm{harm}}}$ for all $O \in O(d)$.

If f and g can be approximated to arbitrary accuracy by elements of $\mathcal{M}_\sigma^{\mathrm{harm}}$, the same is true for $f+g$ and λf with $\lambda \in \mathbb{R}$, so $\overline{\mathcal{M}_\sigma^{\mathrm{harm}}}$ is a linear space. In the terminology of Lemma 1, this means that $V_{\tilde{p}_j} \subseteq \overline{\mathcal{M}_\sigma^{\mathrm{harm}}}$ and thus $\mathcal{HP}_j^d \subseteq \overline{\mathcal{M}_\sigma^{\mathrm{harm}}}$. If $j = m$, then $q \in \overline{\mathcal{M}_\sigma^{\mathrm{harm}}}$ and the proof is concluded.

If $j > m \geq 0$, we observe that there exists $1 \leq \ell \leq k$ such that $\partial_{x^\ell} p(x)$ is not the zero polynomial, since a nontrivial polynomial has a nontrivial gradient. By Königsberger [49, Section 2.2], we have $\Delta \partial_{x^\ell} p = \partial_{x^\ell} \Delta p = 0$. In particular $\partial_{x^\ell} p(x)$ is a homogeneous harmonic polynomial of degree $j-1$ and

$$\partial_{x^\ell} p(x) = \lim_{h \to 0} \frac{p(x + he_\ell) - p(x)}{h} \in \overline{\mathcal{M}_\sigma^{\mathrm{harm}}}$$

by almost the same construction as before. The main difference is that earlier we took the derivative in the scaling factor ρ, where now we take the derivative in the bias b. After $j-m$ steps, we find $\tilde{p} \in \mathcal{HP}_m^d \cap \overline{\mathcal{M}_\sigma^{\mathrm{harm}}}$, and the proof can be concluded as before.

Step 2. We have seen that, if σ is not a harmonic polynomial of degree at most $m-1$, then every harmonic homogeneous polynomial of degree at most m can be approximated by elements of $\mathcal{M}_\sigma$ in the uniform topology on $\overline{D}$. By linearity, this is also true for every harmonic polynomial of degree at most m. We now distinguish two cases:

1. σ is a polynomial of degree m. Then $\mathcal{M}_\sigma^{\mathrm{harm}}$ is contained in the space of harmonic polynomials of degree at most m, so $\mathcal{M}_\sigma^{\mathrm{harm}} = \overline{\mathcal{M}_\sigma^{\mathrm{harm}}} = \oplus_{j=0}^{m} \mathcal{HP}_j^d$ is the space of harmonic polynomials of degree at most m.

2. σ is not a polynomial. Then $\oplus_{j=0}^{\infty}\mathcal{HP}_j^d \subseteq \overline{\mathcal{M}_\sigma^{\text{harm}}}$, where the $\oplus$ denotes the direct sum in which at most finitely many terms are nonzero. By density $\overline{\oplus_{j=0}^{\infty}\mathcal{HP}_j^d} \subseteq \overline{\mathcal{M}_\sigma^{\text{harm}}}$.

Step 3. We now show that $\overline{\mathcal{M}_\sigma^{\text{harm}}} \subseteq \overline{\mathcal{HP}}$. Since σ is a harmonic function on the entire space $\mathbb{R}^k$, it can be represented by a globally convergent power series in many variables:

$$\sigma(x) = \sum_{j=0}^{\infty} p_j(x)$$

due to [27, Section 2.2.e], where p_j is a harmonic homogeneous polynomial of degree j, much like in Step 1. In particular, for any $P \in O(d,k)$ and $\rho, b \in \mathbb{R}$, we see that $x \mapsto \sigma(\rho P x + b)$ is an analytic function. Since power series converge locally uniformly, we can truncate the series at a finite index depending on ρ, b, and $\overline{D}$ to obtain a harmonic polynomial in d variables which uniformly approximates $\sigma(\rho P x + b)$. By linearity, any element in $\mathcal{M}_\sigma^{\text{harm}}$ can be approximated uniformly by harmonic polynomials. By density and selection of a diagonal sequence, this extends to any element in the closure $\overline{\mathcal{M}_\sigma^{\text{harm}}}$. □

Remark 1 The restrictive class of linear maps is crucial in this result, since the harmonicity is only preserved due to the orthogonality constraint. If $\widetilde{M}_{\sigma,n}$ is the class of all maps

$$f(x) = \sum_{i=1}^{n} a_i\, \sigma\big(W_i x + b_i\big)$$

for general linear maps $W_i : \mathbb{R}^n \to \mathbb{R}^k$ and $\widetilde{\mathcal{M}}_\sigma = \bigcup_{n=1}^{\infty} \widetilde{\mathcal{M}}_{\sigma,n}$, then:

1. If σ is a polynomial of degree at most m, then $\widetilde{\mathcal{M}}_\sigma = \mathcal{P}_m$ is the space of polynomials of degree m.
2. If σ is not a polynomial, then $\widetilde{\mathcal{M}}_\sigma$ is dense in the space of continuous functions.

This holds even for harmonic activation functions. An easy way to see this is to fix a direction $v \in \mathbb{R}^k$ such that $t \mapsto \sigma(tv)$ is not a polynomial and consider the uniform approximation theorem in one variable, e.g., Theorem 1 for maps W_i which project to the line tv. To see that such a direction exists, observe the following:

$$\sigma(x) = \sum_{j=0}^{\infty} p_j(x)$$

where p_j is a homogeneous harmonic polynomial of degree j. For all j, the following dichotomy holds: Either $p_j \equiv 0$ or the set

$$N_j = \{v \in \mathbb{R}^k : p_j(v) = 0\}$$

has Lebesgue measure 0.[2] In particular, the set

$$Y = \bigcap_{N_j \neq \mathbb{R}^k} \left(\mathbb{R}^k \setminus N_j\right)$$

has full measure. For any $v \in Y$, $\sigma_v(t) = \sigma(tv)$ is an analytic function which is a polynomial of the same degree as σ if σ is a polynomial, and not a polynomial if σ is not a polynomial.

Analogously, the restriction to complex linear maps plays a major role in the context of [82].

Remark 2 Let us briefly consider the question of finding an element of $\mathcal{M}_{\sigma,n}^{harm}$ computationally. As noted, the constraint that the linear maps $P_i : \mathbb{R}^d \to \mathbb{R}^k$ of an input-harmonic neural network satisfy $P_i P_i^T = I_{k\times k}$ plays a key role in the analysis.

If P_i is updated, e.g., by a gradient-descent type algorithm, it is necessary after every step to project P_i back onto the set of matrices with orthonormal rows. This closest point projection has been studied in detail when distance is measured by the Frobenius norm as the "orthogonal Procrustes problem." The unique solution is given by

$$\tilde{P}_i = P_i\left(P_i P_i^T\right)^{-1/2}$$

as long as P_i is full rank—see, e.g., [72], where an alternative approach using the singular value decomposition is also developed. If $k = 2$ or more generally $k \ll d$, the projection $\tilde{P}_i$ can be computed efficiently.

Projected gradient algorithms have been studied extensively in the context of constrained optimization for machine learning [13, 36]. We do not pursue this topic further here in order to return to the approximation-theoretic core subject of this article.

Remark 3 As in the complex case, the question whether all harmonic functions in D which extend continuously to $\overline{D}$ can be approximated by harmonic polynomials depends on the topology of D. For a general compact set $K \subseteq \mathbb{R}^d$, the following are equivalent [32, Theorem 1.3]:

1. Every continuous function on K which is harmonic in the interior K° of K can be approximated uniformly by harmonic polynomials.
2. $\mathbb{R}^d \setminus K$ and $\mathbb{R}^d \setminus K^\circ$ are *thin* at the same points of K.

For a review of thin sets, see, e.g., [32, Section 0] or [39, Section 5.6]. In particular, if D is a bounded open set with C^2-boundary and $d \geq 3$, every harmonic function

[2] If $n = 1$, this is the fundamental theorem of calculus. In the general case, this can be proved by Fubini's theorem and induction on n. See, e.g., [44] for an alternative elegant proof.

on D which extends continuously to $\overline{D}$ can be approximated uniformly by harmonic polynomials.

4 Proof of Theorem 3: Deep Residual Networks

In this section, we establish the approximation properties of deep residual neural networks. The proof of the first claim in Theorem 3 is a classical technique of "turning a neural network on its side."

Proof **First Claim** Let

$$f_\Theta \in \mathcal{M}_{\sigma,n}, \qquad f_\Theta(x) = \sum_{i=1}^n a_i\,\sigma(\langle w_i, x\rangle + b_i) = \sum_{i=1}^n a_i\,\sigma\left(\sum_{j=1}^d \overline{w}_{i,j}x_j + b_i\right)$$

be a shallow neural network. We represent f_Θ by a residual neural network as a concatenation of linear and residual operations:

$$x \xrightarrow{lin.} \begin{pmatrix} x \\ 0 \end{pmatrix} \xrightarrow{res.} \begin{pmatrix} x + 0\cdot\sigma(0) \\ 0 + \sum_{i=1}^{D_1} a_i\,\sigma(\langle w_i, x\rangle + b_i) \end{pmatrix} = \begin{pmatrix} x \\ \sum_{i=1}^{D_1} a_i\,\sigma(\langle w_i, x\rangle + b_i) \end{pmatrix}$$

$$\xrightarrow{res.} \begin{pmatrix} x \\ \sum_{i=1}^{D_1+D_2} a_i\,\sigma(\langle w_i, x\rangle + b_i) \end{pmatrix} \xrightarrow{res.} \dots \xrightarrow{res.} \begin{pmatrix} x \\ \sum_{i=1}^{n} a_i\,\sigma(\langle w_i, x\rangle + b_i) \end{pmatrix}$$

$$\xrightarrow{linear} \sum_{i=1}^{D_1} a_i\,\sigma(\langle w_i, x\rangle + b_i)$$

with weights

$$A^0 = \begin{pmatrix} 1 & 0 & \dots & 0 \\ 0 & 1 & & \\ \vdots & & \ddots & \vdots \\ 0 & \dots & \dots & 1 \\ 0 & \dots & \dots & 0 \end{pmatrix}, \qquad b^0 = \begin{pmatrix} 0 \\ \vdots \\ 0 \end{pmatrix}, \qquad A^L = (0 \dots 0\ 1)$$

and, for $1 \le \ell \le L-1$,

$$A^\ell = \begin{pmatrix} 0 & \dots & 0 \\ \vdots & \ddots & \vdots \\ 0 & \dots & 0 \\ a_{N_\ell+1} & \dots & a_{N_\ell+D_\ell} \end{pmatrix} \in \mathbb{K}^{(d+1)\times D_\ell}$$

$$W^\ell = \begin{pmatrix} \overline{w}_{N_\ell+1,1} & \dots & \overline{w}_{N_\ell+1,d} & 0 \\ \vdots & \ddots & \vdots & \vdots \\ \overline{w}_{N_\ell+D_\ell,1} & \dots & \overline{w}_{N_\ell+D_\ell,d} & 0 \end{pmatrix} \in \mathbb{K}^{D_\ell \times (d+1)}$$

where $N_\ell = \sum_{i=1}^{\ell-1} D_i$. The biases are chosen accordingly as $b^\ell = (b_{N_\ell+1}, \dots, b_{N_\ell+D_\ell})$. Of course if $\mathbb{K} = \mathbb{R}$, the complex conjugation has no effect.

In particular $\overline{\mathcal{P}} = \overline{\mathcal{M}_\sigma} \subseteq \overline{\mathcal{N}_1}$ since $\mathcal{M}_\sigma \subseteq \mathcal{N}_1$, and similarly for $\mathcal{N}_2$. On the other hand, $f \in \mathcal{R}_{L,\sigma}(d_0, D_1, \dots, D_{L-1})$ is a composition of analytic (vector-valued) functions with globally convergent power series. Consequently, also f can be represented by a globally convergent power series [50, Section 14.2], and by truncating $\mathcal{R}_{L,\sigma}(d_0, D_1, \dots, D_{L-1}) \subseteq \overline{\mathcal{P}}$ for any choice of architecture $d_0, D_1, \dots, D_{L-1}$. Therefore in particular $\overline{\mathcal{N}_1}, \overline{\mathcal{N}_2} \subseteq \overline{\mathcal{P}}$.

Second Claim If σ is not a polynomial, the second claim follows from the first. In the case of polynomial activation, the approximation properties of deep networks are strictly greater than those of networks of fixed depths. We provide a direct proof for the second claim which does not use the first.

First, consider the case that that $\sigma(z) = z^2$. Again, we construct the ResNet such that the vector $z = (z_1, \dots, z_d)$ is available at all layers. We recall that

$$z_i z_j = \frac{\sigma(z_i + z_j) - \sigma(z_i - z_j)}{4} \quad \text{and} \quad z_i = \frac{\sigma(z_i + 1) - \sigma(z_i - 1)}{4}. \tag{1}$$

Let $P(z) = \sum_{j_1+\dots+j_d \le m, j_i \in \mathbb{N}_0} a_{j_1\dots j_d} z_1^{j_1} \dots z_d^{j_d}$. If $d_0 \ge d+2$ and $D_\ell \ge 2$, we construct the residual representation:

$$z \xrightarrow{linear} \begin{pmatrix} a_{0,\dots,0} + \sum_{j=1}^d a_{0,\dots,1,\dots,0} z_j \\ 0 \\ z \end{pmatrix}$$

$$\xrightarrow{residual} \begin{pmatrix} a_{0,\dots,0} + \sum_{j=1}^d a_{0,\dots,1,\dots,0} z_j + a_{1,1,0,\dots,0} z_1^2 \\ 0 \\ z \end{pmatrix}$$

$$\xrightarrow{res.} \begin{pmatrix} a_{0,\dots,0} + \sum_{j=1}^d a_{0,\dots,1,\dots,0} z_j + a_{2,0,\dots,0} z_1^2 + a_{1,1,0,\dots,0} z_1 z_2 \\ 0 \\ z \end{pmatrix}$$

$$\xrightarrow{res.} \dots \xrightarrow{residual} \begin{pmatrix} a_{0,\dots,0} + \sum_{j=1}^d a_{0,\dots,1,\dots,0} z_j + \sum_{i,j=1}^d a_{0,\dots,1,\dots,1,\dots,0} z_i z_j \\ 0 \\ z \end{pmatrix}$$

$$\xrightarrow{res.} \begin{pmatrix} a_{0,\dots,0} + \sum_{j=1}^{d} a_{0,\dots,1,\dots,0} z_j + \sum_{i,j=1}^{d} a_{0,\dots,1,\dots,1,\dots,0} z_i z_j \\ z_1^2 \\ z \end{pmatrix}$$

$$\xrightarrow{res.} \begin{pmatrix} a_{0,\dots,0} + \sum_{j=1}^{d} a_{0,\dots,1,\dots,0} z_j + \sum_{i,j=1}^{d} a_{0,\dots,1,\dots,1,\dots,0} z_i z_j \\ z_1^3 \\ z \end{pmatrix}$$

$$\xrightarrow{res.} \begin{pmatrix} a_{0,\dots,0} + \sum_{j=1}^{d} a_{0,\dots,1,\dots,0} z_j + \sum_{i,j=1}^{d} a_{0,\dots,1,\dots,1,\dots,0} z_i z_j + a_{3,0,\dots,0} z_1^3 \\ 0 \\ z \end{pmatrix}$$

$$\xrightarrow{res.} \dots \begin{pmatrix} P(z) \\ 0 \\ z \end{pmatrix} \xrightarrow{linear} P(z).$$

The additional zero component is needed to "build up" higher powers of z by using σ and

$$z_1^{j_1+1} \dots z_d^{j_d} = z_1^{j_1} \dots z_d^{j_d} \cdot z_1$$

together with (1), before adding them to the polynomial which is assembled over many layers in the first component. We note that the requirement $D_\ell \geq 2$ was needed in order to execute multiplication—if multiplication were stretched over multiple layers, an additional "storage" space would be required. Wider networks with wider residual blocks could fit multiple terms of the polynomial at the same time.

Now consider the case that σ is a general analytic function which is not affine-linear. In this case, there exists $z^* \in \mathbb{K}$ such that $\sigma''(z^*) \neq 0$. As in the proof of Theorem 1, we can approximate $z \mapsto z^2$ to arbitrary accuracy by

$$z^2 = \lim_{h \to 0} \frac{\sigma(z^* + hz) - 2\sigma(z^*) + \sigma(z^* - hz)}{h^2 \sigma''(z^*)},$$

so we note that we can approximate squares to arbitrary accuracy using three evaluations of σ and products to arbitrary accuracy using six evaluations of σ. However, as we only require squares or the differences of squares for which middle function value $\sigma(z^*)$ cancels out, only $m = 4$ neurons are needed rather than $m = 6$.

We thus see that $\mathcal{P} \subseteq \overline{\bigcup_{L=1}^{\infty} \mathcal{R}_{L,\sigma}(d+2, 4, \dots, 4)}$. The opposite inclusion holds since every element of the space on the right is analytic on $\mathbb{K}^d$ and can therefore be approximated uniformly by polynomials on compact sets. □

While every shallow neural network can be expressed as a deep residual network, the number of parameters required to represent the network increases roughly twofold, since we list $D_\ell \cdot d$ zeros explicitly in the weight matrices A^ℓ, W^ℓ in every step which are implicit in the shallow neural network. While a shallow neural network is described by $(d+2)n$ parameters, the same function represented as a deep residual network has

$$(d+1) + \sum_{\ell=1}^{L-1} \big(2D_\ell(d+1) + D_\ell\big) + d(d+1) + (d+1) = 2(n+1)(d+1) + n$$

parameters. On the other hand, deep residual networks have a much larger expressive power for this number of parameters, since a single input can pass through multiple nonlinear activations σ.

A similar argument can be made for traditional fully connected feedforward networks and DenseNets. More details can be found in Appendix "Appendix: Classical Multilayer Perceptra and DenseNets". In the real case, the final proof remains valid under the weaker assumption that σ is not affine linear and $\sigma \in C^2$.

5 Conclusion and Further Directions

We showed that neural networks with (real or complex) analytic activation function can approximate any function which can be approximated by polynomials, and vice versa. While we focused on the C^0-topology in our presentation, the result holds in any C^k- or L^p- topology. The proofs are simple and only require the approximation of derivatives by difference quotients. In the real case, we reprove the universal approximation theorem by elementary means, reducing it to the Stone-Weierstrass theorem. In the complex case, the situation is more complicated, and results may depend on the topology of the domain of approximation.

If a function is known to be holomorphic in $d \geq 2$ complex variables, or the domain of approximation does not satisfy the hypotheses of Mergelyan's theorem for $d = 1$, then also approximation by neural networks is generally impossible. Similar results are obtained for shallow real "harmonic" neural networks, which utilize orthogonal projections of the data onto a lower-dimensional (but not one-dimensional) linear space.

Finally, we showed that polynomial activation functions are admissible from the perspective of approximation theory for residual networks if the network has a certain minimal width and the depth may be taken arbitrarily large.

The results as presented above are not as sharp as would be desirable in some ways:

1. The results presented here are purely qualitative. No rates of approximation are established under stronger assumptions on the target function in terms of, e.g., the number of parameters. Some quantitative rates can be found in [77, 87] for target functions which lie either in Sobolev or Hölder spaces, and for approximation in different topologies. However, we emphasize that *any* function class which has desirable properties from the perspective of statistical learning theory faces

the curse of dimensionality when approximating some function in a too general function class, even in a weak topology [24].

2. The proofs above involved the approximation of derivatives by difference quotients. Consequently, the coefficients are large, and the representation of the neural network depends critically on cancellation between possibly very large terms. For gradient-based optimizers, such subtle coefficients are hard to find.
3. The proofs cannot be directly modified to include the possibly most popular activation function in practice, the rectified linear united (ReLU) activation $\sigma(x) = \max\{x, 0\}$. It is, however, easy to prove by different means that any polynomial in one real dimension can be approximated arbitrarily well by a piecewise linear function/shallow ReLU network in this specific concrete case. The remainder of the proof goes through as before, allowing us to pass from the one-dimensional to the higher-dimensional situation.

From a negative perspective, our results can be interpreted as a statement that neural networks with entire activation functions can approximate exactly the same functions as polynomials. In the real case, this is not surprising since polynomials are dense in the space of continuous functions. Nevertheless, approximation by polynomials of high degree performs poorly even in *one-dimensional* interpolation problems, as interpolation with poorly chosen data points leads to high-amplitude oscillations at the domain boundary known as the *Runge phenomenon* [79, Chapter 13]. Universal approximation theorems are therefore unable to explain the superiority of neural networks over other parametrized function classes.

Approximation results for neural networks with bounded weights in a suitable sense can be obtained in suitable model classes adapted to neural networks [5, 6, 21, 25, 64, 73, 74], in which it can also be demonstrated that neural networks significantly outperform any linear method in spaces of high dimension [6, Theorem 6].

Appendix: On the Uniform Convergence of Difference Quotients

In this appendix, we prove that difference quotients converge to derivatives uniformly on compact sets in the form needed in the main text. While we believe that this result is well-known to the experts, we have not been able to find a reference. Recall that we consider the difference quotients:

$$\Delta_h^m f(x) = \frac{1}{h^m} \sum_{\ell=0}^{m} (-1)^\ell \binom{m}{\ell} f(\ell h z).$$

Lemma 2 *Let $\sigma(z) = \sum_{k=0} \alpha_k z^k$ be an analytic function with a convergent power series on a disk $B_R(0)$. Then for any compact set $K \subseteq \mathbb{K}$ and any $M \in \mathbb{N}$, there* *exists $\bar{h} > 0$ such that*

$$\left|\Delta_h \sigma(z) - m!\alpha_m z^m\right| \leq C_K h \qquad \forall\ |h| < \bar{h} \text{ and } m \leq M$$

with a constant C_K which depends only on K, M, and σ, but not h or m.

Proof Let $\sigma(z) = \sum_{k=0}^{\infty} \alpha_k z^k$ be given by a convergent power series on a disk $B_R(0)$. We note that the power series for $\sigma(h\ell z)$ converges on $B_{R/(\ell h)}(0)$, so for any compact subset in $\mathbb{K}$ and any $m \in \mathbb{N}$, there is $\bar{h}$ such that the power series for $\sigma(h\ell z)$ converges absolutely and uniformly for $z \in \overline{B_R(0)}$. We combine the series without reordering the terms:

$$\begin{aligned}\frac{1}{h^m}\sum_{\ell=0}^{m}(-1)^{\ell}\binom{m}{\ell}\sigma(\ell h z) &= \frac{1}{h^m}\sum_{\ell=0}^{m}(-1)^{\ell}\binom{m}{\ell}\sum_{k=0}^{\infty}\alpha_k(\ell h z)^k \\ &= \sum_{k=0}^{\infty}\alpha_k z^k\left(h^{k-m}\sum_{\ell=0}^{m}(-1)^{\ell}\binom{m}{\ell}\ell^k\right).\end{aligned}$$

We note three important facts:

1. The coefficients in the expansion vanish for $k < m$:

$$k < m \quad \Rightarrow \quad \sum_{\ell=0}^{m}\binom{m}{\ell}(-1)^{\ell}\ell^k = 0.$$

2. For $k = m$, we have:

$$\sum_{\ell=0}^{m}\binom{m}{\ell}(-1)^{\ell}\ell^m = m!.$$

3. For $k > m$, the coefficients satisfy the bound:

$$\left|\sum_{\ell=0}^{m}(-1)^{\ell}\binom{m}{\ell}\ell^k\right| \leq m!\sum_{\ell=0}^{m}\ell^k \leq m^{k+1}m!$$

Indeed, the first two properties are the guiding principle for how we choose the coefficients in the difference quotient Δ_h^m. For a reference, see [8].

In the third one, we only require that the bound for higher-order terms is not such that it alters the radius of convergence. To obtain convergence on all compact sets, we separate a finite number of intermediate terms and truly higher-order terms which are damped significantly for small h. In total

$$\left|\frac{1}{h^m}\sum_{\ell=0}^{m}(-1)^\ell\binom{m}{\ell}\sigma(\ell hz) - m!\,\alpha_m z^m\right|$$
$$\leq h\left(m!\sum_{\ell=m+1}^{2m}|\alpha_k|\,|mz|^k + m!\sum_{\ell=2m+1}^{\infty}|\alpha_k|\,\left|m\bar{h}^{1/2}z\right|^k\right).$$

For any fixed m and z in a compact disk, the bound on the right is uniform. $\bar{h}$ must be chosen to make the series on the right converge on the given compact set K, depending on M and the convergence radius of the original series. □

Difference quotients in $h_1, \ldots, h_d$ for a power function, as in the proof of Theorem 1, can be considered in exactly the same way. The same is true for difference quotients of the type:

$$\Delta_h^m f(z) = \frac{1}{h^m}\sum_{i=1}^{m}\binom{m}{i}(-1)^i f(z+ih)$$

or their combination.

Appendix: Proof of Lemma 1

We present a simple proof which was pointed out to us by an anonymous reviewer.

Proof (Proof of Lemma 1) In [83, Section 4.1, in particular Theorem 4.1], the authors show that if one defines the class of complex-valued harmonic homogeneous polynomials $p : \mathbb{R}^d \to \mathbb{C}$

$$\mathcal{HP}_m^d\mathbb{C} := \{p \in \mathbb{C}[x_1, \ldots, x_d] : p \text{ harmonic, homogeneous of degree } m\},$$

then the representation

$$\pi_{d,m} : O(d) \to U(\mathcal{HP}_{d,m}\mathbb{C}) \qquad \text{given by } \pi_{d,m}(O)p = p \circ O^{-1}$$

is irreducible, meaning that whenever $p \in HP_{d,m}\mathbb{C}$, $p \not\equiv 0$, it holds that

$$\mathcal{HP}_m^d\mathbb{C} = \text{span}_\mathbb{C}\{p(Ox) : O \in O(d)\}$$

where $\text{span}_\mathbb{C}$ denotes the span over the complex numbers.

Now, let $0 \neq p \in \mathcal{HP}_d^j$ be arbitrary. Given any $q \in \mathcal{HP}_d^j \subset \mathcal{HP}_d^j\mathbb{C}$, there then exist $N \in \mathbb{N}$, $\alpha_1, \ldots, \alpha_N \in \mathbb{C}$, and $O_1, \ldots, O_N \in O(d)$ satisfying

$$q(x) = \sum_{j=1}^{N} \alpha_j p(O_j x) \qquad \text{and hence } q(x) = \operatorname{Re} q(x) = \sum_{j=1}^{N} \operatorname{Re}(\alpha_j) p(O_j x)$$

since p is real-valued. Hence, $\mathcal{HP}_d^j \in \operatorname{span}_{\mathbb{R}}\{p(Ox) : O \in O(d)\}$. □

Appendix: Locally Uniform Approximation and Approximation on Compact Sets

There are (at least) two natural ways to study the closure of the function class $\mathcal{M}_\sigma$:

1. Fix a compact set K and take the closure $\overline{\mathcal{M}_\sigma}^K$ of $\mathcal{M}_\sigma$ in $C^0(K)$.
2. Consider the more global closure of $\mathcal{M}_\sigma$ in the *compact-open* topology or *topology of compact convergence* [60, §46]:

$$\overline{\mathcal{M}_\sigma}^{cc} = \big\{ f \in C^0(\mathbb{K}^d) : \exists\, (f_n)_{n\in\mathbb{N}} \in \mathcal{M}_\sigma \text{ s.t. } f_n \to f \text{ in } C^0(K) \\ \text{for all compact } K \subseteq \mathbb{K}^d \big\}.$$

In this appendix, we compare the different closures. By the nature of the subject, this appendix is more technical than the main text and requires some familiarity with topology, Baire categories, and elliptic partial differential equations. Its content is not needed for the main results of this article, but illustrates and justifies the conceptual difference to the approach taken in [82].

The two notions of closure are related as follows.

Lemma 3 *A function $f \in C^0(\mathbb{K}^d)$ satisfies $f \in \overline{\mathcal{M}_\sigma}^{cc}$ if and only if $f|_K \in \overline{\mathcal{M}_\sigma}^K$ for all compact sets K.*

Proof The implication $f \in \overline{\mathcal{M}_\sigma}^{cc} \Rightarrow f|_K \in \overline{\mathcal{M}_\sigma}^K$ is trivial. For the opposite implication, note that by definition for all $n \in \mathbb{N}$ there exists $f_n \in \mathcal{M}_\sigma$ such that

$$\|f_n - f\|_{C^0(\overline{B_n(0)})} < \frac{1}{n}.$$

Since every compact set K is contained in $B_n(0)$ for sufficiently large n, we find that $f_n \to f$ locally uniformly, i.e., $f \in \overline{\mathcal{M}_\sigma}^{cc}$. □

In particular, if $f \in \overline{\mathcal{M}_\sigma}^{cc}$ and $K \subseteq \mathbb{K}^d$ is a compact set, then $f|_K \in \overline{\mathcal{M}_\sigma}^K \subseteq C^0(K)$. The opposite question is more subtle: if $f \in \overline{\mathcal{M}_\sigma}^K$ for a fixed compact set $K \subseteq \mathbb{K}^d$, is there $F \in \overline{\mathcal{M}_\sigma}^{cc}(\mathbb{K}^d)$ such that $f = F|_K$?

First consider neural networks with analytic activation function $\sigma : \mathbb{K} \to \mathbb{K}$:

1. $\mathbb{K} = \mathbb{R}$. In this case, if σ is not a polynomial, $\overline{\mathcal{M}_\sigma}^K = \overline{\mathcal{P}}^K = C^0(K)$ by the Stone-Weierstrass theorem [50, Section 15.7]. Thus $\overline{\mathcal{M}_\sigma}^{cc}$ is the set of functions $f : \mathbb{K}^d \to \mathbb{K}$ such that f is continuous everywhere, the Fréchet space $C^0(\mathbb{K}^d)$.

   We can answer the question in the affirmative in the real case due to the Tietze-Urysohn theorem [70, Theorem 20.4].

   Theorem (Tietze-Urysohn Extension Theorem) Let $K \subseteq \mathbb{K}^d$ be compact and $f \in C^0(K)$. Then there exists $F \in C_c(\mathbb{K}^d)$ such that $F|_K = f$.

2. $\mathbb{K} = \mathbb{C}$. In this case, we claim that $\overline{\mathcal{M}_\sigma}^{cc}$ is the space of holomorphic functions on $\mathbb{C}^d$. To see this, note that a holomorphic function on $\mathbb{C}^d$ can be expanded into a globally convergent power series. By truncating the series, we see that f can be approximated by polynomials in the uniform topology on every bounded set; hence, $f \in \overline{\mathcal{P}} \subseteq M_\sigma^{cc}$.

   On the other hand, assume that $f \in \overline{\mathcal{M}_\sigma}^{cc}$, i.e., $f \in \overline{\mathcal{M}_\sigma}^K$ for all compact subsets of $\mathbb{C}^d$. Then for every $R > 0$, f can be approximated by holomorphic functions arbitrarily well in $C^0(\overline{B_R(0)})$, so f is holomorphic in the interior $B_R(0)$. Consequently f is holomorphic on the entire space $\mathbb{C}^d$.

   It is well-known that even in one complex dimension, there are holomorphic functions which cannot be extended to the entire complex plane. As an example consider $f : \mathbb{C} \setminus \{0\} \to \mathbb{C}$, $f(z) = z^{-1}$. On the other hand, f can be approximated uniformly on any compact set K for which $\mathbb{C} \setminus K$ is connected by Mergelyan's theorem.

   Thus in the complex case, the answer to our question is *negative*.

A statement in the same spirit holds also if $\sigma : \mathbb{R}^k \to \mathbb{R}$ is harmonic. While the closure of $\mathcal{M}_\sigma$ in the topology of locally uniform convergence coincides with the space of harmonic functions on $\mathbb{R}^d$, the closure of $\mathcal{M}_\sigma$ in the topology of uniform convergence on a fixed set $\overline{D}$ contains functions which are harmonic in D, but merely continuous on $\overline{D}$. In particular, there exists no harmonic extension of f to $\mathbb{R}^d$, and we may miss possible limiting functions if we only consider the topology of locally uniform convergence.

To see that this is true, let D be a bounded C^2-domain in $\mathbb{R}^d$. By the Perron method [33, Section 2.8], the Dirichlet problem

$$\begin{cases} -\Delta u = 0 \text{ in } D \\ \quad\ u = f \text{ on } \partial D \end{cases}$$

has a solution for every $f \in C^0(D)$ which is continuous on $\overline{D}$. By Remark 3, if $d \geq 3$, then u can be approximated by harmonic polynomials. On the other hand, if f is merely continuous, then u cannot be extended to a harmonic function on $\mathbb{R}^d$, so u is not the restriction of a function $U \in \overline{\mathcal{M}_\sigma}^{cc}$ to $\overline{D}$, where the closure is taken with respect to the compact-open topology.

In fact, we can define a continuous linear map:

$$A : \overline{\mathcal{M}_\sigma}^{cc} \to \overline{\mathcal{M}_\sigma}^{D}, \qquad u \mapsto u|_{\overline{D}}$$

and compose further with the trace map $B : C^0(\overline{D}) \to C^0(\partial D)$. Then $B \circ A(\overline{\mathcal{M}_\sigma}^{cc}) \subseteq C^2(\partial D)$, while $B(\overline{\mathcal{M}_\sigma}^{D}) = C^0(\partial D)$. It follows from the well-known Banach-Mazurkiewicz theorem [4, Theorem 9] that C^2 is of first Baire category in C^0, i.e., it is the union of countably many closed sets which all have empty interior. In this sense, we miss "most" functions which can be approximated in $\overline{D}$ by studying only the global limiting objects in the topology of locally uniform convergence. A measure-theoretic extension can be found, e.g., in [42].

Appendix: Classical Multilayer Perceptra and DenseNets

Fully Connected Neural Networks

For the sake of completeness, we prove an approximation theorem for classical fully connected neural networks with multiple layers. While the depth of residual networks can reach dozens or hundreds and in extreme cases thousands of layers, the number of layers in a deep neural network is typically more manageable. Nevertheless, both statements in Theorem 3 have analogues for deep fully connected networks.

A fully connected deep neural network is defined as follows:

- For a given input $z \in \overline{D} \subseteq \mathbb{K}^d$, designate $\hat{z}^0 = z$ and $d_0 = d$.
- For $\ell \in \{1, \dots, L\}$, let $d_\ell \in \mathbb{N}$ and set

$$z^\ell := A^\ell \hat{z}^{\ell-1} + b^\ell \quad \text{and} \quad \hat{z}^\ell = \sigma(z^\ell),$$

where $A^\ell \in \mathbb{K}^{d_\ell \times d_{\ell-1}}$ is a linear map which takes $\hat{z}^{\ell-1} \in \mathbb{K}^{d_{\ell-1}}$ to $z^\ell \in \mathbb{K}^{d_\ell}$, and $b^\ell \in \mathbb{K}^{d_\ell}$. The function σ is applied to the vector z^ℓ coordinate-wise.

We designate the parameters (or "weights") of the deep neural network by

$$\Theta = \left(A^1, b^1, \dots, A^L, b^L\right) \in \mathbb{K}^{d_1 \times d} \times \mathbb{K}^{d_1} \times \cdots \times \mathbb{K}^{1 \times d_{\ell-1}} \times \mathbb{K}$$

and set $f_\Theta(z) = z^L \in \mathbb{K}^{d_L} = \mathbb{K}$. A fully connected neural network is described by the choice of activation function σ, depth L, and width of the layers $d_0, \dots, d_L$ (where $d_0 = d$ and $d_L = 1$ in our case). We designate the class of fully connected neural networks with such architecture by $\mathcal{FNN}_{\sigma,L}(d_0, \dots, d_L)$.

The following is the analogue of Theorem 3 for deep fully connected neural networks.

Theorem 4 *Let $d_\ell = d+1+m_\ell$ for $m_\ell \in \mathbb{N}$ and $1 \leq \ell \leq L-1$. Set $n := \sum_{\ell=1}^{L-1} m_\ell$. Assume that σ is an entire function, not constant. Then for every compact set $K \subset$*

$\mathbb{K}^d$, *we have:*

$$\mathcal{M}_{\sigma,n} \subseteq \overline{\mathcal{FNN}_{\sigma,L}(d, d_1, \ldots, d_{L-1}, d_L)}$$

for the closure in $C^0(K)$.

Proof Since σ is not a constant function, there exists z^* in $\mathbb{K}$ such that $\sigma'(z^*) \neq 0$. We write $\sigma(z^* + \varepsilon z) = c_0 + c_1\varepsilon z + O\big((\varepsilon z)^2\big)$ for small ε and $c_0 = \sigma(z^*)$ and $c_1 = \sigma'(z^*)$. Since $z \in K$, there exists $R > 0$ such that $|z| \leq R$, so the error term is uniformly small in z. We initially ignore the quadratic error term and note that any affine function of z can be written as an affine function of $c_0 + c_1\varepsilon z$. By an abuse of notation, we denote by c_0, z^* also the vectors with identical entries in $\mathbb{K}^d$. With the notations

$$N_\ell = \sum_{i=1}^{\ell-1} m_i, \qquad W_\ell = \begin{pmatrix} w_{N_\ell+1,1} & \cdots & w_{N_\ell+1,d} \\ \vdots & \ddots & \vdots \\ w_{N_\ell+m_\ell,1} & \cdots & w_{N_\ell+m_\ell,d} \end{pmatrix} \in \mathbb{K}^{m_\ell \times d}$$

and $a_\ell = (a_{N_\ell+1}, \ldots, a_{N_\ell+m_\ell})$, we represent:

$$z \xrightarrow{linear} \begin{pmatrix} 0 \\ W_1 z + b_1 \\ z^* + \varepsilon z \end{pmatrix} \underset{\approx}{\overset{\sigma}{\to}} \begin{pmatrix} \sigma(0) \\ \sigma(W_1 z + b_1) \\ c_0 + c_1 \varepsilon z \end{pmatrix} \underset{\approx}{\xrightarrow{linear}} \begin{pmatrix} z^* + \varepsilon\, a_1^T \sigma(W_1 z + b_1) \\ W_2 z + b_2 \\ z^* + \varepsilon z \end{pmatrix}$$

$$\cdots \underset{\approx}{\overset{\sigma}{\to}} \begin{pmatrix} c_0 + c_1\varepsilon \sum_{i=1}^{n-m_{L-1}} a_i \sigma(\langle w_i, z\rangle + b_i) \\ \sigma(W_{L-1} z + b_{L-1}) \\ z^* + \varepsilon z \end{pmatrix} \underset{\approx}{\xrightarrow{linear}} \sum_{i=1}^{n} a_i \sigma(\langle w_i, z\rangle + b_i).$$

The approximation error can be made arbitrarily small without increasing the number of parameters, by taking ε sufficiently small. When writing an affine function of z as an affine function of $c_0 + c_1\varepsilon z$, this can lead to potentially poorly conditioned linear maps, which may cause numerical instability in practice. Taking the limit $\varepsilon \to 0$, we see that any shallow neural network can be approximated arbitrarily well by a deeper neural network with sufficient width and length. □

Remark 4 The number of weights of a deep neural network with architecture $d_1, \ldots, d_L$ is

$$N = \sum_{\ell=1}^{L} \big(d_{\ell-1} \cdot d_\ell + d_\ell\big).$$

Assume that $L := n/d$ is an integer. Then, we can construct two networks to approximate $f \in \mathcal{M}_{\sigma,n}$:

1. A network of depth n and width $d + 2$. The number of weights is $N \sim nd^2$.

2. A network of depth L and width $2d+1$. The number of weights is $N \sim L(2d)^2 \sim 4nd$.

Thus, the number of weights increases by a factor d if we approximate f by a thin deep network and by a factor 4 if we use a network which is both wide and deep instead.

Thus in a sense, we can argue that anything which can be achieved by a shallow neural network can also be achieved by a deep neural network with a comparable number of parameters, if the width of the deep network is not too small. Since deep neural networks with analytic activation are analytic, we find the following.

Corollary 2 *Assume that σ is entire, not constant. Consider the classes of neural networks:*

$$\mathcal{NN}_1 = \bigcup_{L=1}^{\infty} \mathcal{FNN}_{\sigma,L}(d, d+1+m, \dots, d+1+m, 1)$$

for any fixed $m \in \mathbb{N}$ and

$$\mathcal{NN}_2 = \bigcup_{m=1}^{\infty} \mathcal{FNN}_{\sigma,L}(d, d+1+m, \dots, d+1+m, 1)$$

for fixed L. Then for any compact set $K \subseteq \mathbb{K}^d$, we have $\overline{\mathcal{NN}_1} = \overline{\mathcal{NN}_2} = \overline{\mathcal{P}}$ in $C^0(K)$.

DenseNets

A DenseNet [43] is a modified neural network structure in which the internal state of the ℓ-th layer is computed linearly from the state of the network at all previous layers $0, \dots, \ell - 1$, rather than just the previous state. Namely, consider the following structure:

- For a given input $z \in \overline{D} \subseteq \mathbb{K}^d$, designate $\hat{z}^0 = z$ and $d_0 = d$.
- For $\ell \in \{1, \dots, L\}$, let $d_\ell \in \mathbb{N}$ the width of the ℓ-th state:

$$n_\ell = \sum_{i=0}^{\ell-1} d_i, \qquad A^\ell \in \mathbb{K}^{d_\ell \times n_\ell}, \quad b^\ell \in \mathbb{K}^{d_\ell}$$

and

$$z^\ell = A^\ell \begin{pmatrix} \hat z^0 \\ \vdots \\ \hat z^{\ell-1} \end{pmatrix} + b^\ell, \qquad \hat z^\ell = \sigma(z^\ell).$$

The function σ is applied to the vector z^ℓ coordinate-wise.

- The output of the network is $f(z) = z^L$, where we suppressed the dependence on the input z and the weights $(A^1, b^1, \dots, A^L, b^L)$ in the notation.

We denote the class of DenseNets with activation σ and widths $(d_0, d_1, \dots, d_{L-1}, d_L)$ by $\mathcal{D}_{\sigma,L}(d_0, d_1, \dots, d_{L-1}, d_L)$. As usual, $d_0 = d$ and $d_L = 1$ are fixed by the problem statement. Since a DenseNet can access all previous states (including the input) at all layers, every neural network with a single hidden layer and activation function σ can be represented exactly by a sufficiently large DenseNet. No restriction to compact sets or closure operation is required, and there is no need to invert potentially ill-conditioned matrices. The same result holds for deep neural networks, since the previous layer can be accessed.

Theorem 5 *Set* $n = \sum_{\ell=1}^{L-1} d_\ell = n_L - d$. *Then, we have:*

$$\mathcal{M}_{\sigma,n} \cup \mathcal{FNN}_{\sigma,L}(d, d_1, \dots, d_{L-1}, 1) \quad \subseteq \quad \mathcal{D}_{\sigma,L}(d, d_1, \dots, d_{L-1}, 1).$$

In particular, if σ is analytic and not polynomial/affine, an analogue of Corollary 2 holds for DenseNets. Note, however, that the number of parameters for DenseNets with comparable widths is significantly larger, since the weights $A^\ell \in \mathbb{K}^{d_\ell \times n_\ell}$ used in the construction of Theorem 5 have block structures which list many zeros explicitly, as only the first/previous layer is accessed.

Acknowledgments The authors would like to thank Ron DeVore, Guergana Petrova, and Peter Binev for inspiring conversations. JP would also like to acknowledge helpful conversations with Dmitriy Bilyk, Alexey Glazyrin, and Oleksandr Vlasiuk at the SIAM TX-LA Conference 2021.

References

1. Monica Alderighi, Sergio D'Angelo, Francesco d'Ovidio, E Gummati, and Giacomo R Sechi. An advanced neuron model for optimizing the siren network architecture. In *Proceedings of 1996 IEEE Second International Conference on Algorithms and Architectures for Parallel Processing, ICA/sup 3/PP'96*, pages 194–200. IEEE, 1996.
2. Brandon Amos, Lei Xu, and J Zico Kolter. Input convex neural networks. In *International Conference on Machine Learning*, pages 146–155. PMLR, 2017.
3. Raman Arora, Amitabh Basu, Poorya Mianjy, and Anirbit Mukherjee. Understanding deep neural networks with rectified linear units. *ICLR Conference paper*, 2018.
4. Pouya Ashraf. Pathological functions and the Baire category theorem. Technical report, Uppsala Universitet, 2017.
5. Francis Bach. Breaking the curse of dimensionality with convex neural networks. *The Journal of Machine Learning Research*, 18(1):629–681, 2017.

6. Andrew R. Barron. Universal approximation bounds for superpositions of a sigmoidal function. *IEEE Trans. Inform. Theory*, 39(3):930–945, 1993.
7. Joshua Bassey, Lijun Qian, and Xianfang Li. A survey of complex-valued neural networks. *arXiv preprint arXiv:2101.12249*, 2021.
8. Khristo N Boyadzhiev. Close encounters with the stirling numbers of the second kind. *Mathematics Magazine*, 85(4):252–266, 2012.
9. Ines Chami, Zhitao Ying, Christopher Ré, and Jure Leskovec. Hyperbolic graph convolutional neural networks. *Advances in neural information processing systems*, 32, 2019.
10. Minshuo Chen, Haoming Jiang, Wenjing Liao, and Tuo Zhao. Efficient approximation of deep ReLU networks for functions on low dimensional manifolds. *Advances in neural information processing systems*, 32, 2019.
11. Ricky TQ Chen, Yulia Rubanova, Jesse Bettencourt, and David K Duvenaud. Neural ordinary differential equations. *Advances in neural information processing systems*, 31, 2018.
12. Yixiao Chen, Linfeng Zhang, Han Wang, and Weinan E. DeePKS: A comprehensive data-driven approach toward chemically accurate density functional theory. *Journal of Chemical Theory and Computation*, 17(1):170–181, dec 2020.
13. Yudong Chen and Martin J Wainwright. Fast low-rank estimation by projected gradient descent: General statistical and algorithmic guarantees. *arXiv preprint arXiv:1509.03025*, 2015.
14. Sanghyun Cho. On the Mergelyan approximation property on pseudoconvex domains in $\mathbb{C}^n$. *Proceedings of the American Mathematical Society*, 126(8):2285–2289, 1998.
15. George Cybenko. Approximation by superpositions of a sigmoidal function. *Mathematics of control, signals and systems*, 2(4):303–314, 1989.
16. Ingrid Daubechies, Ronald DeVore, Nadav Dym, Shira Faigenbaum-Golovin, Shahar Z Kovalsky, Kung-Ching Lin, Josiah Park, Guergana Petrova, and Barak Sober. Neural network approximation of refinable functions. *arXiv preprint arXiv:2107.13191*, 2021.
17. Ingrid Daubechies, Ronald DeVore, Simon Foucart, Boris Hanin, and Guergana Petrova. Nonlinear approximation and (deep) ReLU networks. *Constructive Approximation*, 55(1):127–172, 2022.
18. Ronald DeVore, Boris Hanin, and Guergana Petrova. Neural network approximation. *Acta Numerica*, 30:327–444, 2021.
19. Klas Diederich and John Erik Fornaess. A strange bounded smooth domain of holomorphy. *Bull. Amer. Math. Soc.*, 82(1):74–76, 1976.
20. Weinan E. A proposal on machine learning via dynamical systems. *Communications in Mathematics and Statistics*, 5(1):1–11, 2017.
21. Weinan E, Chao Ma, and Lei Wu. A priori estimates of the population risk for two-layer neural networks. *Comm. Math. Sci.*, 17(5):1407–1425, 2019.
22. Weinan E and Qingcan Wang. Exponential convergence of the deep neural network approximation for analytic functions. *Sci. China Math.*, 61(10):1733–1740, 2018.
23. Weinan E and Stephan Wojtowytsch. On the Banach spaces associated with multi-layer ReLU networks of infinite width. *CSIAM Trans. Appl. Math.*, 1(3):387–440, 2020.
24. Weinan E and Stephan Wojtowytsch. Kolmogorov width decay and poor approximators in machine learning: shallow neural networks, random feature models and neural tangent kernels. *Res. Math. Sci.*, 8(1):Paper No. 5, 28, 2021.
25. Weinan E. and Stephan Wojtowytsch. Representation formulas and pointwise properties for Barron functions. *Calc. Var. Partial Differential Equations*, 61(2):Paper No. 46, 37, 2022.
26. Ronen Eldan and Ohad Shamir. The power of depth for feedforward neural networks. In *Conference on learning theory*, pages 907–940, 2016.
27. Lawrence C. Evans. *Partial differential equations*, volume 19 of *Graduate Studies in Mathematics*. American Mathematical Society, Providence, RI, second edition, 2010.
28. Javier Falcó, Paul M. Gauthier, Myrto Manolaki, and Vassili Nestoridis. A function algebra providing new Mergelyan type theorems in several complex variables. *Adv. Math.*, 381:Paper No. 107649, 31, 2021.

29. John Erik Fornæss, Franc Forstnerič, and Erlend F. Wold. Holomorphic approximation: the legacy of Weierstrass, Runge, Oka-Weil, and Mergelyan. In *Advancements in complex analysis—from theory to practice*, pages 133–192. Springer, Cham, [2020] ©2020.
30. Eberhard Freitag and Rolf Busam. *Funktionentheorie*. Springer-Lehrbuch. [Springer Textbook]. Springer-Verlag, Berlin, 1993.
31. Octavian Ganea, Gary Bécigneul, and Thomas Hofmann. Hyperbolic neural networks. *Advances in neural information processing systems*, 31, 2018.
32. Stephen J. Gardiner. *Harmonic approximation*, volume 221 of *London Mathematical Society Lecture Note Series*. Cambridge University Press, Cambridge, 1995.
33. David Gilbarg and Neil S Trudinger. *Elliptic partial differential equations of second order*, volume 224. Springer, 2015.
34. Rémi Gribonval, Gitta Kutyniok, Morten Nielsen, and Felix Voigtlaender. Approximation spaces of deep neural networks. *Constructive Approximation*, 55(1):259–367, 2022.
35. Steven Gubkin. *L^2-Mergelyan theorems in several complex variables*. ProQuest LLC, Ann Arbor, MI, 2015. Thesis (Ph.D.)–The Ohio State University.
36. Harshit Gupta, Kyong Hwan Jin, Ha Q Nguyen, Michael T McCann, and Michael Unser. CNN-based projected gradient descent for consistent CT image reconstruction. *IEEE transactions on medical imaging*, 37(6):1440–1453, 2018.
37. Jiequn Han, Arnulf Jentzen, and Weinan E. Solving high-dimensional partial differential equations using deep learning. *Proceedings of the National Academy of Sciences*, 115(34):8505–8510, 2018.
38. Kaiming He, Xiangyu Zhang, Shaoqing Ren, and Jian Sun. Deep residual learning for image recognition. In *Proceedings of the IEEE conference on computer vision and pattern recognition*, pages 770–778, 2016.
39. Lester L. Helms. *Potential theory*. Universitext. Springer, London, second edition, 2014.
40. Kurt Hornik. Approximation capabilities of multilayer feedforward networks. *Neural networks*, 4(2):251–257, 1991.
41. John M. Howie. *Complex analysis*. Springer Undergraduate Mathematics Series. Springer-Verlag London, Ltd., London, 2003.
42. Brian R. Hunt. The prevalence of continuous nowhere differentiable functions. *Proc. Amer. Math. Soc.*, 122(3):711–717, 1994.
43. Forrest Iandola, Matt Moskewicz, Sergey Karayev, Ross Girshick, Trevor Darrell, and Kurt Keutzer. Densenet: Implementing efficient convnet descriptor pyramids. *arXiv preprint arXiv:1404.1869*, 2014.
44. Sergei Ivanov. Zariski closed sets in $\mathbb{C}^n$ are of measure 0, https://mathoverflow.net/questions/25513/zariski-closed-sets-in-cn-are-of-measure-0, 2010.
45. Jesse Johnson. Deep, skinny neural networks are not universal approximators. *arXiv preprint arXiv:1810.00393*, 2018.
46. Mark Kac. Can one hear the shape of a drum? *The american mathematical monthly*, 73(4P2):1–23, 1966.
47. Patrick Kidger and Terry Lyons. Universal approximation with deep narrow networks. In *Conference on learning theory*, pages 2306–2327. PMLR, 2020.
48. Jason M Klusowski and Andrew R Barron. Approximation by combinations of ReLU and squared ReLU ridge functions with ℓ^1 and ℓ^0 controls. *IEEE Transactions on Information Theory*, 64(12):7649–7656, 2018.
49. Konrad Königsberger. *Analysis. 2*. Springer-Lehrbuch. [Springer Textbook]. Springer-Verlag, Berlin, 1993. Grundwissen Mathematik. [Basic Knowledge in Mathematics].
50. Konrad Königsberger. *Analysis. 1*. Springer-Lehrbuch. [Springer Textbook]. Springer-Verlag, Berlin, sixth edition, 2004.
51. Moshe Leshno, Vladimir Ya Lin, Allan Pinkus, and Shimon Schocken. Multilayer feedforward networks with a nonpolynomial activation function can approximate any function. *Neural networks*, 6(6):861–867, 1993.
52. Qianxiao Li, Long Chen, Cheng Tai, and E Weinan. Maximum principle based algorithms for deep learning. *The Journal of Machine Learning Research*, 18(1):5998–6026, 2017.

53. Qi Liu, Maximilian Nickel, and Douwe Kiela. Hyperbolic graph neural networks. *Advances in Neural Information Processing Systems*, 32, 2019.
54. Vitaly Maiorov and Allan Pinkus. Lower bounds for approximation by MLP neural networks. *Neurocomputing*, 25(1-3):81–91, 1999.
55. Ashok Makkuva, Amirhossein Taghvaei, Sewoong Oh, and Jason Lee. Optimal transport mapping via input convex neural networks. In *International Conference on Machine Learning*, pages 6672–6681. PMLR, 2020.
56. Y Makovoz. Uniform approximation by neural networks. *Journal of Approximation Theory*, 95(2):215–228, 1998.
57. S. N. Mergelyan. On the representation of functions by series of polynomials on closed sets. *Doklady Akad. Nauk SSSR (N.S.)*, 78:405–408, 1951.
58. H. N. Mhaskar and T. Poggio. Deep vs. shallow networks: an approximation theory perspective. *Anal. Appl. (Singap.)*, 14(6):829–848, 2016.
59. Hrushikesh N Mhaskar. Neural networks for optimal approximation of smooth and analytic functions. *Neural computation*, 8(1):164–177, 1996.
60. James R. Munkres. *Topology*. Prentice Hall, Inc., Upper Saddle River, NJ, 2000. Second edition of [MR0464128].
61. Kiyoshi Oka. *Sur les fonctions analytiques de plusieurs variables*. Iwanami Shoten, Tokyo, 1961.
62. Bernt Øksendal. Stochastic differential equations. In *Stochastic differential equations*, pages 65–84. Springer, 2003.
63. Keiron O'Shea and Ryan Nash. An introduction to convolutional neural networks. *arXiv preprint arXiv:1511.08458*, 2015.
64. Rahul Parhi and Robert D Nowak. Banach space representer theorems for neural networks and ridge splines. *J. Mach. Learn. Res.*, 22(43):1–40, 2021.
65. Rahul Parhi and Robert D Nowak. What kinds of functions do deep neural networks learn? insights from variational spline theory. *arXiv preprint arXiv:2105.03361*, 2021.
66. Wei Peng, Tuomas Varanka, Abdelrahman Mostafa, Henglin Shi, and Guoying Zhao. Hyperbolic deep neural networks: A survey. *arXiv preprint arXiv:2101.04562*, 2021.
67. Philipp Petersen and Felix Voigtlaender. Optimal approximation of piecewise smooth functions using deep ReLU neural networks. *Neural Networks*, 108:296–330, 2018.
68. Allan Pinkus. Approximation theory of the MLP model in neural networks. *Acta numerica*, 8(1):143–195, 1999.
69. Maziar Raissi, Paris Perdikaris, and George E Karniadakis. Physics-informed neural networks: A deep learning framework for solving forward and inverse problems involving nonlinear partial differential equations. *Journal of Computational Physics*, 378:686–707, 2019.
70. Walter Rudin. *Real and complex analysis*. McGraw-Hill Book Co., New York, third edition, 1987.
71. Johannes Schmidt-Hieber. Deep ReLU network approximation of functions on a manifold. *arXiv preprint arXiv:1908.00695*, 2019.
72. Peter H Schönemann. A generalized solution of the orthogonal procrustes problem. *Psychometrika*, 31(1):1–10, 1966.
73. Jonathan W Siegel and Jinchao Xu. On the approximation properties of neural networks. *arXiv preprint arXiv:1904.02311*, 2019.
74. Jonathan W Siegel and Jinchao Xu. Approximation rates for neural networks with general activation functions. *Neural Networks*, 128:313–321, 2020.
75. Jonathan W Siegel and Jinchao Xu. High-order approximation rates for neural networks with ReLU^k activation functions. *arXiv preprint arXiv:2012.07205*, 2020.
76. Vincent Sitzmann, Julien Martel, Alexander Bergman, David Lindell, and Gordon Wetzstein. Implicit neural representations with periodic activation functions. *Advances in Neural Information Processing Systems*, 33:7462–7473, 2020.
77. Taiji Suzuki. Adaptivity of deep ReLU network for learning in Besov and mixed smooth Besov spaces: optimal rate and curse of dimensionality, 2018.

78. C Trabelsi, O Bilaniuk, Y Zhang, D Serdyuk, S Subramanian, JF Santos, S Mehri, N Rostamzadeh, Y Bengio, and CJ Pal. Deep complex networks. *arXiv preprint arXiv:1705.09792*, 2017.
79. Lloyd N. Trefethen. *Approximation theory and approximation practice*. Society for Industrial and Applied Mathematics (SIAM), Philadelphia, PA, 2013.
80. Mark Tygert, Joan Bruna, Soumith Chintala, Yann LeCun, Serkan Piantino, and Arthur Szlam. A mathematical motivation for complex-valued convolutional networks. *Neural computation*, 28(5):815–825, 2016.
81. Patrick Virtue, X Yu Stella, and Michael Lustig. Better than real: Complex-valued neural nets for MRI fingerprinting. In *2017 IEEE international conference on image processing (ICIP)*, pages 3953–3957. IEEE, 2017.
82. Felix Voigtlaender. The universal approximation theorem for complex-valued neural networks. *arXiv:2012.03351 [math.FA]*, 2020.
83. Valery V Volchkov and Vitaly V Volchkov. *Harmonic analysis of mean periodic functions on symmetric spaces and the Heisenberg group*. Springer Science & Business Media, 2009.
84. André Weil. L'intégrale de Cauchy et les fonctions de plusieurs variables. *Math. Ann.*, 111(1):178–182, 1935.
85. Ziwei Xuan and Krishna Narayanan. Deep joint source-channel coding for transmission of correlated sources over AWGN channels. In *ICC 2021-IEEE International Conference on Communications*, pages 1–6. IEEE, 2021.
86. Ziwei Xuan and Krishna Narayanan. Low-delay analog distributed joint source-channel coding using sirens. In *2021 29th European Signal Processing Conference (EUSIPCO)*, pages 1601–1605. IEEE, 2021.
87. Dmitry Yarotsky. Error bounds for approximations with deep ReLU networks. *Neural Networks*, 94:103–114, 2017.
88. Dmitry Yarotsky and Anton Zhevnerchuk. The phase diagram of approximation rates for deep neural networks. *arXiv preprint arXiv:1906.09477*, 2019.
89. Linfeng Zhang. Deep potential molecular dynamics: A scalable model with the accuracy of quantum mechanics. *Physical Review Letters*, 120(14), 2018.

Linearly Embedding Sparse Vectors from ℓ_2 to ℓ_1 via Deterministic Dimension-Reducing Maps

Simon Foucart

Abstract This note is concerned with deterministic constructions of $m \times N$ matrices satisfying a restricted isometry property from ℓ_2 to ℓ_1 on s-sparse vectors. Similarly to the standard (ℓ_2 to ℓ_2) restricted isometry property, such constructions can be found in the regime $m \asymp s^2$, at least in theory. With effectiveness of implementation in mind, two simple constructions are presented in the less pleasing but still relevant regime $m \asymp s^4$. The first one, executing a Las Vegas strategy, is quasideterministic and applies in the real setting. The second one, exploiting Golomb rulers, is explicit and applies to the complex setting. As a stepping stone, an explicit isometric embedding from $\ell_2^n(\mathbb{C})$ to $\ell_4^{cn^2}(\mathbb{C})$ is presented. Finally, the extension of the problem from sparse vectors to low-rank matrices is raised as an open question.

1 Motivation from Sparse Vector Recovery

Almost 20 years ago [1, 2], the realization that high-dimensional but sparse vectors could be efficiently recovered from far fewer linear measurements than expected created a prolific field of research now known as compressive sensing (or compressed sensing). To be specific, vectors $x \in \mathbb{K}^N$, $\mathbb{K} \in \{\mathbb{R}, \mathbb{C}\}$, are called s-sparse if

$$\|x\|_0 := |\mathrm{supp}(x)| \leq s, \qquad \text{where } \mathrm{supp}(x) := \{j \in [1:N] : x_j \neq 0\}.$$

Such vectors can be recovered from compressive measurements $Ax \in \mathbb{K}^m$ with m being of the order of $s\ln(N/s) \ll N$. One refers to [3] for all the nitty-gritty details. One simply mentions here that, on the one hand, the order $s\ln(N/s)$ cannot be lowered if one requires the recovery to be stable and, one the other hand, that random measurement matrices $A \in \mathbb{K}^{m \times N}$ with $m \asymp s\ln(N/s)$ fulfill, with high

S. Foucart (✉)
Department of Mathematics, Texas A&M University, College Station, TX, USA
e-mail: foucart@tamu.edu

S. Foucart, S. Wojtowytsch (eds.), *Explorations in the Mathematics of Data Science*, Applied and Numerical Harmonic Analysis,
https://doi.org/10.1007/978-3-031-66497-7_4

probability, favorable properties that make s-sparse recovery possible. Thus, there is an abundance of matrices suitable for compressive sensing in the optimal regime $m \asymp s\ln(N/s)$, but somehow the mathematical community is unable to pinpoint a single one!

The most popular favorable property—the restricted isometry property (RIP), introduced in [4]—stipulates that the matrix $A \in \mathbb{K}^{m\times N}$ should satisfy:

$$(1-\delta)\|x\|_2^2 \leq \|Ax\|_2^2 \leq (1+\delta)\|x\|_2^2 \qquad \text{for all } s\text{-sparse } x \in \mathbb{K}^N. \tag{1}$$

This standard version of the RIP is ubiquitous in ensuring the success of sparse recovery via a variety of reconstruction algorithms, such as ℓ_1-minimization (aka basis pursuit), orthogonal matching pursuit (OMP), compressive sampling matching pursuit (CoSaMP), iterative hard thresholding (IHT), hard thresholding pursuit (HTP), and to name but a few. It can be interpreted as saying that the linear map $x \mapsto Ax$ provides an embedding from ℓ_2^N to ℓ_2^m with a distortion on s-sparse vectors equal to $\gamma = (1+\delta)/(1-\delta) \geq 1$. The latter can be made arbitrarily close to one by taking $\delta > 0$ small enough. However, the distortion need not be close to one to enable sparse recovery: a requirement $\gamma \leq \gamma^*$ for a fixed threshold $\gamma^* \geq 1$ is enough. Likewise, the embedding need not map into ℓ_2^m: any ℓ_p^m with $0 < p \leq 2$ will be convenient. This note concentrates on the case $p = 1$—interestingly, this version appeared in [2]. Thus, instead of the standard version of the RIP, i.e., (1), one considers an "ℓ_2 to ℓ_1" RIP stipulating that $A \in \mathbb{K}^{m\times N}$ should satisfy:

$$\alpha\|x\|_2 \leq \|Ax\|_1 \leq \beta\|x\|_2 \qquad \text{for all } s\text{-sparse } x \in \mathbb{K}^N \tag{2}$$

with distortion on s-sparse vectors bounded by a fixed threshold, say $\gamma = \beta/\alpha \leq \gamma_*$. I am an advocate of this alternative version, for several reasons:

- the theory of sparse recovery can be built from (2), and not only for basis pursuit as presented in [5, Chapter 14] but also for iterative hard thresholding; see [6];
- the theory of one-bit compressive sensing can be built from (2) as well, as presented [5, Chapter 17], so long as the distortion can be made close to one, which is the case if A is a Gaussian matrix or a partial Gaussian circulant matrix [7];
- Laplace matrices (more generally subexponential random matrices) satisfy (2) in the optimal regime $m \asymp s\ln(N/s)$ (see [8]), while (1) would only hold in the suboptimal regime $m \asymp s\ln^2(N/s)$;
- any ensemble of random matrices $A \in \mathbb{K}^{m\times N}$ yielding (1) with exponentially small failure probability when $m \asymp s\ln(N/s)$ also yields (2) with exponentially small failure probability when $m \asymp s\ln(N/s)$; see [9] for the precise statement.

This last point suggests that the "ℓ_2 to ℓ_1" RIP is easier to fulfill than the standard "ℓ_2 to ℓ_2" RIP, so it is plausible that deterministic constructions of RIP matrices are more accessible through the "ℓ_2 to ℓ_1" avenue. Since deterministic constructions exist for the standard RIP with $m \asymp s^{2-\varepsilon}$ (see next section), the same is expected to hold for the "ℓ_2 to ℓ_1" RIP. This is indeed the case, but through a construction that may be considered inadequate, because it combines two fairly theoretical results that are, in my view, not as explicit as hoped for. The purpose of this note is to exhibit a couple of simple constructions of matrices that fulfill the "ℓ_2 to ℓ_1" RIP in the regime $m \asymp s^4$. This is not the desired regime, for sure, but the advantage here is the simplicity of the constructions. This simplicity is validated by the few-lines MATLAB implementation found in the associated reproducible file (available on the author's webpage).

The rest of this note is organized as follows. Section 2 discusses some known facts about deterministic embeddings. Section 3 presents, in the case $\mathbb{K} = \mathbb{R}$, the first deterministic construction of an embedding of s-sparse vectors from ℓ_2^N to ℓ_1^m with $m \asymp s^4$. Section 4 uncovers, in case $\mathbb{K} = \mathbb{C}$, an explicit isometric embedding from ℓ_2^p to ℓ_4^m with $m \asymp p^2$, p being a prime number. Section 5 exploits this isometric embedding—or rather the argument leading to it—to present, in the case $\mathbb{K} = \mathbb{C}$, a second deterministic, and actually explicit, construction of an embedding of s-sparse vectors from ℓ_2^N to ℓ_1^m with $m \asymp s^4$. Section 6 briefly touches on the mostly uncharted territory of deterministic restricted isometry properties for low-rank matrices. Finally, as an aside, Sect. 7 recalls a connection between (almost) isometric embeddings from ℓ_2^n to ℓ_{2k}^N and (approximate) spherical designs.

2 Known Deterministic Results

The mathematical community's incapability to create nonrandom matrices fulfilling the standard RIP in the optimal regime is vexing. Nonrandom procedures are stuck in the quadratic regime $m \asymp s^2$—in truth, $m \asymp s^2 \mathrm{polylog}(N)$. There are several ways to reach this regime, mostly based on the notion of coherence. The coherence of a matrix $A \in \mathbb{K}^{m \times N}$ with unit ℓ_2-norm columns $a_1, \ldots, a_N \in \mathbb{K}^m$ is defined by

$$\mu(A) := \max_{j \neq \ell} |\langle a_j, a_\ell \rangle|.$$

Indeed, to ensure that $\delta_s(A)$, the smallest $\delta \in (0, 1)$ for which (1) holds, obeys $\delta_s(A) < \delta_*$ as soon as $m \geq Cs^2$, it is sufficient to make the coherence small as per

$$\mu(A) \leq \frac{c}{\sqrt{m}}, \qquad c := \sqrt{C}\, \delta_*,$$

by virtue of the inequality $\delta_s(A) < s\mu(A)$ (see, e.g., [3, Proposition 6.2]). As examples of matrices with small coherence, let me mention:

- an $m \times m^2$ matrix with columns formed by translations and modulations of the Alltop vector ([10], see also [3, Proposition 5.13]): its coherence is $\mu(A) = 1/\sqrt{m}$;
- when p is prime and $d < p$, a $p^2 \times p^{d+1}$ matrix with binary entries in $\{0, 1/\sqrt{p}\}$ ([11], see also [5, Theorem 18.5]): its coherence is $\mu(A) \leq d/p = d/\sqrt{m}$;

- when p is prime and $d < p$, a $p \times p^{d+1}$ matrix with entries

$$A_{k,f} = \frac{1}{\sqrt{p}} \exp\left(i 2\pi \frac{k f(k)}{p}\right) \tag{3}$$

indexed by elements k of $\mathbb{F}_p$ and by polynomials f over $\mathbb{F}_p$ of degree at most d: its coherence is $\mu(A) \leq d/\sqrt{p} = d/\sqrt{m}$. The argument is simple but relies on a deep result known as Weil bound (see, e.g., [12, Proposition 5.3.8]), which says that if f is a nonconstant polynomial over $\mathbb{F}_p$, then

$$\left| \sum_{k \in \mathbb{F}_p} \exp\left(i 2\pi \frac{f(k)}{p}\right) \right| \leq (\deg(f) - 1)\sqrt{p}.$$

There is a notable explicit construction that overcomes, albeit ever so slightly, the quadratic barrier. Indeed, the article [13] uncovered an RIP matrix with $m \asymp s^{2-\varepsilon}$ rows, where $\varepsilon > 0$ was tiny. In a practitioner's mind, this is viewed as an issue, but in fact not as the most critical one: one also has $m \geq N^{1-\varepsilon}$, making the matrix almost square and thus defeating the compressive sensing purpose of taking far fewer measurements than the high ambient dimension. A similar issue occurs in [14]: there, conditionally on a folklore conjecture in number theory, it was shown that the quadratic barrier would be overcome by the Paley matrix fulfilling the "ℓ_2 to ℓ_2" RIP, but this matrix has $m = N/2$ rows.

Turning now to the deterministic "ℓ_2 to ℓ_1" RIP, which has not been explored, at least to the best of my knowledge, one can follow the indirect strategy below:

(i) consider a deterministic linear map $A' : \ell_2^N \to \ell_2^n$ with constant distortion on s-sparse vectors for $n \asymp s^\eta$;
(ii) consider a deterministic linear map $A'' : \ell_2^n \to \ell_1^m$ with constant distortion on arbitrary vectors for $m \asymp n^\theta$.

Then, the deterministic map $A := A'' \circ A' : \ell_2^N \to \ell_1^m$ has constant distortion on s-sparse vectors for $m \asymp s^{\eta\theta}$. For (i), one can take $\eta = 2 - \varepsilon$ according to [13]. For (ii), according to [15] (or [16]),[1] one can take $\theta = 1 + \omega$ for any $\omega > 0$. These choices yield an "ℓ_2 to ℓ_1" RIP in the regime $m \asymp s^{(2-\varepsilon)(1+\omega)}$ for any $\omega > 0$, hence overcoming the quadratic barrier here, too. This is quite compelling until the hands-on stage, because the theoretical results underlying the argument are not easily implementable, despite being deterministic (which often means constructible in polynomial time).

The general strategy is still valid, though, and one can gain in explicitness by giving up on the smallest ν in $m \asymp s^\nu$. This is the spirit of Sect. 5, which obtains

[1] Both [15] and [16] rely on the construction of unbalanced expanders from [17]. This construction, combined with the result of [18], has a direct implication in compressive sensing by providing a deterministic linear embedding of sparse vectors from ℓ_1 (not ℓ_2) to ℓ_1.

$\nu = 4$ by way of taking $\eta = 2$ in (i) and $\theta = 2$ in (ii). For the first step, one can exploit any of the small-coherence matrices listed above. For the second step, one relies on the existence of deterministic constant-distortion linear embeddings from ℓ_2^n to $\ell_1^{cn^2}$. Such an existence result is stated in [19, Lemma 3.3] by citing [20] for a construction of *four*-wise independent families of vectors, which I would qualify as deterministic but not explicit. The article [16] also claims without details that such an existence result can be extracted from [21]. Section 5 will actually provide such an explicit map embedding ℓ_2^n into $\ell_1^{cn^2}$. I suspect this construction to be very close to what should have been extracted from [21]. It will yield an explicit embedding on sparse vectors taking place in the complex setting. In the real setting, the construction presented next in Sect. 3 is more direct, as it bypasses (i)-(ii), but I would qualify it as quasideterministic rather than explicit. Both constructions pertain to the regime $m \asymp s^4$. This should definitely be improved to $m \asymp s^\nu$ with $\nu \leq 2$, but for the moment there is comfort in the simplicity of the constructions.

3 Simple Embedding of s-sparse Vectors from $\ell_2^N(\mathbb{R})$ to $\ell_1^{cs^4}(\mathbb{R})$

The construction proposed in this section is not explicit, but rather (quasi)deterministic, in the sense that it is outputted by the following Las Vegas algorithm:

for $t = 1, 2, \ldots$

- draw $A \in \mathbb{R}^{m \times N}$ populated with independent Rademacher random variables,
- check if conditions (a)-(b) of Theorem 1 below are satisfied for $\kappa = \sqrt{8\ln(N)}$,

and if they are, then return A.

The success of this procedure is guaranteed by the fact that conditions (a)-(b) certify that A yields an embedding on s-sparse vectors from ℓ_2^N to ℓ_1^m (by Theorem 1), together with the fact that each draw of A satisfies (a)-(b) for $\kappa = \sqrt{8\ln(N)}$ with failure probability at most $1/3$, so the failure probability after T independent rounds is at most $(1/3)^T$. This vanishing quantity is never exactly zero; hence, the procedure cannot technically be qualified as deterministic, but quasideterministic sounds like a suitable designation. Note that the verification of (a)-(b) can be performed in polynomial time, precisely in $O(N^4 m)$ multiplications.

Theorem 1 *Let $A \in \mathbb{R}^{m \times N}$ be populated with entries $A_{j,k} = \pm 1$. Assume that*

$$\text{(a)} \quad \left| \sum_{j=1}^{m} A_{j,k} A_{j,k'} \right| \leq \kappa \sqrt{m} \qquad \text{for all distinct } k, k' \in [1:N],$$

(b) $\left| \sum_{j=1}^{m} A_{j,k} A_{j,k'} A_{j,\ell} A_{j,\ell'} \right| \leq \kappa \sqrt{m}$ *for all distinct* $k, k', \ell, \ell' \in [1 : N]$.

Then, for any $\delta \in (0, 1)$*, the linear map* $A : \ell_2^N \to \ell_1^m$ *has the property that*

$$\alpha m \|x\|_2 \leq \|Ax\|_1 \leq \beta m \|x\|_2 \qquad \text{for all } s\text{-sparse } x \in \mathbb{R}^N,$$

with distortion $\gamma = \beta/\alpha \leq \sqrt{3}\big((1+\delta)/(1-\delta)\big)^{3/2}$ *as soon as* $m \geq \kappa^2 \delta^{-2} s^4$.

Before justifying this theorem, it is worth observing that its assumptions (a)-(b) are indeed fulfilled when $A \in \mathbb{R}^{m \times N}$ is a Rademacher random matrix and $\kappa = \sqrt{8 \ln(N)}$. Fixing distinct k, k' in $[1 : N]$, the $A_{j,k} A_{j,k'}$, $j \in [1 : m]$, are independent Rademacher variables, so by Hoeffding inequality (see, e.g., [3, Corollary 8.8])

$$\mathbb{P}\left[\left| \sum_{i=1}^{m} A_{i,j_1} A_{i,j_2} \right| \geq \kappa \sqrt{m} \right] \leq 2 \exp\left(-\frac{\kappa^2}{2} \right).$$

Hence, by a union bound, one derives:

$$\mathbb{P}[\text{(a) fails}] \leq \binom{N}{2} 2 \exp\left(-\frac{\kappa^2}{2} \right) \leq N^2 \exp\left(-\frac{\kappa^2}{2} \right) = \frac{1}{N^2}.$$

Likewise, fixing distinct $k, k', \ell, \ell' \in [1 : N]$, the $A_{j,k} A_{j,k'} A_{j,\ell} A_{i,\ell'}$, $j \in [1 : m]$, are independent Rademacher variables, so Hoeffding inequality followed by a union bound once again yields:

$$\mathbb{P}[\text{(b) fails}] \leq \binom{N}{4} 2 \exp\left(-\frac{\kappa^2}{2} \right) \leq \frac{N^4}{12} \exp\left(-\frac{\kappa^2}{2} \right) = \frac{1}{12}.$$

Consequently, (a) and (b) are indeed both fulfilled with failure probability bounded, for $N \geq 2$, as

$$\mathbb{P}[\text{(a) or (b) fail}] \leq \frac{1}{N^2} + \frac{1}{12} \leq \frac{1}{3}.$$

Before turning the attention to the proof of Theorem 1, it is also worth isolating two key ingredients as separate lemmas, one being a technical calculation to be reused later and the other one being a way to estimate the ℓ_1-norm from below via the ℓ_2- and ℓ_4-norms.

Lemma 1 *Let* $B \in \mathbb{C}^{q \times r}$ *with* $|B_{j,k}| = 1$ *for all* $j \in [1 : q]$ *and* $k \in [1 : r]$*. Then, for any* $x \in \mathbb{C}^r$*, one has*

$$\|Bx\|_2^2 = q\|x\|_2^2 + \sum_{1\le k\ne k'\le r} \left(\sum_{j=1}^{q} \overline{B_{j,k}} B_{j,k'}\right) \overline{x_k} x_{k'}, \tag{4}$$

$$\|Bx\|_4^4 = 2\|x\|_2^2 \|Bx\|_2^2 - q\|x\|_4^4 + \Sigma_1 \tag{5}$$

$$= 2\|x\|_2^2 \|Bx\|_2^2 - q\|x\|_4^4 + \sum_{k\ne k'} \left(\sum_{j=1}^{q} \overline{B_{j,k}}^2 B_{j,k'}^2\right) \overline{x_k}^2 x_{k'}^2 + \Sigma_2, \tag{6}$$

where the quantities Σ_1 *and* Σ_2 *are given by*

$$\Sigma_1 := \sum_{(k\ne k')\ne(\ell\ne\ell')} \left(\sum_{j=1}^{q} \overline{B_{j,k}} B_{j,k'} B_{j,\ell} \overline{B_{j,\ell'}}\right) \overline{x_k} x_{k'} x_\ell \overline{x_{\ell'}},$$

$$\Sigma_2 := \sum_{\substack{(k\ne k')\ne(\ell\ne\ell')\\(k\ne k')\ne(\ell'\ne\ell)}} \left(\sum_{j=1}^{q} \overline{B_{j,k}} B_{j,k'} B_{j,\ell} \overline{B_{j,\ell'}}\right) \overline{x_k} x_{k'} x_\ell \overline{x_{\ell'}}.$$

Proof For the first identity, one writes:

$$\|Bx\|_2^2 = \sum_{j=1}^{q} \overline{(Bx)_j}(Bx)_j = \sum_{j=1}^{q} \left(\sum_{k=1}^{r} \overline{B_{j,k} x_k}\right)\left(\sum_{k'=1}^{r} B_{j,k'} x_{k'}\right)$$

$$= \sum_{j=1}^{q} \left(\sum_{k=1}^{r} \overline{B_{j,k}}\overline{x_k} B_{j,k} x_k + \sum_{k\ne k'} \overline{B_{j,k}}\overline{x_k} B_{j,k'} x_{k'}\right)$$

$$= \sum_{j=1}^{q}\sum_{k=1}^{r} |B_{j,k}|^2 |x_k|^2 + \sum_{j=1}^{q}\sum_{k\ne k'} \overline{B_{j,k}} B_{j,k'} \overline{x_k} x_{k'},$$

which, by virtue of $|B_{j,k}| = 1$, reduces to (4) after changing the order of summation.

For the second identity, one writes:

$$\|Bx\|_4^4 = \sum_{j=1}^{q} \left[\overline{(Bx)_j}(Bx)_j\right]^2 = \sum_{j=1}^{q} \left[\left(\sum_{k=1}^{r} \overline{B_{j,k} x_k}\right)\left(\sum_{k'=1}^{r} B_{j,k'} x_{k'}\right)\right]^2$$

$$= \sum_{j=1}^{q} \left[\sum_{k=1}^{r} |B_{j,k}|^2 |x_k|^2 + S_j\right]^2 \qquad \text{where } S_j := \sum_{k\ne k'} \overline{B_{j,k}} B_{j,k'} \overline{x_k} x_{k'}$$

$$= \sum_{j=1}^{q} \left[\|x\|_2^2 + S_j\right]^2 = q\|x\|_2^4 + 2\|x\|_2^2 \sum_{j=1}^{q} S_j + \sum_{j=1}^{q} S_j^2. \tag{7}$$

Taking (4) into account, one notices that

$$\sum_{j=1}^{q} S_j = \|Bx\|_2^2 - q\|x\|_2^2. \tag{8}$$

Next, exploiting the fact that $S_j \in \mathbb{R}$ and separating the cases $(k \neq k') = (\ell \neq \ell')$ and $(k \neq k') \neq (\ell \neq \ell')$, one obtains:

$$\begin{aligned} S_j^2 &= S_j \overline{S_j} \\ &= \sum_{k\neq k'} |B_{j,k}|^2 |B_{j,k'}|^2 |x_k|^2 |x_{k'}|^2 + \sum_{(k\neq k')\neq(\ell\neq\ell')} \overline{B_{j,k}} B_{j,k'} B_{j,\ell} \overline{B_{j,\ell'}} \overline{x_k} x_{k'} x_\ell \overline{x_{\ell'}}, \end{aligned}$$

from where it follows that

$$\sum_{j=1}^{q} S_j^2 = \sum_{j=1}^{q} \left(\sum_{k,k'} |x_k|^2 |x_{k'}|^2 - \sum_k |x_k|^4 \right) + \Sigma_1 = q\left(\|x\|_2^4 - \|x\|_4^4 \right) + \Sigma_1. \tag{9}$$

It remains to substitute (8) and (9) into (7) to arrive at (5).

For the last identity, one simply writes:

$$\begin{aligned} \Sigma_1 &= \left\{ \sum_{\substack{(k\neq k')\neq(\ell\neq\ell') \\ (k\neq k')=(\ell'\neq\ell)}} + \sum_{\substack{(k\neq k')\neq(\ell\neq\ell') \\ (k\neq k')\neq(\ell'\neq\ell)}} \right\} \left(\sum_{j=1}^{q} \overline{B_{j,k}} B_{j,k'} B_{j,\ell} \overline{B_{j,\ell'}} \right) \overline{x_k} x_{k'} x_\ell \overline{x_{\ell'}} \\ &= \sum_{k\neq k'} \left(\sum_{j=1}^{q} \overline{B_{j,k}}^2 B_{j,k'}^2 \right) \overline{x_k}^2 x_{k'}^2 + \Sigma_2, \end{aligned}$$

and substituting the latter into (5) directly leads to (6). □

As for the lower estimate of the ℓ_1-norm, this is achieved via an upper estimate of the ℓ_4-norm, as stated below.

Lemma 2 *For any* $y \in \mathbb{C}^m$,

$$\|y\|_1 \geq \frac{\|y\|_2^3}{\|y\|_4^2}.$$

Proof This is Hölder inequality in disguise. Namely, one can easily rearrange

$$\|y\|_2^2 = \sum_{i=1}^{m} |y_i|^{1\times\frac{2}{3}+4\times\frac{1}{3}} \leq \left(\sum_{i=1}^{m} |y_i|^1 \right)^{\frac{2}{3}} \left(\sum_{i=1}^{m} |y_i|^4 \right)^{\frac{1}{3}} = \|y\|_1^{\frac{2}{3}} \|y\|_4^{\frac{4}{3}}$$

into the announced inequality. □

One is now ready to justify the main result of this section.

Proof of Theorem 1 Let an s-sparse vector $x \in \mathbb{R}^N$ be fixed throughout the proof. Applying Lemma 1 with the real matrix $A \in \mathbb{R}^{m \times N}$ taking the role of the complex matrix $B \in \mathbb{C}^{q \times r}$, one first observes, from (4), that

$$\|Ax\|_2^2 - m\|x\|_2^2 = \sum_{k \neq k'} \left(\sum_{j=1}^{m} A_{j,k} A_{j,k'} \right) x_k x_{k'}.$$

Taking the bound (a) into consideration, as well as the sparsity of x, yields:

$$\left| \|Ax\|_2^2 - m\|x\|_2^2 \right| \leq \sum_{k \neq k'} \kappa \sqrt{m} |x_k||x_{k'}| \leq \kappa \sqrt{m} \|x\|_1^2 \leq \frac{\kappa s}{\sqrt{m}} m \|x\|_2^2.$$

For $m \geq \kappa^2 \delta^{-2} s^2$ (which is the case since $m \geq \kappa^2 \delta^{-2} s^4$), this implies

$$(1 - \delta)\, m\|x\|_2^2 \leq \|Ax\|_2^2 \leq (1 + \delta)\, m\|x\|_2^2. \tag{10}$$

As a side note, this is the standard RIP for the renormalized matrix $A/\sqrt{m}$. It was derived solely from (a), which is nothing but a coherence assumption for this matrix. Now, using the standard comparison of the ℓ_1-norm and ℓ_2-norm, one arrives at

$$\|Ax\|_1 \leq \sqrt{m}\, \|Ax\|_2 \leq (1 + \delta)^{1/2}\, m\|x\|_2. \tag{11}$$

Next, using (6) in Lemma 1 while taking into account that $(k \neq k') \neq (\ell \neq \ell')$ and $(k \neq k') \neq (\ell' \neq \ell)$ means that k, k', ℓ, ℓ' are all distinct, and using assumption (b) as well, one can write:

$$\begin{aligned}
\|Ax\|_4^4 &\leq 2\|x\|_2^2 \|Ax\|_2^2 + m\|x\|_2^4 \\
&\quad + \sum_{\substack{k,k',\ell,\ell' \\ \text{all distinct}}} \left| \sum_{j=1}^{m} A_{j,k} A_{j,k'} A_{j,\ell} A_{j,\ell'} \right| |x_k||x_{k'}||x_\ell||x_{\ell'}| \\
&\leq (2(1+\delta)+1) m\|x\|_2^4 + \kappa\sqrt{m}\|x\|_1^4 \leq (3 + 2\delta) m\|x\|_2^4 + \kappa\sqrt{m} s^2 \|x\|_2^4 \\
&= \left(3 + 2\delta + \frac{\kappa s^2}{\sqrt{m}} \right) m\|x\|_2^4 \leq (3 + 3\delta) m\|x\|_2^4,
\end{aligned}$$

where the last step exploited $m \geq \kappa^2 \delta^{-2} s^4$. This upper bound on $\|Ax\|_4$, combined with the lower bound on $\|Ax\|_2$ from (10), ensures, according to Lemma 2, that

$$\|Ax\|_1 \geq \frac{\left(\|Ax\|_2^2\right)^{3/2}}{\left(\|Ax\|_4^4\right)^{1/2}} \geq \frac{\left((1-\delta) m\|x\|_2^2\right)^{3/2}}{\left(3(1+\delta) m\|x\|_2^4\right)^{1/2}} = \left(\frac{(1-\delta)^3}{3(1+\delta)} \right)^{1/2} m\|x\|_2. \tag{12}$$

The estimates (11) and (12) together establish the result, noting in particular the expression of the distortion. □

Remark 1 It is unclear if the above argument can be refined to improve the exponent $\nu = 4$ for the regime $m \asymp s^\nu$ where the "ℓ_2 to ℓ_1" RIP provably holds. One point is certain, however: for a matrix $A \in \mathbb{K}^{m \times N}$ whose entries all have modulus/absolute value equal to one, one cannot beat $\nu = 2$ if one estimates the ℓ_1-norm from below via an upper bound on the ℓ_4-norm of the form $\|Ax\|_4^4 \le Cm\|x\|_2^4$ for all s-sparse vectors $x \in \mathbb{K}^N$. Indeed, for a fixed $i \in [1 : m]$, if $x = A_{S,i}$ represents the ith rows of A restricted to a set $S \subseteq [1 : N]$ of size s, then

$$Cms^2 = Cm\|x\|_2^4 \ge \|Ax\|_4^4 = \sum_{j=1}^{m} |\langle A_{:,j}, A_{S,i}\rangle|^4 \ge |\langle A_{:,i}, A_{S,i}\rangle|^4 = s^4,$$

which forces $m \ge C^{-1}s^2$.

4 Explicit Isometric Embedding from $\ell_2^p(\mathbb{C})$ into $\ell_4^{cp^2}(\mathbb{C})$

As mentioned earlier, deterministic linear embeddings from ℓ_2^n to $\ell_1^{cn^2}$ with constant distortion have been claimed to exist in different places without full details, e.g., stating that a construction can be extracted from [21]. In the latter, a central role was played by Sidon sets, which are the same as Golomb rulers. In what follows, I propose a Golomb-ruler argument which likely coincides with what should have been extracted from [21]. Taking a detour via the ℓ_4-norm, the argument actually uncovers a deterministic embedding from ℓ_2^p (p being prime) to $\ell_4^{cp^2}$ whose distortion is exactly equal to one. This isometric embedding is singled out in this section. The main ingredient, due to [22], is reproduced here for completeness.

Lemma 3 *For a prime number $p \ge 3$, the integers*

$$g(k) = 2pk + \left(k^2\right)_p, \qquad k \in [0 : p-1],$$

where $\left(k^2\right)_p$ denotes the integer $t \in [0 : p-1]$ such that $k^2 \equiv t \mod p$, form a Golomb ruler, in the sense that

$$g(k) - g(k') \ne g(\ell) - g(\ell') \qquad \textit{whenever } (k \ne k') \ne (\ell \ne \ell').$$

Proof Let $(k \ne k') \ne (\ell \ne \ell')$ and suppose $g(k) - g(k') = g(\ell) - g(\ell')$. Writing $k' = k + \sigma$ and $\ell' = \ell + \tau$ for some $\sigma, \tau \in [-p+1 : p-1] \setminus \{0\}$, this reads $2pk + \left(k^2\right)_p - 2p(k+\sigma) - \left((k+\sigma)^2\right)_p = 2p\ell + \left(\ell^2\right)_p - 2p(\ell+\tau) - \left((\ell+\tau)^2\right)_p$, i.e.,

$$\left(k^2\right)_p - \left((k+\sigma)^2\right)_p - \left(\ell^2\right)_p + \left((\ell+\tau)^2\right)_p = 2p(\sigma - \tau). \tag{13}$$

Since the left-hand side, in absolute value, is at most $2(p-1) < p$, the right-hand side must be zero, so that $\sigma = \tau$. Taking $\left((k+\sigma)^2\right)_p \equiv \left(k^2\right)_p + \left(2\sigma k\right)_p + \left(\sigma^2\right)_p$ mod p into account, as well as $\left((\ell+\tau)^2\right)_p = \left((\ell+\sigma)^2\right)_p \equiv \left(\ell^2\right)_p + \left(2\sigma\ell\right)_p + \left(\sigma^2\right)_p$ mod p, looking at (13) modulo p yields:

$$\left(2\sigma\ell\right)_p - \left(2\sigma k\right)_p \equiv 0 \mod p, \qquad \text{i.e.,} \qquad 2\sigma(\ell - k) \equiv 0 \mod p.$$

But since $2 \not\equiv 0$ and $\sigma \not\equiv 0 \mod p$, one deduces that $\ell - k \equiv 0 \mod p$ and hence that $k = \ell$. In view of $k' = k + \sigma$ and $\ell' = \ell + \tau$ with $\sigma = \tau$, it follows that $(k \neq k') = (\ell \neq \ell')$. This leads to a contradiction, showing that g indeed generates a Golomb ruler. □

The coveted isometric embedding from ℓ_2^p to $\ell_4^{cp^2}$ will be obtained by combining Lemma 1 with Lemma 3. Before that, it is worth pausing to remark that g maps $[0 : p-1]$ into $[0 : q-1]$, where $q = 3p(p-1)+1$ is quadratic in p, and that such a quadratic order is optimal. Indeed, for any Golomb ruler g from $[0 : p-1]$ into $[0 : q-1]$, all the distinct $g(k) - g(k')$ indexed by ordered pairs $(k \neq k')$ are contained in $[-q+1 : q-1]$, and as such $2q - 1 \geq p(p-1)$.

Theorem 2 *For a prime number $p \geq 3$, let $m = 6p^2 - 6p + 1$. Consider the matrix $A'' \in \mathbb{C}^{m \times p}$ with entries:*

$$A''_{j,k} = \exp\left(i2\pi \frac{jg(k)}{m}\right), \quad j \in [0 : m-1], \ k \in [0 : p-1]. \tag{14}$$

Then the matrix $M \in \mathbb{C}^{(m+p)\times p}$ defined as

$$M := \left[\begin{array}{c} \dfrac{1}{(2m)^{1/4}} A'' \\ \hline \dfrac{1}{2^{1/4}} I_p \end{array}\right] \tag{15}$$

provides an isometric embedding from $\ell_2^p(\mathbb{C})$ into $\ell_2^{m+p}(\mathbb{C})$, i.e.,

$$\|Mx\|_4 = \|x\|_2 \qquad \text{for all } x \in \mathbb{C}^p.$$

Proof Let a vector $x \in \mathbb{C}^p$ be fixed throughout the proof. For any $k \neq k'$, one observes that

$$\sum_{j=0}^{m-1} \overline{A''_{j,k}} A''_{j,k'} = \sum_{j=0}^{m-1} \exp\left(-i2\pi \frac{j(g(k) - g(k'))}{m}\right) = 0,$$

owing to $g(k) - g(k') \in [-q+1 : q-1] \subseteq [-m+1 : m-1]$ being nonzero (otherwise, if $g(k) = g(k')$, choosing $k'' \notin \{k, k'\}$ would yield $g(k) - g(k'') =$

$g(k') - g(k''))$. According to Lemma 1, and specifically to (4), one therefore has

$$\|A''x\|_2^2 = m\|x\|_2^2.$$

Next, for any $(k \neq k') \neq (\ell \neq \ell')$, one observes that

$$\sum_{j=0}^{m-1} \overline{A''_{j,k}} A''_{j,k'} A''_{j,\ell} \overline{A''_{j,\ell'}} = \exp\left(-i2\pi \frac{j\big((g(k) - g(k')) - (g(\ell) - g(\ell'))\big)}{m}\right) = 0,$$

owing to $(g(k) - g(k')) - (g(\ell) - g(\ell')) \in [-2q + 2 : 2q - 2] = [-m + 1 : m - 1]$ being nonzero. According to Lemma 1 again, and specifically to (5), one obtains:

$$\|A''x\|_4^4 = 2\|x\|_2^2\|A''x\|_2^2 - m\|x\|_4^4 = 2m\|x\|_2^4 - m\|x\|_4^4.$$

From here, it easily follows that

$$\|Mx\|_4^4 = \frac{1}{2m}\|A''x\|_4^4 + \frac{1}{2}\|x\|_4^4 = \|x\|_2^4,$$

which is the desired result. □

Remark 2 Isometric embeddings from ℓ_2 to ℓ_4 can alternatively be viewed through the lenses of spherical designs and of tensors. In order not to be diverted from the main goal, this connection will be brought forward much later, in Sect. 7.

5 Explicit Embedding of s-sparse Vectors from $\ell_2^N(\mathbb{C})$ to $\ell_1^{cs^4}(\mathbb{C})$

Based on Theorem 2 in the previous section, a linear embedding from $\ell_2^p(\mathbb{C})$ to $\ell_1^{cp^2}(\mathbb{C})$ with constant distortion can easily be generated. However, instead of using the matrix M from (15), which provided an isometric embedding from $\ell_2^p(\mathbb{C})$ to $\ell_4^{cp^2}(\mathbb{C})$, the matrix A'' from (14) is preferred, as it leads to a nicer expression for the distortion. Recall that the proof of Theorem 2 revealed that, for any $x \in \mathbb{C}^p$,

$$\|A''x\|_2 = \sqrt{m}\|x\|_2 \qquad \text{and} \qquad \|A''x\|_4 \leq (2m)^{1/4}\|x\|_2.$$

Theorem 3 *For a prime number $p \geq 3$, let $m = 6p^2 - 6p + 1$. The matrix $A'' \in \mathbb{C}^{m \times p}$ defined in* (14) *provides an embedding from $\ell_2^p(\mathbb{C})$ into $\ell_1^m(\mathbb{C})$ with distortion at most $\sqrt{2}$. Precisely, one has:*

$$\frac{m}{\sqrt{2}}\|x\|_2 \leq \|A''x\|_1 \leq m\,\|x\|_2 \qquad \textit{for all } x \in \mathbb{C}^p.$$

Proof Let a vector $x \in \mathbb{C}^p$ be fixed throughout this short proof. On the one hand, comparing ℓ_1- and ℓ_2-norms yields:

$$\|A''x\|_1 \leq \sqrt{m}\|A''x\|_2 = m\|x\|_2.$$

On the other hand, according to Lemma 2, one has:

$$\|A''x\|_1 \geq \frac{\|A''x\|_2^3}{\|A''x\|_4^2} \geq \frac{\sqrt{m}^3\|x\|_2^3}{\sqrt{2m}\|x\|_2^2} = \frac{m}{\sqrt{2}}\|x\|_2.$$

These two inequalities together justify the announced embedding. □

It is now time for the main result of this section, namely, the awaited explicit linear embedding of s-sparse vectors from $\ell_2^N(\mathbb{C})$ to $\ell_1^{cs^4}(\mathbb{C})$. It repeats the general strategy (i)-(ii) outlined in Sect. 2.

Theorem 4 *Given integers $N, s \geq 1$ with $N \gg s^4$, let $p \geq 3$ be a prime number between $9s^2\lceil \ln^2(N)\rceil$ and $18s^2\lceil \ln^2(N)\rceil$ and let $m = 6p^2 - 6p + 1 \asymp s^4 \ln^2(N)$. Then the matrix $A \in \mathbb{C}^{m\times N}$ indexed by $[0 : m-1]$ and by an arbitrary N-set $\mathcal{F}$ of polynomials over $\mathbb{F}_p$ of degree at most $d = \lceil \ln(N/p)/\ln(p)\rceil$ and with entries*

$$A_{k,f} = \frac{1}{\sqrt{p}}\sum_{k=0}^{p-1} \exp\left(i2\pi\left(j\frac{2pk+(k^2)_p}{m} + \frac{kf(k)}{p}\right)\right), \quad k \in [0:m-1],\ f \in \mathcal{F}, \tag{16}$$

provides an explicit embedding from $\ell_2^N(\mathbb{C})$ into $\ell_1^m(\mathbb{C})$ with distortion on s-sparse vectors at most 2, namely,

$$\frac{m}{\sqrt{3}}\|x\|_2 \leq \|Ax\|_1 \leq \frac{2m}{\sqrt{3}}\|x\|_2 \qquad \text{for all } s\text{-sparse } x \in \mathbb{C}^N. \tag{17}$$

Proof Picking a prime number p between $9s^2\lceil \ln^2(N)\rceil$ and $18s^2\lceil \ln^2(N)\rceil$ is possible by Bertrand postulate. One considers the smallest integer d such that $p^{d+1} \geq N$, i.e., $d = \lceil \ln(N/p)/\ln(p)\rceil$. Note that $1 < p < d$ since $p^2 \lesssim s^4\text{polylog}(N) < N$ and $p^p \geq e^p \geq e^{\ln(N)} = N$. From $p^d < N$, one also deduces that $d \leq d\ln(p) < \ln(N)$. Let then $A' \in \mathbb{C}^{p\times p^{d+1}}$ be the "Weil" matrix with entries defined in (3), recalling that $\delta_s(A') < s\mu(A) < sd/\sqrt{p} < s\ln(N)/(3s\ln(N)) = 1/3$, so that

$$\|A'x\|_2^2 \begin{cases} \leq (1+\delta_s(A'))\|x\|_2^2 \leq \dfrac{4}{3}\|x\|_2^2 \\ \\ \geq (1-\delta_s(A'))\|x\|_2^2 \geq \dfrac{2}{3}\|x\|_2^2 \end{cases} \qquad \text{for all } s\text{-sparse } x \in \mathbb{C}^{p^{d+1}}.$$

Let also $A'' \in \mathbb{C}^{m\times p}$ be the "Golomb" matrix with entries defined in (14), recalling that

$$\|A''x\|_1 \begin{cases} \le m\|x\|_2 \\ \ge \dfrac{m}{\sqrt{2}}\|x\|_2 \end{cases} \qquad \text{for all } x \in \mathbb{C}^p.$$

As a result, for the matrix $\widetilde{A} := A'' \times A' \in \mathbb{C}^{m\times p^{d+1}}$ and for any s-sparse $x \in \mathbb{C}^{p^{d+1}}$,

$$\|\widetilde{A}x\|_1 = \|A''(A'x)\|_1 \begin{cases} \le m\|A'x\|_2 \le m\sqrt{\dfrac{4}{3}}\|x\|_2 = \dfrac{2m}{\sqrt{3}}\|x\|_2, \\ \ge \dfrac{m}{\sqrt{2}}\|A'x\|_2 \ge \dfrac{m}{\sqrt{2}}\sqrt{\dfrac{2}{3}}\|x\|_2 = \dfrac{m}{\sqrt{3}}\|x\|_2. \end{cases}$$

This strongly resembles the desired inequalities (17); expect that the matrix $\widetilde{A}$ has p^{d+1} columns, while the matrix A should have N columns. But one can simply remove $p^{d+1}-N$ arbitrary columns from $\widetilde{A} \in \mathbb{C}^{m\times p^{d+1}}$ to create a matrix $A \in \mathbb{C}^{m\times N}$ satisfying (17). The entries of this matrix, indexed by j in $[0:m-1]$ and by f in a set $\mathcal{F}$ of polynomials over $\mathbb{F}_p$ of degree at most d with size N, are given by

$$A_{j,f} = \sum_{k=0}^{p-1} A''_{j,k}A'_{k,f} = \sum_{k=0}^{p-1} \exp\left(i2\pi\frac{jg(k)}{m}\right)\frac{1}{\sqrt{p}}\exp\left(i2\pi\frac{kf(k)}{p}\right),$$

which reduces to the expression announced in (16). □

6 Outlook into Low-Rank Recovery

The theory of compressive sensing also deals with objects of nominally high but intrinsically low dimension beyond s-sparse vectors $x \in \mathbb{K}^N$, prototypically with matrices $X \in \mathbb{K}^{n\times n}$ of rank at most r. In this scenario, too, one can recover such objects from their compressive measurements $\mathcal{A}(X) \in \mathbb{K}^m$ with m being of the order of $nr \ll n^2$. This coup can be achieved (see [23]) by nuclear norm minimization when the linear map $\mathcal{A} : \mathbb{K}^{n\times n} \to \mathbb{K}^m$ satisfies an RIP of the form:

$$(1-\delta)\|X\|_F^2 \le \|\mathcal{A}(X)\|_2^2 \le (1+\delta)\|X\|_F^2 \qquad \text{for all rank-}r\ X \in \mathbb{K}^{n\times n}. \tag{18}$$

For $m \asymp nr$, this RIP is fulfilled when $A_1, \dots, A_m \in \mathbb{K}^{n\times n}$ in $\mathcal{A}(Z)_i = \langle A_i, X\rangle_F$ are independent random matrices populated with independent properly normalized Gaussian entries; see [24]. But, as in the vector case, no deterministic linear map $\mathcal{A} : \mathbb{K}^{n\times n} \to \mathbb{K}^m$ is known in this optimal regime $m \asymp nr$. Worst, I am not aware of simple deterministic constructions overcoming the trivial regime $m \asymp n^2$, and I have not seen an effective analog of the notion of coherence.

Interestingly, the theory of low-rank recovery can also be built from a modification of the RIP featuring the ℓ_1-norm as the inner norm (see [5, Chapter 16]), namely, from

$$(1-\delta)\|X\|_F \leq \|\mathcal{A}(X)\|_1 \leq (1+\delta)\|X\|_F \qquad \text{for all rank} - r\ X \in \mathbb{K}^{n\times n}. \tag{19}$$

This alternative version seems even more relevant in the present scenario. Indeed, when $m \asymp nr$, the rank-one measurements

$$\mathcal{A}(Z)_i = \langle b_i, Za_i\rangle = \langle A_i, Z\rangle_F, \qquad A_i := b_i a_i^*,$$

with independent properly normalized Gaussian vectors $a_1, \ldots, a_m, b_1, \ldots, b_m \in \mathbb{K}^n$ do not lead to the RIP (18) but to the RIP (19), while the latter still enables low-rank recovery via nuclear norm minimization; see [25]. It also enables low-rank recovery via iterative-thresholding-type algorithms; see [26]. But the possibility of fulfilling this modified RIP in a regime $m \asymp n^\lambda r^\mu \ll n^2$ with deterministic matrices $A_1, \ldots, A_m \in \mathbb{K}^{n\times n}$—of rank one or even unrestricted—is a wide open question. This question is hereby set as a challenge to the readers.

7 Addendum: Isometric Embeddings, Spherical Designs, and Tensors

As mentioned in Remark 2, isometric embedding from $\ell_2^N(\mathbb{K})$ to $\ell_{2k}(\mathbb{K})$, $\mathbb{K} \in \{\mathbb{R}, \mathbb{C}\}$, has connections with spherical designs and tensors. This is made precise by the following result, found in [27] for the case $\mathbb{K} = \mathbb{R}$. There is a connection with cubature formulas on the sphere, too, as found, e.g., in [28]. I thank G. Paouris who recently brought the latter to my attention—it also contains an explicit isometric embedding from $\ell_2^n(\mathbb{C})$ into $\ell_2^{\Theta(n^4)}(\mathbb{C})$ when n is an odd prime power.

Theorem 5 *Let $2k \geq 2$ be an even integer. The following properties are equivalent:*

(1) There exists $A \in \mathbb{K}^{N\times n}$ providing an isometric embedding from ℓ_2^n into ℓ_{2k}^N, i.e.,

$$\|Ax\|_{2k} = \|x\|_2 \qquad \textit{for all } x \in \mathbb{K}^n;$$

(2) There exist $x_1, \ldots, x_N \in S_2^n$ and $\tau_1, \ldots, \tau_N \geq 0$ with $\tau_1 + \cdots + \tau_N = 1$ such that

$$\sum_{i,j=1}^{N} \tau_i\tau_j |\langle x_i, x_j\rangle|^{2k} = \int_{S_2^n\times S_2^n} |\langle x, y\rangle|^{2k} d\sigma(x)d\sigma(y),$$

where σ is the normalized standard measure on the unit sphere S_2^n of ℓ_2^n;

(3) *There exist* $x_1, \ldots, x_N \in S_2^n$ *and* $\tau_1, \ldots, \tau_N \geq 0$ *with* $\tau_1 + \cdots + \tau_N = 1$ *such that*

$$\sum_{i=1}^{N} \tau_i \otimes^k (x_i \otimes \overline{x_i}) = \int_{S_2^n} \otimes^k (x \otimes \overline{x}) d\sigma(x).$$

Since RIPs are not genuine isometric embeddings, but almost isometric ones, this result will be established in a slightly stronger form for the sake of completeness. Towards this end, some pieces of notations and some identities are brought forth as a preamble. First, one considers the quantity δ (independent of $x \in S_2^n$) defined by

$$\delta = \delta_{n,2k} := \int_{S_2^n} |\langle x, y\rangle|^{2k} d\sigma(y) = \begin{cases} \dfrac{(2k-1)\cdots 3\cdot 1}{(n+2k-2)\cdots(n+2)\cdot n} & \text{for } \mathbb{K} = \mathbb{R}, \\ \dfrac{k!}{(n+k-1)\cdots(n+1)\cdot n} & \text{for } \mathbb{K} = \mathbb{C}, \end{cases}$$

whose numerical value is given in [27] for $\mathbb{K} = \mathbb{R}$ and in [29] for $\mathbb{K} = \mathbb{C}$. Note that δ coincides with the double integral appearing in (2). As for the integral appearing in (3), called distribution $2k$-tensor (when $\mathbb{K} = \mathbb{R}$), it shall be denoted by D. Thus,

$$D = D_{n,2k} := \int_{S_2^n} \otimes^k (y \otimes \overline{y}) d\sigma(y).$$

Using the standard notion of inner product on tensor spaces, one verifies below that, for any $x \in S_2^n$,

$$\langle \otimes^k (x \otimes \overline{x}), D\rangle = \delta \qquad \text{and} \qquad \langle D, D\rangle = \delta. \tag{20}$$

To justify the leftmost identity of (20), it suffices to write:

$$\begin{aligned} \langle \otimes^k (x \otimes \overline{x}), D\rangle &= \left\langle \otimes^k (x \otimes \overline{x}), \int_{S_2^n} \otimes^k (y \otimes \overline{y}) d\sigma(y) \right\rangle \\ &= \int_{S_2^n} \left\langle \otimes^k (x \otimes \overline{x}), \otimes^k (y \otimes \overline{y}) \right\rangle d\sigma(y) \\ &= \int_{S_2^n} \langle x \otimes \overline{x}, y \otimes \overline{y}\rangle^k \, d\sigma(y) = \int_{S_2^n} \left(|\langle x, y\rangle|^2\right)^k d\sigma(y) \\ &= \delta. \end{aligned}$$

To justify the rightmost identity of (20), it suffices to write

$$\begin{aligned} \langle D, D\rangle &= \left\langle \int_{S_2^n} \otimes^k (x \otimes \overline{x}) d\sigma(x), D \right\rangle = \int_{S_2^n} \left\langle \otimes^k (x \otimes \overline{x}), D \right\rangle d\sigma(x) = \int_{S_2^n} \delta \, d\sigma(x) \\ &= \delta. \end{aligned}$$

From here, an important identity follows easily, namely: for any $x_1, \ldots, x_N \in S_2^n$ and any $\tau_1, \ldots, \tau_N \geq 0$ with $\tau_1 + \cdots + \tau_N = 1$,

$$\left\| \sum_{i=1}^{N} \tau_i \otimes^k (x_i \otimes \overline{x_i}) - D \right\|_2^2 = \sum_{i,j=1}^{N} \tau_i \tau_j |\langle x_i, x_j \rangle|^{2k} - \delta. \tag{21}$$

Indeed, the justification of (21) simply reads:

$$\begin{aligned}
&\left\| \sum_{i=1}^{N} \tau_i \otimes^k (x_i \otimes \overline{x_i}) - D \right\|_2^2 \\
&= \sum_{i,j=1}^{N} \tau_i \tau_j \langle \otimes^k (x_i \otimes \overline{x_i}), \otimes^k (x_j \otimes \overline{x_j}) \rangle - 2\,\mathrm{Re} \left\langle \sum_{i=1}^{N} \tau_i \otimes^k (x_i \otimes \overline{x_i}), D \right\rangle + \langle D, D \rangle \\
&= \sum_{i,j=1}^{N} \tau_i \tau_j \langle x_i \otimes \overline{x_i}, x_j \otimes \overline{x_j} \rangle^k - 2 \sum_{i=1}^{N} \tau_i \,\mathrm{Re} \left\langle \otimes^k (x_i \otimes \overline{x_i}), D \right\rangle + \langle D, D \rangle \\
&= \sum_{i,j=1}^{N} \tau_i \tau_j \left(|\langle x_i, x_j \rangle|^2 \right)^k - 2 \sum_{i=1}^{N} \tau_i \,\mathrm{Re}(\delta) + \delta,
\end{aligned}$$

where both identities of (20) were used in the last step. Taking $\sum_{i=1}^{N} \tau_i = 1$ into account then leads to the desired identity (21). Note that it implies the so-called Sidelnikov inequality, namely:

$$\sum_{i,j=1}^{N} \tau_i \tau_j |\langle x_i, x_j \rangle|^{2k} \geq \delta \quad \text{whenever } x_1, \ldots, x_N \in S_2^n, \ \tau_1, \ldots, \tau_N \geq 0, \ \sum_{i=1}^{N} \tau_i = 1.$$

After this preparatory work, one can now state and prove the slight generalization of Theorem 5, which is retrieved as the special case $\varepsilon_1 = \varepsilon_2 = \varepsilon_3 = 0$.

Theorem 6 *Let $2k \geq 2$ be an even integer. For $\varepsilon_1, \varepsilon_2, \varepsilon_3 \geq 0$ that depend on each other, the following properties are equivalent:*

(1') There exists a matrix $A \in \mathbb{K}^{N \times n}$ for which

$$(1 - \varepsilon_1) \|x\|_2^{2k} \leq \|Ax\|_{2k}^{2k} \leq (1 + \varepsilon_1) \|x\|_2^{2k} \qquad \text{for all } x \in \mathbb{K}^n;$$

(2') There exist $x_1, \ldots, x_N \in S_2^n$ and $\tau_1, \ldots, \tau_N \geq 0$ with $\tau_1 + \cdots + \tau_N = 1$ such that

$$\sum_{i,j=1}^{N} \tau_i \tau_j |\langle x_i, x_j \rangle|^{2k} \leq \delta + \varepsilon_2;$$

(3') *There exist* $x_1, \dots, x_N \in S_2^n$ *and* $\tau_1, \dots, \tau_N \geq 0$ *with* $\tau_1 + \cdots + \tau_N = 1$ *such that*

$$\left\| \sum_{i=1}^{N} \tau_i \otimes^k (x_i \otimes \overline{x_i}) - D \right\|_2 \leq \varepsilon_3.$$

Proof (2') $\Leftrightarrow$ (3') with $\varepsilon_3 = \sqrt{\varepsilon_2}$. This equivalence results from identity (21).

(3') $\Rightarrow$ (1') with $\varepsilon_1 = \varepsilon_3/\delta$. For any $x \in \mathbb{K}^n$, one observes that the leftmost identity of (20) yields $\langle D, \otimes^k(x \otimes \overline{x})\rangle = \delta \|x\|_2^{2k}$ by homogeneity. One also observes that

$$\begin{aligned}\left\langle \sum_{i=1}^{N} \tau_i \otimes^k (x_i \otimes \overline{x_i}), \otimes^k(x \otimes \overline{x}) \right\rangle &= \sum_{i=1}^{N} \tau_i \langle \otimes^k(x_i \otimes \overline{x_i}), \otimes^k(x \otimes \overline{x})\rangle \\ &= \sum_{i=1}^{N} \tau_i \left(|\langle x_i, x\rangle|^2\right)^k.\end{aligned}$$

From both these observations, it follows that

$$\begin{aligned}\left| \sum_{i=1}^{N} \tau_i |\langle x_i, x\rangle|^{2k} - \delta \|x\|_2^{2k} \right| &= \left| \left\langle \sum_{i=1}^{N} \tau_i \otimes^k (x_i \otimes \overline{x_i}) - D, \otimes^k(x \otimes \overline{x}) \right\rangle \right| \\ &\leq \left\| \sum_{i=1}^{N} \tau_i \otimes^k (x_i \otimes \overline{x_i}) - D \right\|_2 \left\| \otimes^k (x \otimes \overline{x}) \right\|_2 \\ &\leq \varepsilon_3 \|x\|_2^{2k}.\end{aligned}$$

Therefore, setting $a_i := (\tau_i/\delta)^{\frac{1}{2k}} x_i$ for each $i \in [1:N]$, the above becomes:

$$\left| \sum_{i=1}^{N} |\langle a_i, x\rangle|^{2k} - \|x\|_2^{2k} \right| \leq \varepsilon_1 \|x\|_2^{2k},$$

which reduces to (1') for the matrix $A \in \mathbb{K}^{N \times n}$ with rows $a_1^*, \dots, a_N^*$.

(1') $\Rightarrow$(2') with $\varepsilon_2 = 4\varepsilon_1\delta$ when $\varepsilon_1 \leq 1/2$. With $a_1^*, \dots, a_N^*$ denoting the rows of the matrix $A \in \mathbb{K}^{N \times n}$, the almost isometric embedding takes the form:

$$(1 - \varepsilon_1)\|x\|_2^{2k} \leq \sum_{i=1}^{N} |\langle a_i, x\rangle|^{2k} \leq (1 + \varepsilon_1)\|x\|_2^{2k} \qquad \text{for all } x \in \mathbb{K}^n. \tag{22}$$

First, integrating (22) over $x \in S_2^n$ implies that $1 - \varepsilon_1 \le \sum_{i=1}^N \|a_i\|_2^{2k}\delta \le 1 + \varepsilon_1$. Setting $S := \sum_{i=1}^N \|a_i\|_2^{2k}$, this means that

$$\frac{\delta}{1+\varepsilon_1} \le \frac{1}{S} \le \frac{\delta}{1-\varepsilon_1}. \tag{23}$$

Second, setting $x_i := a_i/\|a_i\|_2 \in S_2^n$ for each $i \in [1 : N]$ and selecting $x = x_j$ in (22) yields

$$1 - \varepsilon_1 \le \sum_{i=1}^N \|a_i\|_2^{2k} |\langle x_i, x_j\rangle|^{2k} \le 1 + \varepsilon_1.$$

Defining $\tau_i := \|a_i\|_2^{2k}/S$, multiplying the latter by τ_j and summing over $j \in [1 : N]$ leads to

$$\frac{1-\varepsilon_1}{S} \le \sum_{i,j=1}^N \tau_i \tau_j |\langle x_i, x_j\rangle|^{2k} \le \frac{1+\varepsilon_1}{S}. \tag{24}$$

From the estimates (23) and (24), one finally concludes that

$$\sum_{i,j=1}^N \tau_i \tau_j |\langle x_i, x_j\rangle|^{2k} \le \frac{1+\varepsilon_1}{1-\varepsilon_1}\delta \le (1 + 4\varepsilon_1)\delta,$$

where the last step used $\varepsilon_1 \le 1/2$. This is the desired inequality with $\varepsilon_2 = 4\varepsilon_1\delta$. □

Acknowledgments S. F. is partially supported by grants from the NSF (DMS-2053172) and from the ONR (N00014-20-1-2787).

References

1. Candès, E., Romberg, J., Tao, T.: Robust uncertainty principles: exact signal reconstruction from highly incomplete frequency information. IEEE Transactions on Information Theory **52**, 489–509 (2006).
2. Donoho, D.: For most large underdetermined systems of linear equations the minimal ℓ_1 solution is also the sparsest solution. Communications on Pure and Applied Mathematics **59**, 797–829 (2006).
3. Foucart, S., Rauhut, H.: A Mathematical Introduction to Compressive Sensing. Birkhäuser (2013).
4. Candès, E., Tao, T.: Decoding by linear programming. IEEE Transactions on Information Theory **51**, 4203–4215 (2005).
5. Foucart, S.: Mathematical Pictures at a Data Science Exhibition. Cambridge University Press (2022).

6. Foucart, S. Lecué, G.: An IHT algorithm for sparse recovery from subexponential measurements. IEEE Signal Processing Letters **24**, 1280–1283 (2017).
7. Dirksen, S., Jung, H. C., Rauhut, H.: One-bit compressed sensing with partial Gaussian circulant matrices. Information and Inference **9**, 601–626 (2020).
8. Foucart, S., Lai, M.-J.: Sparse recovery with pre-Gaussian random matrices. Studia Mathematica **200**, 91–102 (2010).
9. Foucart, S.: The sparsity of LASSO-type minimizers. Applied and Computational Harmonic Analysis **62**, 441–452 (2023).
10. Strohmer, T., Heath Jr., R. W.: Grassmannian frames with applications to coding and communication. Applied and Computational Harmonic Analysis **14**, 257–275 (2003).
11. DeVore, R. A.: Deterministic constructions of compressed sensing matrices. Journal of Complexity **23**, 918–925 (2007).
12. Niederreiter, H., Winterhof, A.: Applied Number Theory. Springer (2015).
13. Bourgain, J., Dilworth, S., Ford, K., Konyagin, S., Kutzarova, D.: Explicit constructions of RIP matrices and related problems. Duke Mathematical Journal **159**, 145–185 (2011).
14. Bandeira, A. S., Mixon, D. G., Moreira, J.: A conditional construction of restricted isometries. International Mathematics Research Notices **2017**, 372–381 (2017).
15. Indyk, P.: Uncertainty principles, extractors, and explicit embeddings of ℓ_2 into ℓ_1. Proceedings of the thirty-ninth annual ACM symposium on Theory of Computing (2007).
16. Guruswami, V., Lee, J. R., Razborov, A.: Almost Euclidean subspaces of ℓ_1^N via expander codes. Combinatorica **30**, 47–68 (2010).
17. Guruswami, V., Umans, C., Vadhan, S.: Unbalanced expanders and randomness extractors from Parvaresh–Vardy codes. Journal of the ACM **56**, 20 (2009).
18. Berinde, R., Gilbert, A. C., Indyk, P., Karloff. H., Strauss, M. J.: Combining geometry and combinatorics: a unified approach to sparse signal recovery. Proceedings of the forty-sixth annual Allerton conference on Communication, Control, and Computing, 798–805 (2008).
19. Linial, N., London, E., and Rabinovich, Y.: The geometry of graphs and some of its algorithmic applications. Combinatorica **15**, 215–245 (1995).
20. Berger, B.: The fourth moment method. SIAM Journal on Computing **26**, 1188–1207 (1997).
21. Rudin, W.: Trigonometric series with gaps. Journal of Mathematics and Mechanics **9**, 203–227 (1960).
22. Erdős, P., Turán, P.: On a problem of Sidon in additive number theory, and on some related problems. Journal of the London Mathematical Society **16**, 212–215 (1941).
23. Recht, B., Fazel, M., Parrilo, P. A.: Guaranteed minimum-rank solutions of linear matrix equations via nuclear norm minimization. SIAM Review **52**, 471–501 (2010).
24. Candès, E., Plan, Y.: Tight oracle inequalities for low-rank matrix recovery from a minimal number of noisy random measurements. IEEE Transactions on Information Theory **57**, 2342–2359 (2011).
25. Cai, T., Zhang, A.: ROP: Matrix recovery via rank-one projections. The Annals of Statistics **43**, 102–138 (2015).
26. Foucart, S., Subramanian, S.: Iterative hard thresholding for low-rank recovery from rank-one projections. Linear Algebra and its Applications **572**, 117–134 (2019).
27. Seidel, J. J.: Spherical designs and tensors. In: Progress in Algebraic Combinatorics **24**, 309–322. Mathematical Society of Japan (1996).
28. König, H.: Isometric imbeddings of Euclidean spaces into finite dimensional ℓ_p-spaces. Banach Center Publications **34**, 79–87 (1995).
29. Kotelina, N. O., Pevnyi, A. B.: Complex spherical semi-designs. Russian Mathematics **61**, 46–51 (2017).

Ridge Function Machines

David E. Stewart

Abstract Ridge function machines implement a computationally tractable and theoretically sound means of creating approximations to functions given by data. A ridge function is a function $\mathbb{R}^n \to \mathbb{R}$ given by $\boldsymbol{x} \mapsto g(\boldsymbol{w}^T\boldsymbol{x})$ where $\boldsymbol{w} \in \mathbb{R}^n$ and $g\colon \mathbb{R} \to \mathbb{R}$. A ridge function machine is a sum of ridge functions $\sum_{j=1}^{m} g_j(\boldsymbol{w}_j^T\boldsymbol{x})$. While it is computationally infeasible to allow for arbitrary continuous functions g_j in this sum, we can approximate such functions by means of linear combinations of B-splines, for example. Given data points $(\boldsymbol{x}_i, y_i)$, $i = 1, 2, \ldots, m$, the regularized ridge function approximation problem $\min_{g\in\mathcal{G}} N^{-1}\sum_{i=1}^{N}(g(\boldsymbol{x}_i) - y_i)^2 + \alpha\,\|g\|_{L^2}^2$ can be solved in $O(N+p)$ time where $\mathcal{G}$ is the span of p uniformly spaced B-splines. For approximation by sums of ridge functions, a block Gauss–Seidel method can be used. The use of a limited number of weight vectors $\boldsymbol{w}_j$ means that the method effectively performs an orthogonal projection onto a low or modest-dimensional subspace, an example of dimension reduction. Gradients with respect to the weight vectors $\boldsymbol{w}_j$ can also be efficiently computed.

1 Introduction

A *ridge function* is a function $f\colon \mathbb{R}^n \to \mathbb{R}$ of the form $f(\boldsymbol{x}) = \varphi(\boldsymbol{w}^T\boldsymbol{x})$ for some continuous function $\varphi\colon \mathbb{R} \to \mathbb{R}$. These can be used at least as theoretical tools for understanding shallow neural networks which typically have the form $\boldsymbol{x} \mapsto \sum_{i=1}^{m} z_i\sigma(\boldsymbol{w}_i^T\boldsymbol{x} + b_i)$ where m is the number of hidden units in the neural network, $\sigma\colon \mathbb{R} \to \mathbb{R}$ is a fixed activation function, and z_i are the output weights. Clearly, if each occurrence of $\boldsymbol{x} \mapsto \sigma(\boldsymbol{w}_i^T\boldsymbol{x} + b_i)$ can be replaced by a general continuous function $\boldsymbol{x} \mapsto \varphi_i(\boldsymbol{w}_i^T\boldsymbol{x})$, the functions that can be approximated becomes much

D. E. Stewart (✉)
Department of Mathematics, University of Iowa, Iowa, IA, USA

S. Foucart, S. Wojtowytsch (eds.), *Explorations in the Mathematics of Data Science*, Applied and Numerical Harmonic Analysis,
https://doi.org/10.1007/978-3-031-66497-7_5

larger:

$$\boldsymbol{x} \mapsto \sum_{i=1}^{m} \varphi_i(\boldsymbol{w}_i^T \boldsymbol{x}) \tag{1}$$

with the output weight z_i absorbed into φ_i.

Ridge functions are useful as a theoretical tool for understanding the approximation power of shallow neural networks, as discussed in [4, 7–10]. To the best of the author's knowledge, this is the first time that ridge functions have been proposed as a practical method of creating functions from data, with efficient computational methods. These methods proposed here have certain desirable properties compared with standard neural network formulations, with respect to training complexity and dimension reduction.

Extreme learning machines (ELMs) [6, 10] inspired some of this work, in that the "training" of an ELM amounts to solving a least squares problem by focusing on the output weights of a shallow neural network: the coefficients z_i in $\boldsymbol{x} \mapsto \sum_{i=1}^{m} z_i \sigma(\boldsymbol{w}_i^T \boldsymbol{x} + b_i)$. While ELMs have found some success [5], we seek to expand the role of the "output" coefficients while limiting the number m of weight vectors $\boldsymbol{w}_i$ used. Note that the function in (1) is a function of the orthogonal projection of $\boldsymbol{x}$ onto span $\{\boldsymbol{w}_i \mid i = 1, \ldots, m\}$, which is a form of dimension reduction. Ridge function machines also enable the optimization over the $\boldsymbol{w}_i$'s to better identify the subspace onto which the input should be projected.

Arguments have been made that conventional training techniques such as limited time stochastic gradient descent (SGD) "freeze" the weight vectors $\boldsymbol{w}_i$ if the weights z_i over-parameterize the function needed to fit the data. See, for example, [3].

2 Basic Properties

$L^2(\mathbb{R}^d)$ is the standard Hilbert space of square-integrable functions $\mathbb{R}^d \to \mathbb{R}$. For a probability measure π over a set $\mathcal{X}$, let $L^2(\pi)$ be the Hilbert space of functions $\mathcal{X} \to \mathbb{R}$ where $\int_{\mathcal{X}} f(x)^2 \, d\pi(x)$ is finite, with inner product $(f, g)_\pi = \int_{\mathcal{X}} f(x)\, g(x)\, d\pi(x)$. This can also be understood in terms of random variables: $(f, g)_\pi = \mathbb{E}_{X\sim\pi}[f(X)\, g(X)]$. Note that in $L^2(\pi)$, we identify $f = g$ if $\pi(\{x \mid f(x) \neq g(x)\}) = 0$.

2.1 Single Ridge Function

We aim to minimize the error in a least squares sense. That is, we want to find a function $g \colon \mathbb{R} \to \mathbb{R}$ and a weight vector $\boldsymbol{w} \in \mathbb{R}^d$ that minimizes

$$\mathbb{E}_{\boldsymbol{X}\sim\pi}\left[\left(g(\boldsymbol{w}^T \boldsymbol{X}) - f(\boldsymbol{X})\right)^2\right]$$

where π is the probability distribution for $\boldsymbol{X}$. That is, we wish to find the minimizer for

$$\min_{\boldsymbol{w}} \min_{g} \int_{\mathbb{R}^d} \left(g(\boldsymbol{w}^T \boldsymbol{x}) - f(\boldsymbol{x})\right)^2 \, d\pi(\boldsymbol{x}). \tag{2}$$

We can split the minimization into two parts: the minimization over g and the minimization over $\boldsymbol{w} \in \mathbb{R}^d$. While the first is an infinite-dimensional minimization, because of the affine character of the map $g \mapsto f(\boldsymbol{x}) - g(\boldsymbol{w}^T \boldsymbol{x})$, this minimization is essentially a linear least squares problem. Also note that replacing $\boldsymbol{w}$ with a multiple $\beta \boldsymbol{w}$ with $\beta \neq 0$ does not essentially change the problem as the same value of the objective function is obtained by replacing $g(u)$ with $g(u/\beta)$. It is not necessary, but can be practically useful to enforce a constraint on $\boldsymbol{w}$ such as $\|\boldsymbol{w}\|_2 = 1$.

Using the marginal probability measure on measurable subsets $E \subseteq \mathbb{R}$

$$\pi_{\boldsymbol{w}}(E) = \pi\left(\left\{\boldsymbol{x} \in \mathbb{R}^d \mid \boldsymbol{w}^T \boldsymbol{x} \in E\right\}\right),$$

we can write:

$$\begin{aligned}
&\mathbb{E}_{\boldsymbol{X}\sim\pi}\left[\left(g(\boldsymbol{w}^T \boldsymbol{X}) - f(\boldsymbol{X})\right)^2\right] \\
&\quad = \int_{\mathbb{R}} \mathbb{E}\left[\left(g(\boldsymbol{w}^T \boldsymbol{X}) - f(\boldsymbol{X})\right)^2 \mid \boldsymbol{w}^T \boldsymbol{X} = u\right] d\pi_{\boldsymbol{w}}(u).
\end{aligned}$$

The linearized perturbation in the loss due to perturbing g to $g + \delta g$ is then

$$\begin{aligned}
&\delta\left[\mathbb{E}_{\boldsymbol{X}\sim\pi}\left[\left(g(\boldsymbol{w}^T \boldsymbol{X}) - f(\boldsymbol{X})\right)^2\right]\right] \\
&\quad = 2\,\mathbb{E}_{\boldsymbol{X}\sim\pi}\left[\left(g(\boldsymbol{w}^T \boldsymbol{X}) - f(\boldsymbol{X})\right) \delta g(\boldsymbol{w}^T \boldsymbol{X})\right] \\
&\quad = 2 \int_{\mathbb{R}} \mathbb{E}_{\boldsymbol{X}\sim\pi}\left[\left(g(\boldsymbol{w}^T \boldsymbol{X}) - f(\boldsymbol{X})\right) \delta g(\boldsymbol{w}^T \boldsymbol{X}) \mid \boldsymbol{w}^T \boldsymbol{X} = u\right] d\pi_{\boldsymbol{w}}(u) \\
&\quad = 2 \int_{\mathbb{R}} \mathbb{E}_{\boldsymbol{X}\sim\pi}\left[(g(u) - f(\boldsymbol{X}))\, \delta g(u) \mid \boldsymbol{w}^T \boldsymbol{X} = u\right] d\pi_{\boldsymbol{w}}(u) \\
&\quad = 2 \int_{\mathbb{R}} \left(g(u) - \mathbb{E}_{\boldsymbol{X}\sim\pi}\left[f(\boldsymbol{X}) \mid \boldsymbol{w}^T \boldsymbol{X} = u\right]\right) \delta g(u)\, d\pi_{\boldsymbol{w}}(u).
\end{aligned}$$

Setting this to zero for any δg gives the solution:

$$g^*_{\boldsymbol{w}}(u) = \mathbb{E}_{\boldsymbol{X}\sim\pi}\left[f(\boldsymbol{X}) \mid \boldsymbol{w}^T \boldsymbol{X} = u\right]. \tag{3}$$

Provided f is continuous and π has a continuous probability density function with bounded support, then g^* is defined and continuous on $\operatorname{supp} \pi_{\boldsymbol{w}}$ for $\boldsymbol{w} \neq \boldsymbol{0}$.

The regularized problem of minimizing

$$\mathbb{E}_{X\sim\pi}\left[\left(g(\boldsymbol{w}^T\boldsymbol{X}) - f(\boldsymbol{X})\right)^2\right] + \alpha\,\|g\|_{L^2(\mathbb{R})}^2 \tag{4}$$

has an easily described solution provided $\pi_{\boldsymbol{w}}$ is absolutely continuous with respect to the Lebesgue measure λ: Let $p_{\boldsymbol{w}} = d\pi_{\boldsymbol{w}}/d\lambda$ be the probability density function for $\pi_{\boldsymbol{w}}$; then the minimizing g is

$$g_{\boldsymbol{w}}^{(\alpha)}(u) = \frac{p_{\boldsymbol{w}}(u)}{\alpha + p_{\boldsymbol{w}}(u)}\mathbb{E}_{X\sim\pi}\left[f(\boldsymbol{X}) \mid \boldsymbol{w}^T\boldsymbol{X} = u\right]. \tag{5}$$

Regularizing using an $L^2(\mathbb{R})$ regularizer for g avoids using data- or distribution-dependent regularizers, while using an $L^2(\pi)$ regularizer would become problematic near the boundary of $\operatorname{supp}\pi_{\boldsymbol{w}}$. The regularized $g_{\boldsymbol{w}}^{(\alpha)}$ in (5) goes to zero outside $\operatorname{supp}\pi_{\boldsymbol{w}}$. When we have finite data sets, it will be especially important for the regularizer to prevent bad behavior for u beyond the values of $\boldsymbol{w}^T\boldsymbol{x}_i$ where $\boldsymbol{x}_i$ is in the data set. In addition, the use of an L^2-based regularizer results in linear systems of equations, making the method easier to implement.

The optimization problem for $\boldsymbol{w}$ is, however, more difficult. Let

$$\begin{aligned}\psi(\boldsymbol{w}) &= \min_g \mathbb{E}\left[\left(g(\boldsymbol{w}^T\boldsymbol{X}) - f(\boldsymbol{X})\right)^2\right]\\ &= \mathbb{E}\left[\left(g_{\boldsymbol{w}}^*(\boldsymbol{w}^T\boldsymbol{X}) - f(\boldsymbol{X})\right)^2\right]\end{aligned}$$

for the unregularized problem, and

$$\begin{aligned}\psi_\alpha(\boldsymbol{w}) &= \min_g \mathbb{E}\left[\left(g(\boldsymbol{w}^T\boldsymbol{X}) - f(\boldsymbol{X})\right)^2\right] + \alpha\,\|g\|_{L^2}^2\\ &= \mathbb{E}\left[\left(g_{\boldsymbol{w}}^{(\alpha)}(\boldsymbol{w}^T\boldsymbol{X}) - f(\boldsymbol{X})\right)^2\right] + \alpha\left\|g_{\boldsymbol{w}}^{(\alpha)}\right\|_{L^2}^2.\end{aligned}$$

We can compute gradients of ψ provided $g_{\boldsymbol{w}}^*$ is sufficiently smooth. Since $g_{\boldsymbol{w}}^*$ minimizes $\mathbb{E}\left[\left(g(\boldsymbol{w}^T\boldsymbol{X}) - f(\boldsymbol{X})\right)^2\right]$ over all $g \in L^2(\pi_{\boldsymbol{w}})$ say, then for any $\delta g \in L^2(\pi_{\boldsymbol{w}})$, $\mathbb{E}\left[\left(g(\boldsymbol{w}^T\boldsymbol{X}) - f(\boldsymbol{X})\right)\delta g(\boldsymbol{w}^T\boldsymbol{X})\right] = 0$ if $g = g_{\boldsymbol{w}}^*$. Now consider the linearized perturbation:

$$\begin{aligned}\delta\left[\psi(\boldsymbol{w})\right] &= \delta\left[\mathbb{E}\left[\left(g_{\boldsymbol{w}}^*(\boldsymbol{w}^T\boldsymbol{X}) - f(\boldsymbol{X})\right)^2\right]\right]\\ &= 2\,\mathbb{E}\left[\left(g_{\boldsymbol{w}}^*(\boldsymbol{w}^T\boldsymbol{X}) - f(\boldsymbol{X})\right)\left(\delta g_{\boldsymbol{w}}^*(\boldsymbol{w}^T\boldsymbol{X}) + (g_{\boldsymbol{w}}^*)'(\boldsymbol{w}^T\boldsymbol{X})\boldsymbol{X}^T\delta\boldsymbol{w}\right)\right].\end{aligned}$$

From the previous computation

$$\mathbb{E}\left[\left(g_{\boldsymbol{w}}^{*}(\boldsymbol{w}^{T}\boldsymbol{X})-f(\boldsymbol{X})\right)\delta g_{\boldsymbol{w}}^{*}(\boldsymbol{w}^{T}\boldsymbol{X})\right]=0,$$

and so

$$\nabla\psi(\boldsymbol{w})=2\,\mathbb{E}\left[\left(g_{\boldsymbol{w}}^{*}(\boldsymbol{w}^{T}\boldsymbol{X})-f(\boldsymbol{X})\right)(g_{\boldsymbol{w}}^{*})'(\boldsymbol{w}^{T}\boldsymbol{X})\boldsymbol{X}\right].$$

Similarly, for the regularized problem

$$\begin{aligned}\delta\left[\psi_{\alpha}(\boldsymbol{w})\right]&=2\,\mathbb{E}\left[\left(g_{\boldsymbol{w}}^{(\alpha)}(\boldsymbol{w}^{T}\boldsymbol{X})-f(\boldsymbol{X})\right)\left(\delta g_{\boldsymbol{w}}^{(\alpha)}(\boldsymbol{w}^{T}\boldsymbol{X})+(g_{\boldsymbol{w}}^{(\alpha)})'(\boldsymbol{w}^{T}\boldsymbol{X})\boldsymbol{X}^{T}\delta\boldsymbol{w}\right)\right]\\&\quad+2\alpha\int_{\mathbb{R}}g_{\boldsymbol{w}}^{(\alpha)}(u)\,\delta g_{\boldsymbol{w}}^{(\alpha)}(u)\,du.\end{aligned}$$

Again, from optimality of $g_{\boldsymbol{w}}^{(\alpha)}$ for the regularized problem

$$\mathbb{E}\left[\left(g_{\boldsymbol{w}}^{(\alpha)}(\boldsymbol{w}^{T}\boldsymbol{X})-f(\boldsymbol{X})\right)\delta g_{\boldsymbol{w}}^{(\alpha)}(\boldsymbol{w}^{T}\boldsymbol{X})\right]+\alpha\int_{\mathbb{R}}g_{\boldsymbol{w}}^{(\alpha)}(u)\,\delta g_{\boldsymbol{w}}^{(\alpha)}(u)\,du=0,\qquad\text{so}$$

$$\nabla\psi_{\alpha}(\boldsymbol{w})=2\,\mathbb{E}\left[\left(g_{\boldsymbol{w}}^{(\alpha)}(\boldsymbol{w}^{T}\boldsymbol{X})-f(\boldsymbol{X})\right)(g_{\boldsymbol{w}}^{(\alpha)})'(\boldsymbol{w}^{T}\boldsymbol{X})\boldsymbol{X}\right].$$

These give efficient formulas for a single ridge function, in both the regularized and unregularized cases, which can be used for optimization algorithms such as gradient descent and other gradient-based optimization methods such as those in [11].

Note that both $g_{\boldsymbol{w}}^{*}$ and $g_{\boldsymbol{w}}^{(\alpha)}$ are C^1 provided f and $p_{\boldsymbol{w}}$ are C^1; it is sufficient that f and p are C^1.

Note that if we restrict the optimization of g over a subspace $\mathcal{G}\subseteq L^2(\pi_{\boldsymbol{w}})$, then if $g_{\boldsymbol{w},\mathcal{G}}^{*}\in\mathcal{G}$ is the minimizer over $\mathcal{G}$, then for any $\delta g\in\mathcal{G}$,

$$\mathbb{E}\left[\left(g_{\boldsymbol{w},\mathcal{G}}^{*}(\boldsymbol{w}^{T}\boldsymbol{X})-f(\boldsymbol{X})\right)\delta g(\boldsymbol{w}^{T}\boldsymbol{X})\right]=0\qquad\text{in the unregularized case, and}$$

$$\mathbb{E}\left[\left(g_{\boldsymbol{w},\mathcal{G}}^{(\alpha)}(\boldsymbol{w}^{T}\boldsymbol{X})-f(\boldsymbol{X})\right)\delta g(\boldsymbol{w}^{T}\boldsymbol{X})\right]+\alpha\int_{\mathbb{R}}g_{\boldsymbol{w},\mathcal{G}}^{(\alpha)}(u)\,\delta g(u)\,du=0$$

in the regularized case. Thus if $\psi_{\mathcal{G}}(\boldsymbol{w})=\min_{g\in\mathcal{G}}\mathbb{E}\left[\left(g(\boldsymbol{w}^{T}\boldsymbol{X})-f(\boldsymbol{X})\right)^2\right]$ and $\psi_{\alpha,\mathcal{G}}(\boldsymbol{w})=\min_{g\in\mathcal{G}}\mathbb{E}\left[\left(g(\boldsymbol{w}^{T}\boldsymbol{X})-f(\boldsymbol{X})\right)^2\right]+\alpha\left\|g\right\|_{L^2}^2$, we have:

$$\nabla\psi_{\mathcal{G}}(\boldsymbol{w})=2\,\mathbb{E}\left[\left(g_{\boldsymbol{w},\mathcal{G}}^{*}(\boldsymbol{w}^{T}\boldsymbol{X})-f(\boldsymbol{X})\right)(g_{\boldsymbol{w},\mathcal{G}}^{*})'(\boldsymbol{w}^{T}\boldsymbol{X})\boldsymbol{X}\right],\qquad\text{and}\qquad(6)$$

$$\nabla\psi_{\alpha,\mathcal{G}}(\boldsymbol{w})=2\,\mathbb{E}\left[\left(g_{\boldsymbol{w},\mathcal{G}}^{(\alpha)}(\boldsymbol{w}^{T}\boldsymbol{X})-f(\boldsymbol{X})\right)(g_{\boldsymbol{w},\mathcal{G}}^{(\alpha)})'(\boldsymbol{w}^{T}\boldsymbol{X})\boldsymbol{X}\right].\qquad(7)$$

Thus even if $\mathcal{G}$ is restricted to a computationally convenient subspace of $L^2(\pi_{\boldsymbol{w}})$, gradient information with respect to $\boldsymbol{w}$ is available.

The version of (6) for data given by pairs $(\boldsymbol{x}_i, y_i)$, $i = 1, 2, \ldots, N$ is

$$\nabla\psi_{\mathcal{G}}(\boldsymbol{w}) = 2\,N^{-1}\sum_{i=1}^{N}\left(g^*_{\boldsymbol{w},\mathcal{G}}(\boldsymbol{w}^T\boldsymbol{x}_i) - y_i\right)(g^*_{\boldsymbol{w},\mathcal{G}})'(\boldsymbol{w}^T\boldsymbol{x}_i)\boldsymbol{x}_i, \qquad \text{and} \tag{8}$$

$$\nabla\psi_{\alpha,\mathcal{G}}(\boldsymbol{w}) = 2\,N^{-1}\sum_{i=1}^{N}\left(g^{(\alpha)}_{\boldsymbol{w},\mathcal{G}}(\boldsymbol{w}^T\boldsymbol{x}_i) - y_i\right)(g^{(\alpha)}_{\boldsymbol{w},\mathcal{G}})'(\boldsymbol{w}^T\boldsymbol{x}_i)\boldsymbol{x}_i, \tag{9}$$

where $g^*_{\boldsymbol{w}}$ respectively $(g^{(\alpha)}_{\boldsymbol{w}})$ is the optimal choice of $g \in \mathcal{G}$ for the given weight vector $\boldsymbol{w}$ in the unregularized (respectively regularized) case.

It should be noted that $\psi(\boldsymbol{w})$ is typically a highly non-convex function of $\boldsymbol{w}$. First, $\psi(\beta\boldsymbol{w}) = \psi(\boldsymbol{w})$ for any $0 \neq \beta \in \mathbb{R}$. This means, for example, that $\psi(-\boldsymbol{w}) = \psi(\boldsymbol{w})$. We show that if ψ is convex satisfying $\psi(\beta\boldsymbol{w}) = \psi(\boldsymbol{w})$ for any $0 \neq \beta \in \mathbb{R}$, then ψ is constant. Consider any two vectors $\boldsymbol{u}$ and $\boldsymbol{v}$ in $\mathbb{R}^n$. Then by convexity, $\psi(\boldsymbol{u}+\boldsymbol{v}) = \psi(\frac{1}{2}\boldsymbol{u}+\frac{1}{2}\boldsymbol{v}) \leq \frac{1}{2}\psi(\boldsymbol{u})+\frac{1}{2}\psi(\boldsymbol{v})$; replacing $\boldsymbol{v}$ by $-\boldsymbol{v}$ gives $\psi(\boldsymbol{u}-\boldsymbol{v}) = \psi(\frac{1}{2}\boldsymbol{u}-\frac{1}{2}\boldsymbol{v}) \leq \frac{1}{2}\psi(\boldsymbol{u})+\frac{1}{2}\psi(-\boldsymbol{v}) = \frac{1}{2}\psi(\boldsymbol{u})+\frac{1}{2}\psi(\boldsymbol{v})$. Thus

$$\begin{aligned}
\psi(\boldsymbol{u}) &= \psi(2\,\boldsymbol{u}) = \psi((\boldsymbol{u}+\boldsymbol{v})+(\boldsymbol{u}-\boldsymbol{v})) \\
&\leq \frac{1}{2}\psi(\boldsymbol{u}+\boldsymbol{v}) + \frac{1}{2}\psi(\boldsymbol{u}-\boldsymbol{v}) \\
&\leq \frac{1}{2}\left(\frac{1}{2}\psi(\boldsymbol{u})+\frac{1}{2}\psi(\boldsymbol{v})\right) + \frac{1}{2}\left(\frac{1}{2}\psi(\boldsymbol{u})+\frac{1}{2}\psi(\boldsymbol{v})\right) \\
&= \frac{1}{2}\psi(\boldsymbol{u}) + \frac{1}{2}\psi(\boldsymbol{v}), \qquad \text{and so}
\end{aligned}$$

$$\frac{1}{2}\psi(\boldsymbol{u}) \leq \frac{1}{2}\psi(\boldsymbol{v}).$$

Since this is true for any $\boldsymbol{u}$ and $\boldsymbol{v}$, we have ψ constant.

2.2 Sums of Ridge Functions

Here we consider approximations of the form:

$$f(\boldsymbol{x}) \approx \sum_{j=1}^{m} g_j(\boldsymbol{w}_j^T\boldsymbol{x})$$

where $g_i : \mathbb{R} \to \mathbb{R}$ and $\boldsymbol{w}_i \in \mathbb{R}^d$ for $i = 1, 2, \ldots, m$. We choose to use the loss function:

$$\mathbb{E}_{\boldsymbol{X}\sim\pi}\left[\left(\sum_{j=1}^{m} g_j(\boldsymbol{w}_j^T\boldsymbol{X}) - f(\boldsymbol{X})\right)^2\right], \tag{10}$$

which we minimize over the $g_i \in L^2(\mathbb{R})$ and $\boldsymbol{w}_i \in \mathbb{R}^d$.

This is the natural loss function as the expectation in (10) is the almost sure limit of

$$\frac{1}{N}\sum_{i=1}^{N}\left(\sum_{j=1}^{m} g_j(\boldsymbol{w}_j^T\boldsymbol{x}_i) - y_i\right)^2 \tag{11}$$

as $N \to \infty$ where the $\boldsymbol{x}_i$ are iid samples from probability distribution π and $y_i = f(\boldsymbol{x}_i)$.

For sums of ridge functions, we can use the formula for a single ridge function (3), to show that the optimal function g_j^* is given by

$$g_j^*(u) = \mathbb{E}_{\boldsymbol{X}\sim\pi}\left[f(\boldsymbol{X}) - \sum_{k=1;k\neq j}^{m} g_k^*(\boldsymbol{w}^T\boldsymbol{X}) \mid \boldsymbol{w}_j^T\boldsymbol{X} = u\right].$$

We can use this to create block Gauss–Seidel methods [12, Sec. 2.4] for solving the least squares problem. Consider the problem of minimizing (10) plus a regularization term: let $\boldsymbol{g} = [g_1, g_2, \ldots, g_m]^T \in L^2(\mathbb{R})^m$ and $\mathcal{A}_j : L^2(\mathbb{R}) \to L^2(\pi)$ be given by $\mathcal{A}_j(g_j) = g_j(\boldsymbol{w}_j^T\cdot)$. Given an $f \in L^2(\pi)$, we seek the functions $g_1, g_2, \ldots, g_m \in L^2(\mathbb{R})$ that minimizes

$$\left\| f - \sum_j \mathcal{A}_j(g_j)\right\|_\pi^2 + \alpha\sum_{j=1}^{m}\|g_j\|_{L^2}^2 = \|f - \mathcal{A}(\boldsymbol{g})\|_\pi^2 + \alpha\,\|\boldsymbol{g}\|_{L^2}^2,$$

where $\mathcal{A} = [\mathcal{A}_1, \mathcal{A}_2, \ldots, \mathcal{A}_m]$ and $\mathcal{A}(\boldsymbol{g}) = \sum_{j=1}^m \mathcal{A}_j(g_j)$. If $\mathcal{A}_j^*$ is the adjoint of $\mathcal{A}_j$, then the modified normal equations are

$$(\mathcal{A}^*\mathcal{A} + \alpha I)\boldsymbol{g} = \mathcal{A}^*(f).$$

Expanding this gives the system of equations:

$$\begin{bmatrix} \mathcal{A}_1^*\mathcal{A}_1 + \alpha I & \mathcal{A}_1^*\mathcal{A}_2 & \cdots & \mathcal{A}_1^*\mathcal{A}_m \\ \mathcal{A}_2^*\mathcal{A}_1 & \mathcal{A}_2^*\mathcal{A}_2 + \alpha I & \cdots & \mathcal{A}_2^*\mathcal{A}_m \\ \vdots & \vdots & \ddots & \vdots \\ \mathcal{A}_m^*\mathcal{A}_1 & \mathcal{A}_m^*\mathcal{A}_2 & \cdots & \mathcal{A}_m^*\mathcal{A}_m + \alpha I \end{bmatrix}\begin{bmatrix} g_1 \\ g_2 \\ \vdots \\ g_m \end{bmatrix} = \begin{bmatrix} \mathcal{A}_1^*(f) \\ \mathcal{A}_2^*(f) \\ \vdots \\ \mathcal{A}_m^*(f) \end{bmatrix}. \tag{12}$$

The block Gauss–Seidel method involves solving a sequence of regularized normal equations:

$$(\mathcal{A}_j^*\mathcal{A}_j+\alpha I)g_j^{(k+1)} = \mathcal{A}_j^*(f-\sum_{i=1}^{j-1}\mathcal{A}_i g_i^{(k+1)}-\sum_{i=j+1}^{m}\mathcal{A}_i g_i^{(k)}), \qquad j=1,2,\ldots,m. \tag{13}$$

The block Gauss–Seidel method converges for $\alpha > 0$ by an extension of the Ostrowski–Reich theorem to Hilbert spaces [2]. Each of these systems (13) can be solved easily using (5).

It should be noted that if $\boldsymbol{w}_i \neq \boldsymbol{w}_j$ and $\operatorname{supp}\pi$ contains an open set, then $\operatorname{range}\mathcal{A}_i \cap \operatorname{range}\mathcal{A}_j \subseteq \operatorname{span}\{\mathbf{1}\}$ where $\mathbf{1}\colon \mathbb{R}^n \to \mathbb{R}$ is the constant function $\mathbf{1}(\boldsymbol{x}) = 1$ for all $\boldsymbol{x}$.

In certain cases, the matrix $\mathcal{A}^*\mathcal{A}$ is nearly block diagonal. Consider, for example, the case where the vectors $\boldsymbol{w}_i$ and $\boldsymbol{w}_j$ are orthogonal, and $\boldsymbol{X} \sim \operatorname{Normal}(\mathbf{0}, I_{n\times n})$. In that case, $\boldsymbol{w}_i^T\boldsymbol{X}$ and $\boldsymbol{w}_j^T\boldsymbol{X}$ are independent random variables, and so also are $g_i(\boldsymbol{w}_i^T\boldsymbol{X})$ and $g_j(\boldsymbol{w}_j^T\boldsymbol{X})$. Hence $\mathbb{E}\left[g_i(\boldsymbol{w}_i^T\boldsymbol{X})\, g_j(\boldsymbol{w}_j^T\boldsymbol{X})\right] = \mathbb{E}\left[g_i(\boldsymbol{w}_i^T\boldsymbol{X})\right]\, \mathbb{E}\left[g_j(\boldsymbol{w}_j^T\boldsymbol{X})\right]$. From this, we deduce that $\operatorname{range}(\mathcal{A}_i^*\mathcal{A}_j) \subseteq \operatorname{span}\{\mathbf{1}\}$.

Gradients with respect to the weight vectors $\boldsymbol{w}_j$ can be computed similarly to (6) and (8). If

$$\psi_{\mathcal{G}}(\boldsymbol{w}_1, \boldsymbol{w}_2, \ldots, \boldsymbol{w}_m) = \min_{g_1,g_2,\ldots,g_m\in\mathcal{G}} \mathbb{E}_{\boldsymbol{X}\sim\pi}\left[\left(\sum_{j=1}^{m} g_j(\boldsymbol{w}_j^T\boldsymbol{X}) - f(\boldsymbol{X})\right)^2\right],$$

then (14)

$$\nabla_{\boldsymbol{w}_k}\psi_{\mathcal{G}}(\boldsymbol{w}_1, \boldsymbol{w}_2, \ldots, \boldsymbol{w}_m) = 2\,\mathbb{E}_{\boldsymbol{X}\sim\pi}\left[\left(\sum_{j=1}^{m} g_j(\boldsymbol{w}_j^T\boldsymbol{X}) - f(\boldsymbol{X})\right) g_k'(\boldsymbol{w}^T\boldsymbol{X})\boldsymbol{X}\right]. \tag{15}$$

For the case of the objective function for given data

$$\psi_{\mathcal{G}}(\boldsymbol{w}_1, \boldsymbol{w}_2, \ldots, \boldsymbol{w}_m) = \min_{g_1,g_2,\ldots,g_m\in\mathcal{G}} N^{-1}\sum_{i=1}^{N}\left(\sum_{j=1}^{m} g_j(\boldsymbol{w}_j^T\boldsymbol{x}_i) - y_i\right)^2,$$

then (16)

$$\nabla_{\boldsymbol{w}_k}\psi_{\mathcal{G}}(\boldsymbol{w}_1, \boldsymbol{w}_2, \ldots, \boldsymbol{w}_m) = 2\,N^{-1}\sum_{i=1}^{N}\left(\sum_{j=1}^{m} g_j(\boldsymbol{w}_j^T\boldsymbol{x}_i) - y_i\right) g_k'(\boldsymbol{w}^T\boldsymbol{x}_i)\boldsymbol{x}_i, \tag{17}$$

where $g_1, \ldots, g_m \in \mathcal{G}$ are the optimal choices. Essentially the same formulas can be used for the regularized problem: if

$$\psi_{\alpha,\mathcal{G}}(\boldsymbol{w}_1, \boldsymbol{w}_2, \ldots, \boldsymbol{w}_m) = \min_{g_1, g_2, \ldots, g_m \in \mathcal{G}} \mathbb{E}_{\boldsymbol{X}\sim\pi}\left[\left(\sum_{j=1}^{m} g_j(\boldsymbol{w}_j^T \boldsymbol{X}) - f(\boldsymbol{X})\right)^2\right]$$

$$+\alpha \sum_{j=1}^{m} \|g_j\|_{L^2}^2, \qquad \text{then}$$

$$\nabla_{\boldsymbol{w}_k} \psi_{\alpha,\mathcal{G}}(\boldsymbol{w}_1, \boldsymbol{w}_2, \ldots, \boldsymbol{w}_m) = 2\,\mathbb{E}_{\boldsymbol{X}\sim\pi}\left[\left(\sum_{j=1}^{m} g_j(\boldsymbol{w}_j^T \boldsymbol{X}) - f(\boldsymbol{X})\right) g_k'(\boldsymbol{w}^T \boldsymbol{X})\boldsymbol{X}\right],$$

where $g_1, \ldots, g_m \in \mathcal{G}$ are the optimal choices.

2.3 *Multidimensional Ridge Functions*

Standard ridge functions have the form $f(\boldsymbol{x}) = g(\boldsymbol{w}^T \boldsymbol{x})$ where $g\colon \mathbb{R} \to \mathbb{R}$. A multidimensional ridge function however has the form $f(\boldsymbol{x}) = g(W^T \boldsymbol{x})$ where $g\colon \mathbb{R}^m \to \mathbb{R}$ and W is $d \times m$, typically with $m \ll d$. These suffer from the curse of dimensionality if $m \gg 1$ as they require approximating a general function in m variables, and are probably practical only for $m \le 4$. Nevertheless, the formulas of Sect. 2.1 can be generalized to this case.

Assuming that we wish to minimize the expected least squares loss

$$\mathbb{E}_{\boldsymbol{X}\sim\pi}\left[\left(g(W^T \boldsymbol{X}) - f(\boldsymbol{X})\right)^2\right],$$

over $g\colon \mathbb{R}^m \to \mathbb{R}$ in $L^2(\mathbb{R}^m)$ or some other suitable subspace of $L^2(\mathbb{R}^m)$, and $W \in \mathbb{R}^{d\times m}$. Again, the solution of this linear least squares problem in g is

$$g_W^*(\boldsymbol{u}) = \mathbb{E}_{\boldsymbol{X}\sim\pi}\left[f(\boldsymbol{X}) \mid W^T \boldsymbol{X} = \boldsymbol{u}\right].$$

Since the unknown function g is a map $\mathbb{R}^m \to \mathbb{R}$, we expect the number of basis functions needed for an accuracy of ϵ to be asymptotically $O(\epsilon^{-m})$. For this reason we expect the method to be impractical for large and even modest m, such as $m > 4$.

The effect of changing W on

$$\psi(W) = \min_{g} \mathbb{E}\left[\left(g(W^T \boldsymbol{X}) - f(\boldsymbol{X})\right)^2\right]$$

can be inferred from

$$\nabla\psi(W) = 2\,\mathbb{E}\left[\left(f(X) - g_W^*(W^T X)\right) X \nabla g_W^*(W^T X)^T\right]$$

with the understanding that $\psi(W+\delta W) - \psi(W) = \mathrm{trace}(\delta W^T \nabla\psi(W)) + o(\|\delta W\|)$ as $\|\delta W\| \to 0$.

3 Practical Construction of Ridge Function Approximations

3.1 *Discretization*

Since the computation of the optimal $g \in L^2(\mathbb{R})$ to minimize $\mathbb{E}\left[\left(g(\boldsymbol{w}^T X) - f(X)\right)^2\right]$ is an infinite-dimensional computation, we need to select a finite-dimensional (or locally finite-dimensional) subspace $\mathcal{G} \subset L^2(\mathbb{R})$ in which we can perform finite-dimensional computations. We have chosen $\mathcal{G} = \mathrm{span}\,\{u \mapsto \phi(u/h - k) \mid k \in \mathbb{Z}\}$ where $\phi\colon \mathbb{R} \to \mathbb{R}$ is a fixed function and $h > 0$ is the spacing parameter. The function ϕ obviously should not be one where translates can form linearly dependent sets of functions. This rules out exponentials, polynomials, and linear combinations of products of exponentials and polynomials. However, suitable functions include B-spline functions such as

$$\phi(u) = \begin{cases} 1 - \frac{3}{2}u^2 + \frac{3}{4}|u|^3, & |u| \le 1, \\ \frac{1}{4}(2 - |u|)^3, & 1 \le |u| \le 2, \\ 0, & 2 \le |u|, \end{cases} \tag{18}$$

which is C^2 and nonnegative.

3.2 *Overfitting*

Overfitting can occur when the number of parameters vastly exceeds the number of data points, and ridge function machines are not an exception. In the original undiscretized version, the minimization of $\sum_{i=1}^N \left(g(\boldsymbol{w}^T \boldsymbol{x}_i) - y_i\right)^2$ over g is an infinite-dimensional optimization problem, and is vastly underdetermined. In fact, as long as each of the values $\boldsymbol{w}^T \boldsymbol{x}_i$ is distinct, there is a function g where $g(\boldsymbol{w}^T \boldsymbol{x}_i) = y_i$ for all i: any interpolant of the data points $(\boldsymbol{w}^T \boldsymbol{x}_i, y_i)$ will do.

In the discretized case, the number of data points needed to uniquely determine the least squares solution is $O(\|\boldsymbol{w}\|_2\, L/h)$ where L is a measure of the spread of the distribution of data points, such as $\mathbb{E}\left[\|X\|_2\right]$ or the norm of the variance–covariance matrix of $\boldsymbol{X}$. This does not take into account the problems of ill-conditioning. Regularization, as discussed in the next section, should be incorporated into the method.

It should be clear that if $h \to 0$, then we expect overfitting to become a problem, and the method would not produce good results. We therefore need to balance the value of L/h with the number of data points.

An even better approach to overfitting is to adapt the B-splines so that each B-spline covers a minimal number of data points. Consider the ridge function approximation problem:

$$\min_{g \in \mathcal{G}} \frac{1}{N} \sum_{i=1}^{N} \left(g(\boldsymbol{w}^T \boldsymbol{x}_i) - y_i \right)^2 + \alpha \, \|g\|_{L^2}^2$$

where $\mathcal{G}$ is the span of shifted and scaled B-splines,

$$\mathcal{G} = \text{span}\left\{ u \mapsto \phi(s_k u - t_k) \mid k = 1, 2, \ldots, p \right\},$$

while fixing the length of $\boldsymbol{w}$: $\|\boldsymbol{w}\|_2 = 1$. We can choose the scales s_k and shifts t_k to balance the number of coefficients with the number of data points. This can be done by assigning each basis function $u \mapsto \phi(s_k u - t_k)$ with a measure of the amount of data used to compute its coefficient. For example, we can require that $\sum_{i=1}^{N} \left|\phi(s_k(\boldsymbol{w}^T \boldsymbol{x}_i) - t_k)\right| \geq \beta$ for each k where β is a user-specified threshold. If we use a sum of m ridge functions, then we will need more data points per coefficient in a single ridge function, so the condition may be strengthened to requiring $\sum_{i=1}^{N} \left|\phi(s_k(\boldsymbol{w}^T \boldsymbol{x}_i) - t_k)\right| \geq m\beta$ for each k.

To make this approach computationally efficient, we can restrict the scales s_k to powers of two times $1/h_0$ where h_0 is a user-specified basic scaling parameter, and the shifts t_k integers. This structure will disrupt the convenient banded structure of the system of equations to solve in the nonadaptive case, but may be more effective as an approximation and machine learning tool.

3.3 *Regularization*

There is a problem in minimizing $\mathbb{E}\left[\left(g(\boldsymbol{w}^T \boldsymbol{X}) - f(\boldsymbol{X})\right)^2\right]$ over g is that as the probability density function $p(\boldsymbol{x})$ goes to zero as $\|\boldsymbol{x}\| \to \infty$ in a suitable sense, such as $\int_{\{\boldsymbol{x} \mid \|\boldsymbol{x}\| \geq R\}} p(\boldsymbol{x})\, d\boldsymbol{x} \to 0$ as $R \to \infty$, so the penalty for incorrect $g(u)$ for large $|u|$ also goes to zero. We should therefore regularize $\mathbb{E}\left[\left(g(\boldsymbol{w}^T \boldsymbol{X}) - f(\boldsymbol{X})\right)^2\right]$ with a penalty on $g(u)$ that does not decay as $|u| \to \infty$. The regularized minimization problem should therefore be modified to something like

$$\min_{g \in \mathcal{G}} \mathbb{E}\left[\left(g(\boldsymbol{w}^T \boldsymbol{X}) - f(\boldsymbol{X})\right)^2\right] + \alpha \, \|g\|_{L^2(\mathbb{R})}^2 \tag{19}$$

for some $\alpha > 0$.

The other issue is that $\mathbb{E}\left[\left(g(\boldsymbol{w}^T \boldsymbol{X}) - f(\boldsymbol{X})\right)^2\right]$ is uncomputable given f, g and $\boldsymbol{w}$ if $p(\cdot)$ is not known. Instead, we have data points $(\boldsymbol{x}_i, y_i)$, $i = 1, 2, \ldots, N$. Replacing the expectation with the empirical values using the data set with $y_i = f(\boldsymbol{x}_i)$ gives the problem:

$$\min_{g \in \mathcal{G}} \frac{1}{N} \sum_{i=1}^{N} \left(g(\boldsymbol{w}^T \boldsymbol{x}_i) - y_i\right)^2 + \alpha \int_{\mathbb{R}} g(u)^2 \, du.$$

This is a linear least squares problem in g. The normal equations matrix that arises from this formulation is sparse if the function ϕ chosen has bounded support—in fact, the normal equation matrix that arises is then banded.

There are two hyperparameters in this framework: the regularization parameter $\alpha > 0$ and the spacing parameter $h > 0$. It should be noted that if all the values $\boldsymbol{w}^T \boldsymbol{x}_i$, $i = 1, 2, \ldots, N$ are distinct, then overfitting can easily occur with sufficiently small $h > 0$ (take $0 < h < \frac{1}{2} \min_{i,j} \left|\boldsymbol{w}^T (\boldsymbol{x}_i - \boldsymbol{x}_j)\right|$) regardless of the value of $\alpha > 0$ provided α is fixed. Thus we wish to keep h from going to zero.

3.4 Practical Construction of Multidimensional Ridge Functions

For practical computations we need to create a finite-dimensional subspace $\mathcal{G}$ of $L^2(\mathbb{R}^m)$ over which to minimize $\mathbb{E}\left[\left(g(W^T \boldsymbol{X}) - f(\boldsymbol{X})\right)^2\right]$ over g. This can be done using, for example, tensor products of B-splines (18):

$$g(\boldsymbol{u}) \approx \sum_{\boldsymbol{k} \in \mathbb{Z}^m} c_{\boldsymbol{k}} \prod_{j=1}^{m} \phi(u_j/h - k_j).$$

This approach again raises issues of the "curse of dimensionality" if m is large, but could be practical for $m \leq 4$, for example. The linear system will again be sparse for $m > 1$, but will no longer be banded with a fixed small bandwidth. Direct methods can be applied to the resulting system of equations for the least squares fit, but the computational complexity can increase fairly rapidly in the number of nonzero entries. Iterative methods can be developed for solving the normal equations using algebraic multigrid techniques [1], for example.

4 Performance

The basic methods were implemented for a simple two-dimensional problem. The test basic problem was to identify the function $f(\boldsymbol{x}) = f_1(\boldsymbol{w}_1^T\boldsymbol{x}) + f_2(\boldsymbol{w}_2^T\boldsymbol{x})$ from a data set of pairs $(\boldsymbol{x}_i, f(\boldsymbol{x}_i))$. The points $\boldsymbol{x}_i$ are chosen pseudorandomly, and independently, from a uniform distribution over $[0, 1] \times [0, 1]$, plus the four corners of the square.[1] Two tests were done: one with 44 data points as a proof-of-concept and to represent low-data situations and another with 10,004 data points. The final 4 points were the vertices of the unit square. The functions of one variable are $f_1(u) = e^{-u}\cos(u)/\sqrt{1+u^2}$ and $f_2(u) = 2(2u^2 - 1)/(1 + 0.2\,u^3)$, and the weight vectors are $\boldsymbol{w}_1 = [1, 0.7]^T$ and $\boldsymbol{w}_2 = [-0.5, 0.6]$. The normalized weight vectors are $\widehat{\boldsymbol{w}}_1 \approx [0.8166, 0.5772]$ and $\widehat{\boldsymbol{w}}_2 \approx [-0.6402, 0.7682]^T$.

The tests were performed in Julia v. 1.9.2 on a Dell Optiplex 7090 running Linux. Code and notebook for the tests are available on request.

The block Gauss–Seidel method used was tested to see how quickly the coefficients of the B-splines would converge both for the case where the weight vectors are known and where they were not. The initial coefficients in all cases were taken to be zero. The spacing used for the B-splines was $h = 0.4$. Regularization simply involves adding αI to the linear system for the coefficients of the B-spline. Thus it is not an L^2 regularization, which would involve α times the mass matrix of the B-splines. However, its effects will be similar. Convergence of the block Gauss–Seidel method is shown in Figs. 1 and 2. Both use logarithmic scaling of the vertical axis.

Each inner iteration updates the coefficients for one of the weight vectors; since there are two weight vectors, two inner iterations form one complete iteration of the block Gauss–Seidel method. Both figures show the reduction in the unregularized objective function $N^{-1}\sum_{i=1}^{N}(\sum_{j=1}^{m} g_j(\boldsymbol{w}_j^T\boldsymbol{x}_i) - f(\boldsymbol{x}_i))^2$ with the number of inner iterations. Figure 1 shows the reduction of the objective function for the case where the weight vectors used are the correct weight vectors. Figure 2 shows the reduction in the unregularized objective for the incorrect weight vectors $\boldsymbol{w}_1 = [1, -0.2]^T$ and $\boldsymbol{w}_2 = [-0.7, 0.4]^T$. Note that reducing $\alpha > 0$ appears to improve the performance of the algorithm for both correct and incorrect weight vectors. However, reducing $\alpha > 0$ to near machine epsilon (which is $\approx 2 \times 10^{-16}$ for double-precision floating point arithmetic which is used here) will result in large roundoff errors and poor performance. Finding the optimal regularization parameter is a subject for further investigation.

Clearly, in actual use, the correct weight vectors will be unknown. Computing the correct weight vectors can be done using a gradient descent algorithm using gradients with respect to $\boldsymbol{w}$ computed using the formula (17). For this test, ten steps of the block Gauss–Seidel method were used per gradient descent step. The gradient descent step parameter was chosen to be $s = 0.4$. In the infinite-

[1] The points were computed using Julia's MersenneTwister pseudorandom number generator, with seed 87656324.

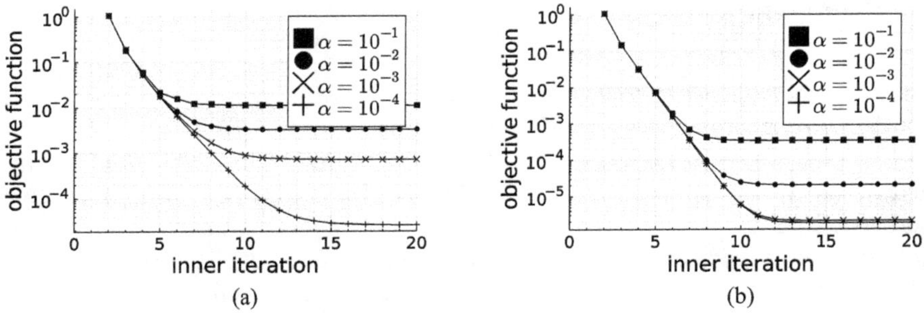

Fig. 1 Objective reduction by block Gauss–Seidel for correct weight vectors. (**a**) 44 data points. (**b**) 10,004 data points

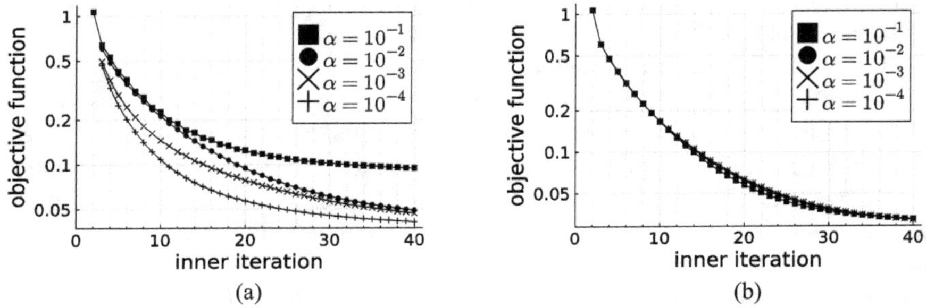

Fig. 2 Objective reduction by block Gauss–Seidel for incorrect weight vectors. (**a**) 44 data points. (**b**) 10,004 data points

dimensional case, scaling of the weight vector is irrelevant as $\boldsymbol{w}_j \mapsto \alpha \boldsymbol{w}_j$ can be compensated for with $g_j(u) \mapsto g_j(u/\alpha)$. However, because of the potential for numerical problems with $\boldsymbol{w}_j$ becoming large, especially with the use of B-splines, the constraint that $\|\boldsymbol{w}_j\|_2 = 1$ for all j was imposed: first, the gradient vector for $\boldsymbol{w}_j$ was orthogonalized against $\boldsymbol{w}_j$, and second, the $\boldsymbol{w}_j$ after the gradient step was normalized to have length one. Both weight vectors were updated in each gradient descent step. The gradient of the unregularized objective function is plotted against gradient step number in Fig. 3. The regularization parameter used throughout this test was $\alpha = 10^{-4}$.

The curious behavior of the convergence curve in Fig. 3a actually shows an important aspect of this approach: the gradient norm has a local minimum at the 48th gradient descent step, and a local maximum at the 94th gradient descent step. The reason for this is the non-convex nature of the optimization problem over the weight vectors: the local minimum indicates passage near to a saddle point, which the method ultimately avoids. The weight vectors after 300 gradient descent steps are $\boldsymbol{w}_1^{(300)} \approx [0.817651, 0.575714]^T$ and $\boldsymbol{w}_2^{(300)} \approx [-0.640822, 0.767690]^T$, compared to the normalized weight vectors that generated the data $\boldsymbol{w}_1^* \approx [0.819232, 0.573462]^T$ and $\boldsymbol{w}_2^* \approx [-0.640184, 0.768221]^T$. Furthermore, the

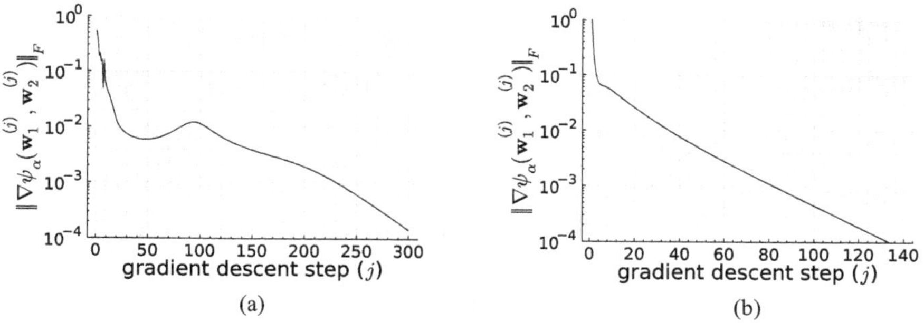

Fig. 3 Norm of gradients against gradient step number. (**a**) 44 data points. (**b**) 10,004 data points

unregularized objective function after 300 iterations of $\approx 7.99808 \times 10^{-5}$. Both indicate that the method has converged to the global minimum. Figure 3b shows a much smaller inflection in the decrease in the norm of the gradient. This is probably also due to the weights passing by a saddle point. That the inflection is much smaller when there are many more data points is an interesting observation, and worthy of future investigation.

5 Conclusion

Ridge function machines as described in here are a potentially useful method for approximating multivariate functions. The use of block Gauss–Seidel to reduce the computational costs for finding the coefficients of B-splines and the efficient computation of gradients with respect to weight vectors makes this a practical method.

More tests are needed on high-dimensional data sets, to check its practicality for such problems. Even if there are a large number of data points, the use of taking significant but modest sized subsets of the data can keep the computational costs modest even for large data sets. Furthermore, ridge function machines may provide a type of dimension reduction that can avoid some of the problems with high- dimensional data sets and other kinds of machine learning approaches.

The optimization methods over the weight vectors can be improved. In particular, methods that are better at handling saddle points can be applied. Trust region methods [11, Chap. 4], perhaps combined with a symmetric rank-1 quasi-Newton method [11, Sec. 6.2], might work well for this.

Further work needs to be done to adapt the method so as to avoid overfitting, and improving generalization performance. Regularization is needed in practice for solving the linear systems, but the overall method still performs well with quite small regularization parameters ($\alpha = 10^{-4}$).

References

1. A. Brandt, S. McCormick, and J. Ruge. Algebraic multigrid (AMG) for sparse matrix equations. In *Sparsity and its applications (Loughborough, 1983)*, pages 257–284. Cambridge Univ. Press, Cambridge, 1985.
2. John de Pillis. Gauss-Seidel convergence for operators on Hilbert space. *SIAM J. Numer. Anal.*, 10:112–122, 1973.
3. Weinan E, Chao Ma, and Lei Wu. A comparative analysis of optimization and generalization properties of two-layer neural network and random feature models under gradient descent dynamics. *Science China Mathematics*, 63(7):1235–1258, jan 2020.
4. Y. Gordon, V. Maiorov, M. Meyer, and S. Reisner. On the best approximation by ridge functions in the uniform norm. *Constr. Approx.*, 18(1):61–85, 2002.
5. Renjie Hu, Edward Ratner, David Stewart, Kaj-Mikael Björk, and Amaury Lendasse. A modified lanczos algorithm for fast regularization of extreme learning machines. *Neurocomputing*, 414:172–181, November 2020.
6. Guang-Bin Huang, Qin-Yu Zhu, and Chee-Kheong Siew. Extreme learning machine: Theory and applications. *Neurocomputing*, 70(1-3):489–501, December 2006.
7. V. È. Ismailov. Approximation by sums of ridge functions with fixed directions. *Algebra i Analiz*, 28(6):20–69, 2016.
8. Vugar E. Ismailov. Approximation by neural networks with weights varying on a finite set of directions. *J. Math. Anal. Appl.*, 389(1):72–83, 2012.
9. Vugar E. Ismailov. *Ridge functions and applications in neural networks*, volume 263 of *Mathematical Surveys and Monographs*. American Mathematical Society, Providence, RI, [2021] ©2021.
10. Palle Jorgensen and David E. Stewart. Approximation properties of ridge functions and extreme learning machines. *SIAM J. Math. Data Sci.*, 3(3):815–832, 2021.
11. Jorge Nocedal and Stephen J. Wright. *Numerical optimization*. Springer Series in Operations Research and Financial Engineering. Springer, New York, second edition, 2006.
12. David E. Stewart. *Numerical Analysis: A Graduate Course*. Springer International Publishing, 2022.

Learning Collective Behaviors from Observation

Jinchao Feng and Ming Zhong

Abstract We present a comprehensive examination of learning methodologies employed for the structural identification of dynamical systems. These techniques are designed to elucidate emergent phenomena within intricate systems of interacting agents. Our approach not only ensures theoretical convergence guarantees but also exhibits computational efficiency when handling high-dimensional observational data. The methods adeptly reconstruct both first- and second-order dynamical systems, accommodating observation and stochastic noise, intricate interaction rules, absent interaction features, and real-world observations in agent systems. The foundational aspect of our learning methodologies resides in the formulation of tailored loss functions using the variational inverse problem approach, inherently equipping our methods with dimension reduction capabilities.

1 Introduction

Data-driven modeling of governing differential equations for biological and physical systems from observations can be traced back to at least Gauss, Lagrange, and Laplace [1]. Kepler's laws of basic planetary motion, Newton's universal law of gravitation, and Einstein's general relativity provide excellent examples for discovering physical laws to understand motion from data. The capture and subsequent description of such biological and physical laws have always been important research topics. We are particularly interested in data-driven modeling of collective behaviors, also known as self-organization. These behaviors can manifest as clustering [2], flocking [3–5], milling [6], swarming [7], and synchronization [8,

J. Feng
Great Bay University, Dongguan, China
e-mail: jcfeng@gbu.edu.cn

M. Zhong (✉)
Illinois Institute of Technology, Chicago, IL, USA
e-mail: mzhong3@iit.edu

S. Foucart, S. Wojtowytsch (eds.), *Explorations in the Mathematics of Data Science*,
Applied and Numerical Harmonic Analysis,
https://doi.org/10.1007/978-3-031-66497-7_6

9]. These behaviors can be observed in spontaneous magnetization, Bose-Einstein condensation, molecular self-assembly, eusocial behaviors in insects, herd behavior, groupthink, and many others (see detailed references in [10–12]). Proper mathematical modeling of these behaviors can provide predictions and control of large systems. Hence, we review a series of learning methods specifically designed to provide mathematical insight from data observed in systems demonstrating collective behaviors.

There are many different kinds of mathematical models for collective behaviors; we focus on the agent-based aspect of these models. Therefore, we assume that the observational data follows certain dynamical systems. To simplify our discussion, we use the following autonomous dynamical system:

$$\dot{\mathbf{Y}} = \mathbf{f}(\mathbf{Y}), \quad \mathbf{Y} \in \mathbb{R}^D,$$

where the time-dependent variable $\mathbf{Y}$ describes a certain state in the system (such as position, velocity, temperature, phase, opinion, or a combination of these) and the function $\mathbf{f} : \mathbb{R}^D \to \mathbb{R}^D$ provides the change needed to update $\mathbf{Y}$. Given the observation of $\mathbf{Y}(t)_{t\in[0,T]}$, is it possible to identify $\mathbf{f}$ to explain the data within the given time interval $[0, T]$? Many system identification methods have been developed, such as learning parameterized systems via the maximum likelihood approach, which includes parameter estimation [13–17], and nonparametric estimation of drift in the stochastic McKean-Vlasov equation [18–20], Sparse Identification of Nonlinear Dynamics (SINDy, [21]), Weak SINDy that leverages the weak form of the differential equation and sparse parametric regression [22, 23], Physics-Informed Neural Network (PINN, [24]), NeuralODE [25], and many more. However, in the case of high-dimensional data, i.e., $D \gg 1$, the identification task becomes computationally prohibitive and time-consuming.

To overcome the curse of dimensionality, we employ a special and effective dimension reduction technique by constructing our learning methods based on the unique structure of the right-hand side function, $\mathbf{f}$. This unique structure arises from the fact that these dynamical systems are used to model collective behaviors, from which global patterns can emerge via only local interaction between pairs of agents.[1] To simplify the discussion, we can focus on first-order systems given as follows:

$$\dot{\mathbf{y}}_i = \frac{1}{N} \sum_{i'=1, i'\neq i}^{N} \phi(||\mathbf{y}_{i'} - \mathbf{y}_i||)(\mathbf{y}_{i'} - \mathbf{y}_i), \quad i = 1, \cdots, N.$$

Here, $\mathbf{y}_i \in \mathbb{R}^d$ describes the state of the ith agent, and $\phi : \mathbb{R}^+ \to \mathbb{R}$ gives the interaction law on how agent i' influences the change of state for agent i. This system is a gradient flow of a certain system energy with rotation/permutation invariances and symmetry. Such a first-order system can be used to model opinion dynamics (formation of consensus), crystal structure, and other pattern formations

[1] Agents here can be referred to as particles, cells, bacteria, robots, UAV, etc.

in skin pigmentation [26]. Using vector notation, i.e., letting

$$\mathbf{Y} = \begin{bmatrix} \vdots \\ \mathbf{y}_i \\ \vdots \end{bmatrix} \in \mathbb{R}^{D=Nd} \text{ and } \mathbf{f}_\phi(\mathbf{Y}) = \begin{bmatrix} \vdots \\ \frac{1}{N}\sum_{i'=1,i'\neq i}^{N} \phi(||\mathbf{y}_{i'} - \mathbf{y}_i||)(\mathbf{y}_{i'} - \mathbf{y}_i) \\ \vdots \end{bmatrix} \in \mathbb{R}^D,$$

we end up with the aforementioned autonomous dynamical system in the form of $\dot{\mathbf{Y}} = \mathbf{f}_\phi(\mathbf{Y})$ with $\mathbf{Y} \in \mathbb{R}^D$ and $D = Nd \gg 1$. For example, even for a simple system with 10 agents where $\mathbf{y}_i \in \mathbb{R}^2$, the system state variable lives in $D = 20$ dimensions. How to efficiently deal with such high dimensionality has become a crucial aspect of developing our learning algorithms [27–35]. Our methods are not only proven to have theoretical guarantees but are also verified through rigorous numerical testing. Even when learning from real observational data, our method can offer valuable insights into how the framework of collective behaviors can be used to explain complex systems.

The remaining sections of the paper are organized as follows: in Sect. 2, we discuss the general model form used for our learning paradigm. In Sect. 3, we go through various learning scenarios, from first-order to second-order, from homogeneous agents to heterogeneous agents, and with Gaussian process priors. We compare our methods to other well-established methods in Sect. 3.1.1. Finally, in Sect. 4, we conclude our review with a highlight of several future directions.

2 Model Equation

In order to provide a more unified picture of modeling collective dynamics, we consider the following state variable $\mathbf{y}_i$ for the ith agent in a system of N interacting agents. Moreover, $\mathbf{y}_i$ is expressed in the form of $\mathbf{y}_i = \begin{bmatrix}\mathbf{x}_i \; \mathbf{v}_i \; \zeta_i\end{bmatrix}$, where $\mathbf{x}_i \in \mathbb{R}^d$ represents the position of agent i, $\mathbf{v}_i \in \mathbb{R}^d$ is its corresponding velocity (hence $\mathbf{v}_i = \dot{\mathbf{x}}_i$), and $\zeta_i \in \mathbb{R}$ is an additional state that can be used to describe opinions, phases, excitation levels (e.g., towards light sources), emotions, etc. The coupled state variable $\mathbf{y}_i \in \mathbb{R}^{2d+1}$ satisfies the following coupled dynamical system (for $i = 1, \cdots, N$):

$$\begin{cases} \dot{\mathbf{x}}_i &= \mathbf{v}_i \\ m_i\dot{\mathbf{v}}_i &= F^{\mathbf{v}}(\mathbf{y}_i) + \sum_{i'=1,i'\neq i}^{N} \frac{1}{N_{k_i}}\Big[\Phi^E_{k_i,k_{i'}}(\mathbf{y}_i, \mathbf{y}_{i'})\mathbf{w}^E(\mathbf{x}_i, \mathbf{x}_{i'}) \\ &\quad +\Phi^A_{k_i,k_{i'}}(\mathbf{y}_i, \mathbf{y}_{i'})\mathbf{w}^A(\mathbf{v}_i, \mathbf{v}_{i'})\Big] \\ \dot{\zeta}_i &= F^\zeta(\mathbf{y}_i) + \sum_{i'=1,i'\neq i}^{N} \frac{1}{N_{k_i}}\Phi^\zeta_{k_i,k_{i'}}(\mathbf{y}_i, \mathbf{y}_{i'})(\zeta_{i'} - \zeta_i) \end{cases}. \tag{1}$$

Here, we consider the system to be partitioned into K types, i.e., $C_{k_1} \cap C_{k_2} = \emptyset$ for $1 \leq k_1 \neq k_2 = K$, $\cup_{k=1}^{K} C_K = [N] = 1, \cdots, N$, and N_k is the set of agents in type

k. Moreover, the types of the agents do not change over time,[2] and there is a type function $k.: \mathbb{N} \to \mathbb{N}$, which returns the type index for an agent index. Furthermore, the external force $F^{\mathbf{v}} : \mathbb{R}^{2d+1} \to \mathbb{R}^d$ gives the environmental effect on the velocity of the agent i, similarly $F^{\zeta} : \mathbb{R}^{2d+1} \to \mathbb{R}$ gives the environmental effect on ξ_i, $\Phi^E : \mathbb{R}^{4d+2} \to \mathbb{R}$ gives the energy-based interaction, $\Phi^A : \mathbb{R}^{4d+2} \to \mathbb{R}$ gives the alignment-based interaction, $\Phi^{\zeta} : \mathbb{R}^{4d+2} \to \mathbb{R}$ gives the ζ-based interaction, and $\mathbf{w}^E, \mathbf{w}^A : \mathbb{R}^{2d} \to \mathbb{R}^d$ give the directions for Φ^E, Φ^A, respectively.

Remark 1 The inclusion of $(\mathbf{x}_i, \mathbf{v}_i)$ into the system also makes it a second-order system (second-order time derivative for $\mathbf{x}_i$). We include the additional state ζ_i to provide a more realistic modeling of emergent behaviors. This formulation can also include first-order systems as special cases. When $F^{\mathbf{v}}(\mathbf{y}_i) = -\nu_i \mathbf{v}_i$ (the usual friction) with $\nu_i \gg m_i$, $\Phi^A = 0$, and Φ^E depends on $(\mathbf{x}_i, \zeta_i, \mathbf{x}_{i'}, \zeta_{i'})$, the second-order systems become:

$$\mathbf{0} \approx -\mathbf{v}_i + \frac{1}{\nu_i} \sum_{i'=1, i' \neq i}^{N} \frac{1}{N_{k_i}} \Phi^E_{k_i, k_{i'}}(\mathbf{x}_i, \zeta_i, \mathbf{x}_{i'}, \zeta_{i'}) \mathbf{w}^E(\mathbf{x}_i, \mathbf{x}_{i'}).$$

This results in a first-order system of $\mathbf{x}_i$ coupled with ζ_i (see the swarmalator model [8, 9, 36]).

Our second-order coupled formulation (1) can include a rather extensive list of behaviors, such as flocking with external potential [37], anticipated flocking dynamics [38], fish swarming dynamics [39], swarmalator dynamics (concurrent swarming and synchronization) [8, 9, 36, 40], and line alignment dynamics [41], where the interaction kernel depends on pairwise state variables. It is possible to include more state variables; for example, we can also consider $(\mathbf{x}_i, \mathbf{v}_i, \zeta_i, \eta_i)$, where η_i describes a different state than ζ_i. However, learning such a system with four state variables adds mere technical complexity, and we will leave the details for future projects.

Using the vector notation, we can obtain a simplified system denoted as $\dot{\mathbf{Y}} = \mathbf{f}_{\Phi^E, \Phi^A, \Phi^{\zeta}}(\mathbf{Y})$, where $\mathbf{Y} = \begin{bmatrix} \cdots \ \mathbf{y}_i^{\top} \ \cdots \end{bmatrix}^{\top} \in \mathbb{R}^{D=N(2d+1)}$ and

$$\mathbf{f}(\mathbf{Y}) = \begin{bmatrix} \vdots \\ \mathbf{v}_i \\ F^{\mathbf{v}}(\mathbf{y}_i) + \sum_{i'=1, i' \neq i}^{N} \frac{1}{N_{k_i}} \Big[\Phi^E_{k_i, k_{i'}}(\mathbf{y}_i, \mathbf{y}_{i'}) \mathbf{w}^E(\mathbf{x}_i, \mathbf{x}_{i'}) \\ + \Phi^A_{k_i, k_{i'}}(\mathbf{y}_i, \mathbf{y}_{i'}) \mathbf{w}^A(\mathbf{v}_{i'} - \mathbf{v}_i) \Big] \\ F^{\zeta}(\mathbf{y}_i) + \sum_{i'=1, i' \neq i}^{N} \frac{1}{N_{k_i}} \Phi^{\zeta}_{k_i, k_{i'}}(\mathbf{y}_i, \mathbf{y}_{i'})(\zeta_{i'} - \zeta_i) \\ \vdots \end{bmatrix} \in \mathbb{R}^{N(2d+1)}.$$

[2] Such a restriction can be relaxed, as long as we have the type information for the agents at all times.

For such a coupled system, we seek to identify $\mathbf{f} = \mathbf{f}_{\{\Phi^E_{k_1,k_2}, \Phi^A_{k_1,k_2}, \Phi^\zeta_{k_1,k_2}\}_{k_1,k_2=1}^K}$ in terms of the interaction kernels $\{\Phi^E_{k_1,k_2}, \Phi^A_{k_1,k_2}, \Phi^\zeta_{k_1,k_2}\}_{k_1,k_2=1}^K$ from observations. With $\mathbf{Y} \in \mathbb{R}^{D=N(2d+1)}$, the computational complexity of learning $\mathbf{f}$ increases exponentially. We will discuss the details of how to handle such high-dimensional learning in Sect. 3.

3 Learning Framework

We are now prepared to discuss a unified learning framework developed for the efficient learning of such high-dimensional dynamical systems used to model collective behaviors. As mentioned in Sect. 1, we are interested in learning the dynamical system from observation in the form $\dot{\mathbf{y}} = \mathbf{f}(\mathbf{y})$, where $\mathbf{y} \in \mathbb{R}^D$ with $D \gg 1$. However, due to the special structure in $\mathbf{f}$, i.e., $\mathbf{f} = \mathbf{f}_\phi$ where $\phi : \Omega \subset \mathbb{R}^{D'}$ with $D' \ll D$, we can exploit it and reduce the dimension for our learning framework. The framework focuses on constructing a suitable loss function designed for various φ's from the observation data $\mathbf{Y}(t), \dot{\mathbf{Y}}(t) t \in [0, T]$ with $\mathbf{Y}(0) \sim \mu\mathbf{Y}$; i.e., the unified loss takes on the following form:

$$\mathcal{E}(\varphi) = \mathbb{E}_{\mathbf{Y}(0)\sim\mu_\mathbf{Y}}\Big[\int_{t=0}^{T} ||\dot{\mathbf{Y}}(t) - \mathbf{f}_\varphi(\mathbf{Y}(t))||^2_{\mathcal{Y}}\, dt\Big],$$

where the vector norm $||\cdot||_{\mathcal{Y}}$ is designed for the special structure of $\mathbf{Y}$. We will begin our discussion from the simplest case, first-order systems, and then gradually extend the method to more complex scenarios, such as systems with multiple types of agents, stochastic noise, geometry-constrained dynamics, missing feature maps, coupled systems (for high-order systems), and learning with a Gaussian process prior.

Before we dive into the details, we would like to provide a brief summary of the papers that contribute to various aspects of our learning framework. An initial framework on first- and second-order multispecies systems was developed and analyzed in [27], where the major focus is on cases with a fixed number of agents rather than the mean-field limit, i.e., $N \to \infty$, as discussed in [42]. A subsequent inquiry, directed towards the theoretical foundations of estimators applied to multispecies agents, was undertaken as detailed in [30]. A numerical study of the steady-state behavior of our learned estimators was presented in [28] with an extension to a two-dimensional interaction function ϕ. A comprehensive analysis of second-order systems of heterogeneous agents with multidimensional ϕ was presented in [32], compared to learning scenario examined in [30], which involved first-order systems of heterogeneous agents characterized by one-dimensional interaction functions. An extension to dynamics constrained on Riemannian manifolds was given in [29], where the convergence rate similar to the one in Euclidean space was established by preserving the geometric structure in the learning. In [31], a combination of feature map learning and dynamical system learning was presented to effectively reduce the high-dimensional interaction function Φ within the collective dynamics framework. An application to re-establish Newton's framework of the universal law of gravitation from NASA JPL's Horizon database (a highly accurate synthetic database of our solar system used in space exploration) was presented in [33], where our learning methods were able to provide relative errors at the scale of 10^{-8} to capture some of the general relativity effects within the limitation of the collective dynamics scheme. A learning framework with a Gaussian prior in order to reduce the number of learning samples needed and provide uncertainty quantification was presented in [34, 35].

3.1 First Order

We start with a simple first-order system to build our learning framework. The first-order system which we consider is given as follows:

$$\dot{\mathbf{x}}_i = \frac{1}{N} \sum_{i'=1, i' \neq i}^{N} \Phi(\mathbf{x}_i, \mathbf{x}_{i'})(\mathbf{x}_{i'} - \mathbf{x}_i), \quad i = 1, \cdots, N. \tag{2}$$

Here, $\mathbf{x}_i$ is the state variable of the ith agent, and $\Phi = \Phi^E : \mathbb{R}^{2d} \to \mathbb{R}$ gives the energy-based (short-range repulsion and long-range attraction; we omit the superscript E here to streamline the notations) interaction. We only require the interaction kernel Φ to be symmetric, i.e., $\Phi(\mathbf{x}_i, \mathbf{x}_{i'}) = \Phi(\mathbf{x}_{i'}, \mathbf{x}_i)$. We shall further assume that Φ can be written in the following compositional form:

$$\Phi(\mathbf{x}_i, \mathbf{x}_{i'}) = \phi(\boldsymbol{\xi}(\mathbf{x}_i, \mathbf{x}_i)), \quad \phi : \mathbb{R}^{d_\phi} \to \mathbb{R} \quad \boldsymbol{\xi} : \mathbb{R}^{2d} \to \mathbb{R}^{d_\phi}. \tag{3}$$

Here, $d_\phi \ll 2d$ is the number of feature variables for the interaction function, ϕ is the reduced interaction kernel, and $\boldsymbol{\xi}$ is the reduced interaction variable. In most of the collective behavior models, $\boldsymbol{\xi}$ is assumed to be the pairwise distance variable, i.e. (Fig. 1),

$$\boldsymbol{\xi}(\mathbf{x}_i, \mathbf{x}_{i'}) = ||\mathbf{x}_{i'} - \mathbf{x}_i||, \quad ||\mathbf{x}|| = \sqrt{\sum_{j=1}^{d} |(\mathbf{x})_j|^2}, \quad \forall \mathbf{x} \in \mathbb{R}^d.$$

Hence, we arrive at a simpler and more familiar form of (2)

$$\dot{\mathbf{x}}_i = \frac{1}{N} \sum_{i'=1, i' \neq i}^{N} \phi(||\mathbf{x}_{i'} - \mathbf{x}_i||)(\mathbf{x}_{i'} - \mathbf{x}_i), \quad i = 1, \cdots, N. \tag{4}$$

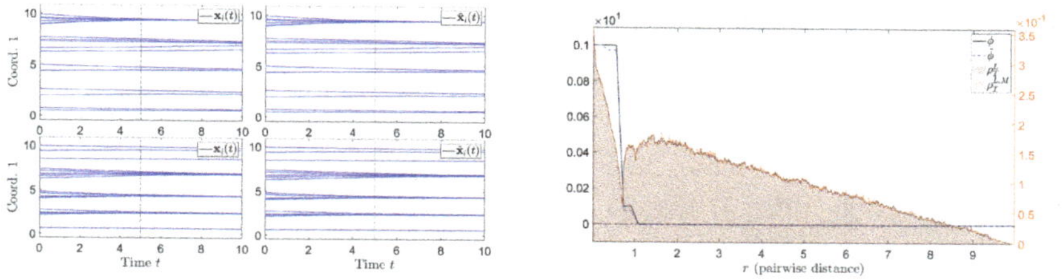

(a) (b)

Fig. 1 Opinion dynamics introduced in [2, 43], with $N = 20$ agents and $\phi(r) = \chi_{[0,\frac{1}{\sqrt{2}})}(r) + 0.1 * \chi_{[\frac{1}{\sqrt{2}},1]}(r)$, learned on [0, 10]; for other parameters see [27]. **(a)** $\mathbf{X}(t)$ vs $\hat{\mathbf{X}}(t)$. **(b)** ϕ vs $\hat{\phi}$

Remark 2 (4) can be viewed as a gradient flow of the following form:

$$\dot{\mathbf{x}}_i = \nabla_{\mathbf{x}_i}\Big(\frac{1}{2N}\sum_{1\le i\ne i\le N} U(||\mathbf{x}_{i'} - \mathbf{x}_i||)\Big), \quad i = 1, \cdots, N.$$

Here $U : \mathbb{R}^+ \to \mathbb{R}$ is an energy potential that depends on pairwise distance.

We are now ready to discuss the learning framework based on the *one*-dimensional ϕ; such a learning method can be easily extended to two- or three-dimensional ϕ as presented in [28, 32]. However, in order to efficiently handle the $2d$-dimensional Φ, we leave the discussion in the feature map learning section. Furthermore, we consider observation data in this particular form, i.e., $\{\mathbf{x}_i^m(t_l), \dot{\mathbf{x}}_i^m(t_l)\}_{i,l,m=1}^{N,L,M}$ for $0 = t_1 < t_2 < \cdots < t_L = T$, and $\mathbf{x}_i^m(0)$s are i.i.d. samples of a certain probability distribution μ_0; where m indicates different initial conditions from the observation data. By using m, we bring back some level of data independence. We now can find the estimator for ϕ by minimizing the following loss functional over all test functions, φ, from a carefully designed convex and compact function/hypothesis space $\mathcal{H}$, i.e.:

$$\mathcal{E}_{L,M,\mathcal{H}}(\varphi) = \frac{1}{LM}\sum_{l,m=1}^{L,M} ||\dot{\mathbf{X}}_l^m - \mathbf{f}_\varphi(\mathbf{X}_l^m)||_{\mathcal{S}}^2, \quad \varphi \in \mathcal{H}. \tag{5}$$

Here $\mathbf{X}_l^m = \big[\cdots (\mathbf{x}_i^m(t_l))^\top \cdots\big]^\top \in \mathbb{R}^{D=Nd}$,

$$\mathbf{f}_\varphi(\mathbf{X}_l^m) = \begin{bmatrix} \vdots \\ \frac{1}{N}\sum_{i'=1,i'\ne i}^N \varphi(||\mathbf{x}_{i'}^m(t_l) - \mathbf{x}_i^m(t_l)||)(\mathbf{x}_{i'}^m(t_l) - \mathbf{x}_i^m(t_l)) \\ \vdots \end{bmatrix} \in \mathbb{R}^D,$$

and the norm $||\cdot||_{\mathcal{S}}$ is defined as follows:

$$||\mathbf{U}-\mathbf{V}||_{\mathcal{S}}^2 = \frac{1}{N}\sum_{i=1}^{N}||\mathbf{u}_i-\mathbf{v}_i||_{\ell^2(\mathbb{R}^d)}^2, \quad \mathbf{U} = \begin{bmatrix} \mathbf{u}_1 \\ \vdots \\ \mathbf{u}_i \\ \vdots \\ \mathbf{u}_N \end{bmatrix}, \quad \mathbf{V} = \begin{bmatrix} \mathbf{v}_1 \\ \vdots \\ \mathbf{v}_i \\ \vdots \\ \mathbf{v}_N \end{bmatrix}, \quad \mathbf{u}_i, \mathbf{v}_i \in \mathbb{R}^d.$$

Since $\mathcal{H}$ is chosen to be convex and compact, the existence and uniqueness of the minimization are guaranteed. We denote the unique minimizer of $\mathcal{E}_{L,M,\mathcal{H}}$ as $\hat{\phi}_{L,M,\mathcal{H}}$, i.e., $\hat{\phi}_{L,M,\mathcal{H}} = \underset{\varphi\in\mathcal{H}}{\arg\min}\, \mathcal{E}_{L,M,\mathcal{H}}(\varphi)$.

Remark 3 By rewriting the dynamical system into $\dot{\mathbf{X}} = \mathbf{f}_\phi(\mathbf{X})$, it seems we might be able to use the regression framework to learn $\mathbf{f}$, simply by letting the input be $\mathbf{X}$ and the response be $\dot{\mathbf{X}}$. However, due to the high dimensionality of $\mathbf{X}$, this approach is computationally prohibitive, not to mention the lack of data independence from our observation data. But by exploiting the structure of $\mathbf{f}_\phi$, i.e., the dependence on ϕ, the learning can be thought of as *one*-dimensional learning. However, it is not a regression learning since we do not have a direct observation of ϕ at various points; instead, we have a linear combination of ϕ. Therefore, we need to cast it as a variational inverse problem to learn ϕ.

In order to demonstrate the convergence of $\hat{\phi}_{L,M,\mathcal{H}}$ to ϕ as $M \to \infty$, we will need to consider the expected version of the loss functional, i.e.:

$$\mathcal{E}_{L,\mathcal{H}}(\varphi) = \frac{1}{L}\mathbb{E}_{\mathbf{X}_0\sim\mu_0}\Big[\sum_{l}^{L}||\dot{\mathbf{X}}_l^m - \mathbf{f}_\varphi(\mathbf{X}_l^m)||_{\mathcal{S}}^2\Big].$$

We denote the minimizer $\hat{\phi}_{L,\mathcal{H}}$ from minimizing $\mathcal{E}_{L,\mathcal{H}}$ over the same hypothesis space $\mathcal{H}$, i.e., $\hat{\phi}_{L,\mathcal{H}} = \underset{\varphi\in\mathcal{H}}{\arg\min}\, \mathcal{E}_{L,\mathcal{H}}(\varphi)$. Using the law of large numbers, $\hat{\phi}_{L,M,\mathcal{H}} \to \hat{\phi}_{L,\mathcal{H}}$ as $M \to \infty$. The only obstacle left is to bound the distance between $\hat{\phi}_{L,\mathcal{H}}$ and ϕ due to the approximation power of $\mathcal{H}$. Before we show the lemma needed to prove convergence, we need to define a weighted $L^2(\rho)$ norm where the weight function ρ is given as follows:

$$\rho(r) = \frac{1}{\binom{N}{2}T}\int_{t=0}^{T}\mathbb{E}_{\mathbf{X}_0\sim\mu_0}\Big[\sum_{i,i'=1,i<i'}^{N}\delta_{r_{i,i'}(t)}(r)\,dt\Big]. \tag{6}$$

Here, $\mathbf{r}_{i,i'} = \mathbf{x}_{i'} - \mathbf{x}_i$ and $r_{i,i'} = ||\mathbf{r}_{i,i'}||$. ρ is used to illustrate the distribution of pairwise distance data through the interaction of agents caused by the dynamics. Such data is also used to learn ϕ; however, we do not have direct access to individual

$\phi(r)$; rather, we have the indirect observation in terms of the linear combination of $\phi(r)$, i.e., $\sum_{j=1,j\neq i}^{N}\phi(r_{i,j})\mathbf{r}_{i,j}$. With this definition of ρ, we can define our weighted $L^2(\rho)$ norm as follows:

$$||\phi_1(\cdot)\cdot - \phi_2(\cdot)\cdot||^2_{L^2(\rho)} = \int_{r\in \text{supp}(\rho)} |\phi_1(r) - \phi_2(r)| r \, d\rho(r), \tag{7}$$

for any $\phi_1, \phi_2 \in L^2(\text{supp}(\rho), \rho)$. Now we are ready to show the lemma needed to prove the convergence theorem.

Definition 1 (Definition 3.1 **in [27])** The dynamical system given by (4) with IC sampled from μ_0 on $\mathbb{R}^{D=Nd}$ satisfies the **coercivity condition** on a set $\mathcal{H}$ if there exists a constant $C_{L,N,\mathcal{H}}$ such that for all $\varphi \in \mathcal{H}$ with $\varphi(\cdot)\cdot \in L^2(\rho)$,

$$C_{L,N,\mathcal{H}}||\varphi(\cdot)\cdot||^2_{L^2(\rho)} \leq \frac{1}{LN}\sum_{i,l=1}^{N,L}\mathbb{E}||\frac{1}{N}\sum_{i'=1,i'\neq i}^{N}\varphi(r_{i,i'}(t_l)\mathbf{r}_{i,i'}(t_l)||^2.$$

The coercivity condition, given by Definition 1, basically states that minimizing $\mathcal{E}_{L,\mathcal{H}}$ would also minimize the distance between $\hat{\phi}_{L,\mathcal{H}}$ and ϕ under the weighted $L^2(\rho)$ norm, i.e.:

$$C_{L,N,\mathcal{H}}||\phi(\cdot)\cdot - \varphi(\cdot)\cdot||^2_{L^2(\rho)} \leq \mathcal{E}_{L,\mathcal{H}}(\varphi)$$

To show this, one simply uses the fact that ϕ is the interaction kernel which gives the observation data, i.e., $\dot{\mathbf{x}}_i = \frac{1}{N}\sum_{i'=1,i'\neq i}^{N}\phi(r_{i,i'})\mathbf{r}_{i,i'}$. With the coercivity condition established, we are ready to show Theorem 1.

Theorem 1 (Theorem 3.1 **in [27])** *Assume that $\phi \in \mathcal{K}_{R,S}$ (an admissible set $\mathcal{K}_{R,S} = \{\phi \in C^1(\mathbb{R}_+) : supp(\phi) \subset [0,R], \sup_{r\in[0,R]}|\phi(r)| + |\phi'(r)| \leq S\}$ for some $R, S > 0$). Let $\{\mathcal{H}_n\}_n$ be a sequence of subspaces of $L^\infty([0,R])$, with $dim(\mathcal{H}_n) \leq c_0 n$ and $\inf_{\varphi\in\mathcal{H}_n}||\varphi - \phi||_{L^\infty([0,R])} \leq c_1 n^{-s}$, for some constants $c_0, c_1, s > 0$. Assume that the coercivity condition holds on $\mathcal{H} := \overline{\cup_{n=1}^{\infty}\mathcal{H}_n}$. Such a sequence exists, for example, if ϕ is s-Hölder regular, and can be chosen so that $\mathcal{H}$ is compact in $L^2(\rho)$. Choose $n_* = (\frac{M}{\log(M)})^{\frac{1}{2s+1}}$. Then, there exists a constant $C = C(c_0, c_1, R, S)$ such that*

$$\mathbb{E}\Big[||\hat{\phi}_{L,M,\mathcal{H}_{n*}}(\cdot)\cdot - \phi(\cdot)\cdot||_{L^2(\rho)}\Big] \leq \frac{C}{C_{L,N,\mathcal{H}}}(\frac{M}{\log(M)})^{\frac{s}{2s+1}}.$$

Our learning rate is as optimal as if it were learned from the regression setting, and the proof is an elegant combination of the proof presented in [42] and [44]. For a more detailed discussion and the actual proof, see [27] and its supplementary information. Furthermore, our method is also robust against observation noise, as shown in Figure 8 in [27].

When we choose a basis for $\mathcal{H}_n$, i.e., $\mathcal{H}_n = \text{span}\{\psi_1, \cdots, \psi_n\}$ and let $\varphi = \sum_{\eta=1}^n \alpha_\eta \psi_\eta$, then the aforementioned minimization problem can be rewritten as a linear system $A\boldsymbol{\alpha} = \mathbf{b}$, where $\boldsymbol{\alpha} = \begin{bmatrix}\alpha_1 & \cdots & \alpha_n\end{bmatrix}^\top$, $A \in \mathbb{R}^{n\times n}$ with

$$A_{i,j} = \frac{1}{LM}\sum_{l,m=1}^{L,M} < \mathbf{f}_{\psi_i}(\mathbf{X}_l^m), \mathbf{f}_{\psi_j}(\mathbf{X}_l^m) >_{\mathcal{S}}, \quad < \mathbf{U}, \mathbf{U} >_{\mathcal{S}} = ||\mathbf{U}||_{\mathcal{S}}^2,$$

and $\mathbf{b} \in \mathbb{R}^n$ with

$$\mathbf{b}_i = \frac{1}{LM}\sum_{l,m=1}^{L,M} < \mathbf{f}_{\psi_i}(\mathbf{X}_l^m), \dot{\mathbf{X}}_l^m >_{\mathcal{S}} .$$

Moreover, we have a theoretical guarantee for the well-conditioning of the system; hence, it is computationally tractable to solve $A\boldsymbol{\alpha} = \mathbf{b}$. Algorithm 1 shows the pseudo-code for implementing the learning framework.

Algorithm 1 Learning ϕ from observations of first-order system

1: Input: $\{\mathbf{x}_i^m(t_l)$ and/or $\dot{\mathbf{x}}_i^m(t_l)$.
2: Output: estimators for the interaction kernels.
3: Find out the maximum/minimum interaction radii $R_{\min}$, $\mathbb{R}_{\max}$.
4: Construct the basis, $\{\psi_\eta\}_{\eta=1}^n$ on $[R_{\min}, R_{\max}]$.
5: Assemble $A\boldsymbol{\alpha} = \mathbf{b}$ (in parallel).
6: Solve for $\boldsymbol{\alpha}$.
7: Assemble $\hat{\phi}(r) = \sum_{\eta=1}^n \alpha_\eta \psi_\eta(r)$.

Furthermore, for massive datasets, one can use parallelization to assemble A and $\mathbf{b}$, which makes the overall process (near-)linear. See [27] for the detailed description of the algorithm, https://github.com/mingjzhong/LearningDynamics for the software package, and https://youtu.be/yc-AIAEtGDc?si=cv9TckRX_grMCOt8 for on how to use the software package.

3.1.1 Other System Identification Methods

There are many different methods for learning dynamical systems, such as using the parametric structure of the right-hand side with Bayesian inference, leveraging the sparse structure of the right-hand side (SINDy [21]), employing random features to approximate the interaction kernel [45], and using neural networks to estimate the right-hand side (PINN [24] and NeuralODE [25]). We will discuss SINDy and the neural network approach and make a direct comparison of these two methods with ours.

SINDy (Sparse Identification of Nonlinear Dynamics) is a data-driven identification method for dynamical systems. It assumes that the dynamical system $\dot{\mathbf{Y}} = \mathbf{f}(\mathbf{Y})$ can be approximated using the following form:

$$\mathbf{f}(\mathbf{Y}) \approx a_1\psi_1(\mathbf{Y}) + \cdots + a_n\psi_n(\mathbf{Y}), \quad \mathbf{Y} = \begin{bmatrix} (\mathbf{Y})_1 \\ \vdots \\ (\mathbf{Y})_D \end{bmatrix} \in \mathbb{R}^D.$$

ψ_i is applied component-wise to $\mathbf{Y}$, i.e.:

$$\psi_i(\mathbf{Y}) = \begin{bmatrix} \psi_i(y_1) \\ \vdots \\ \psi_i(y_D) \end{bmatrix}, \quad \mathbf{Y} = \begin{bmatrix} y_1 \\ \vdots \\ y_D \end{bmatrix}.$$

Furthermore, $\{\psi_1, \cdots, \psi_n\}$ is a set of basis functions from a predetermined dictionary; for example, they can be a set of polynomials, sine/cosine, and/or negative power polynomials. Given the observation data $\{\mathbf{Y}(t_1), \cdots, \mathbf{Y}(t_L)\}$, we assemble the matrix:

$$\mathbf{Y} = \begin{bmatrix} \mathbf{Y}^\top(t_1) \\ \vdots \\ \mathbf{Y}^\top(t_L) \end{bmatrix} = \begin{bmatrix} y_1(t_1) & y_2(t_1) & \cdots & y_D(t_1) \\ y_1(t_2) & y_2(t_2) & \cdots & y_D(t_2) \\ \vdots & \vdots & \ddots & \vdots \\ y_1(t_L) & y_2(t_L) & \cdots & y_D(t_L) \end{bmatrix} \in \mathbb{R}^{L\times D}.$$

We construct a dictionary (or library) $\Theta(\mathbf{Y})$ of nonlinear candidate functions of $\mathbf{Y}$, i.e.:

$$\Theta(\mathbf{Y}) = \begin{bmatrix} | & | & & | \\ \psi_1(\mathbf{Y}) & \psi_2(\mathbf{Y}) & \cdots & \psi_n(\mathbf{Y}) \\ | & | & & | \end{bmatrix} \in \mathbb{R}^{L\times nD}.$$

Again $\psi_i(\mathbf{Y})$ is applied component-wise, i.e.:

$$\psi_i(\mathbf{Y}) = \begin{bmatrix} \psi_i(y_1(t_1)) & \psi_i(y_2(t_1)) & \cdots & \psi_i(y_D(t_1)) \\ \psi_i(y_1(t_2)) & \psi_i(y_2(t_2)) & \cdots & \psi_i(y_D(t_2)) \\ \vdots & \vdots & \ddots & \vdots \\ \psi_i(y_1(t_L)) & \psi_i(y_2(t_L)) & \cdots & \psi_i(y_D(t_L)) \end{bmatrix}.$$

The set of basis functions can be $\{1, y, y^2, y^3, \cdots, \sin(y), \cos(y), \cdots\}$, depending on prior knowledge or computational capacity. We write the dynamical system in the new form:

$$\dot{\mathbf{Y}} = \Theta(\mathbf{Y})\Xi, \quad \Xi = \begin{bmatrix} \boldsymbol{\xi}_1 \cdots \boldsymbol{\xi}_D \end{bmatrix}, \quad \boldsymbol{\xi}_i \in \mathbb{R}^{nD \times D}.$$

SINDy assumes that the expansion of $\Theta(\mathbf{Y})$ is the same for all time; it simplifies the structure of Ξ down to

$$\Xi = \begin{bmatrix} \boldsymbol{\xi} \; \boldsymbol{\xi} \cdots \boldsymbol{\xi} \end{bmatrix} = \boldsymbol{\xi} \cdot \mathbf{1}, \quad \boldsymbol{\xi} \in \mathbb{R}^{nD \times 1} \quad \text{and} \quad \mathbf{1} = \begin{bmatrix} 1 \; 1 \cdots 1 \end{bmatrix} \in \mathbb{R}^{1 \times D}.$$

Furthermore, SINDy assumes that $\mathbf{f}(\mathbf{Y}(t))$ admits a spare representation in $\Theta(\mathbf{Y})$. Moreover, it finds a parsimonious model by performing least squares regression with sparsity-promoting regularization, i.e.:

$$\hat{\boldsymbol{\xi}} = \underset{\boldsymbol{\xi} \in \mathbb{R}^{nD \times 1}}{\arg\min} \, \{||\dot{\mathbf{Y}} - \Theta(\mathbf{Y})(\boldsymbol{\xi} \cdot \mathbf{1})||_{\text{Frob}}^2 + \lambda ||\boldsymbol{\xi}||_1\}.$$

The SINDy approach has been shown to be effective for observational data from various kinds of dynamics, especially for the Lorenz system [21, 46, 47]. The assumption that the right-hand side function can be approximated well by a set of predetermined dictionaries opens up a question on how to optimally choose such a dictionary. In the case of collective dynamics, a direct application is ineffective. Although the form of collective dynamics can be written as polynomials in components of $\mathbf{y}_i$, the coefficients change all the time, and they rarely assume a sparse representation in terms of the coordinate systems (or any coordinate system). However, they might have a sparse representation in terms of the pairwise variables. In order to use it for collective dynamics, one has to rework the approximation scheme; see [48] for an example.

We can also use neural network structures to solve and infer information about dynamical systems. Given the observation data $\{\mathbf{Y}(t_l), \dot{\mathbf{Y}}(t_l)\}_{l=1}^L$, we find the estimator to $\mathbf{f}$ as in $\dot{\mathbf{Y}} = \mathbf{f}(\mathbf{Y})$ from minimizing the following loss functional:

$$\text{Loss}(\tilde{\mathbf{f}}) = \frac{1}{L} \sum_{l=1}^{L} ||\dot{\mathbf{Y}}(t_l) - \tilde{\mathbf{f}}(\mathbf{Y}(t_l))||_{\ell^2(\mathbb{R}^D)}^2, \quad \tilde{\mathbf{f}} \in \mathcal{H}.$$

Here $\tilde{\mathbf{f}} : \mathbb{R}^D \to \mathbb{R}^D$ and $\mathcal{H}$ is a set of neural networks of the same depth, the same neurons on each hidden layer, and the same activation function on each hidden layer. The neural network solution is

$$\hat{\mathbf{f}} = \underset{\tilde{\mathbf{f}} \in \mathcal{H}}{\arg\min} \, \text{Loss}(\tilde{\mathbf{f}}).$$

Neural network approximation has been shown to be effective for high-dimensional function estimation. However, the right-hand side $\mathbf{f}(\mathbf{Y})$ might have a special structure that a usual neural network might fail to capture. In this case, NeuralODE

[25], which uses the ODE solver and recurrent neural network structure for training, can be employed to capture special properties of the ODE system. However, in the case of collective dynamics, when **f** on **Y** is learned directly, we limit ourselves to a fixed system (fixed number of agents). Recall that **Y** is a concatenation of all system state variables. In other words, if we switch to a system where the number of agents is different from the training data, we have to retrain. PINN (Physics-Informed Neural Networks [24]), on the other hand, may be adopted for such special needs, as it can learn both the right-hand side and the ODE solution together. A comprehensive study on the comparison between vanilla SINDy and NeuralODE was conducted in [30]; we are planning a more comprehensive comparison study where both the SINDy and NeuralODE are reformulated to conform to the collective dynamics models.

Another similar method is introduced as a random feature learning in [45], where the loss function combines both the observation loss and regularization on $\varphi \in \mathcal{H}$, i.e.:

$$\mathcal{E}_{\mathcal{H}}(\varphi) = \mathbb{E}_{\mathbf{Y}_0 \sim \mu_{\mathbf{Y}}}\Big[\frac{1}{T}\int_{t=0}^{T} ||\dot{\mathbf{Y}}_t - \mathbf{f}_{\varphi}(\mathbf{Y}_t)||_{\mathcal{S}}^2 + \lambda J(\varphi)\Big].$$

Here λ is a regularization parameter, $J : \mathcal{H} \to \mathbb{R}^+$ is a regularizing function on φ, and the function/hypothesis space $\mathcal{H}$ uses random features as the basis. The addition of regularization is also presented in [35] (Fig. 2).

3.2 *Heterogeneous Agents*

It is natural to consider heterogeneous agents, i.e., agents of different types, for there are many dynamics that have multiple types of agents involved, e.g., leader-follower, predator-prey, pedestrian-vehicle, etc. We consider the scenario that the system of N agents is partitioned into K types, i.e., $C_{k_1} \cap C_{k_2} = \emptyset$ for $1 \leq k_1 \neq k_2 = K$ and $\cup_{k=1}^{K} C_K = [N] = \{1, \cdots, N\}$. Moreover, the type of the agents does not change in

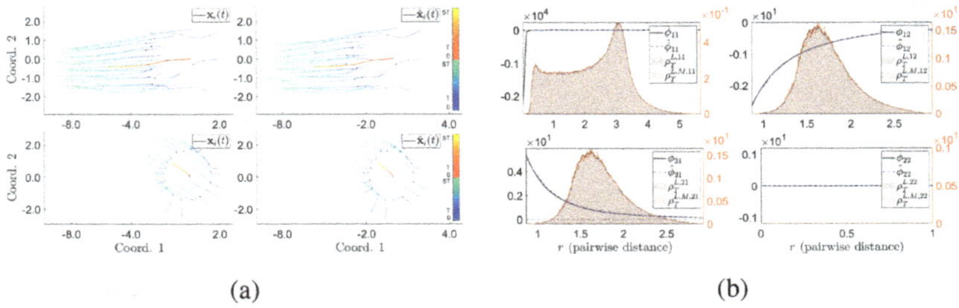

Fig. 2 Predator-prey dynamics introduced in [49], with 1 : prey, 2 : predator, $N_1 = 19$ preys, $N_2 = 1$ predator, learned on [0, 5]; for other parameters see [27]. **(a)** $\mathbf{X}(t)$ vs $\hat{\mathbf{X}}(t)$. **(b)** ϕ_{k_1,k_2} vs $\hat{\phi}_{k_1,k_2}$

time,[3] and there is a type function $k.: \mathbb{N} \to \mathbb{N}$, which returns the type index for an agent index.

Then, we can consider a simple extension of (4) as follows:

$$\dot{\mathbf{x}}_i = \sum_{i'=1, i' \neq i}^{N} \frac{1}{N_{k_i}} \phi_{k_i, k_{i'}}(||\mathbf{x}_{i'} - \mathbf{x}_i||)(\mathbf{x}_{i'} - \mathbf{x}_i), \quad i = 1, \cdots, N. \tag{8}$$

Now, the interaction kernel ϕ_{k_1,k_2} not only depends on the pairwise distance variable r but is also different for agents of different types. Given the same observation data, i.e., $\{\mathbf{x}_i^m(t_l), \dot{\mathbf{x}}_i^m(t_l)\}_{i,l,m=1}^{N,L,M}$, we will learn a total of K^2 interaction kernels, i.e., $\boldsymbol{\phi} = \{\phi_{k_1,k_2}\}_{k_1,k_2=1}^{K}$, all together by minimizing a slightly updated loss functional:

$$\mathcal{E}_{L,M,\mathcal{H}}^{\text{Mul}}(\boldsymbol{\varphi}) = \frac{1}{LM} \sum_{l,m=1}^{L,M} ||\dot{\mathbf{X}}_l^m - \mathbf{f}_{\boldsymbol{\varphi}}(\mathbf{X}_l^m)||_{\mathcal{S}'}^2, \tag{9}$$

Here $\boldsymbol{\varphi} = \{\varphi_{k_1,k_2} \in \mathcal{H}_{k_1,k_2}\}_{k_1,k_2=1}^{K}$, the direct sum space $\mathcal{H} = \oplus_{k_1,k_2=1}^{K} \mathcal{H}_{k_1,k_2}$, $\mathbf{X}_l^m$ is the usual vector notation and

$$\mathbf{f}_{\boldsymbol{\varphi}}(\mathbf{X}_l^m) = \begin{bmatrix} \vdots \\ \sum_{i'=1, i' \neq i}^{N} \frac{1}{N_{k_i}} \varphi_{k_i, k_{i'}}(||\mathbf{x}_{i'}^m(t_l) - \mathbf{x}_i^m(t_l)||)(\mathbf{x}_{i'}^m(t_l) - \mathbf{x}_i^m(t_l)) \\ \vdots \end{bmatrix}.$$

The norm, $|| \cdot ||_{\mathcal{S}'}$, is changed slightly as

$$||\mathbf{U} - \mathbf{V}||_{\mathcal{S}'}^2 = \sum_{i=1}^{N} \frac{1}{N_{k_i}} ||\mathbf{u}_i - \mathbf{v}_i||_{\ell^2(\mathbb{R}^d)}^2, \quad \mathbf{U}, \mathbf{V} \in \mathbb{R}^{D=Nd}.$$

The main difficulty in learning multiple ϕ_{k_1,k_2} at the same time is that there is no way to separate the trajectory data into K^2 groups. However, we are able to capture each individual ϕ_{k_1,k_2} differently from the same set of trajectory data. For proof of convergence, it follows similarly from the proof presented in [27, 30].

3.3 Stochastic Noise

Noise comes in various forms; it might appear in the observation data due to imperfect measurements, or it might manifest in the model as a stochastic noise to

[3] Such restriction can be relaxed, as long as we know the type information for the agents at all time.

reflect the randomness in decision-making of these interacting agents. We consider the second scenario where a stochastic noise term is added to the first-order system (8). It is given as follows (recall that $\mathbf{x}_i \in \mathbb{R}^d$ represents the state of the ith agent):

$$d\mathbf{x}_i = \Big(\sum_{i'=1, i'\neq i}^{N} \frac{1}{N_{k_i}} \phi_{k_i,k_{i'}}(||\mathbf{x}_{i'} - \mathbf{x}_i||)(\mathbf{x}_{i'} - \mathbf{x}_i) \Big) dt + d\mathbf{B}_i, \tag{10}$$

for $i = 1, \cdots, N$. Here $d\mathbf{B}_i$ is a Brownian motion with a state-dependent symmetric and positive-definite covariance matrix $\Xi_i(\mathbf{x})$ ($\Xi : \mathbb{R}^d \to \mathbb{R}^{d\times d}$). Using vector-notation, i.e., let

$$\mathbf{f}_{\varphi}(\mathbf{X}_l^m) = \begin{bmatrix} \vdots \\ \sum_{i'=1, i'\neq i}^{N} \frac{1}{N_{k_i}} \varphi_{k_i,k_{i'}}(||\mathbf{x}_{i'}^m(t_l) - \mathbf{x}_i^m(t_l)||)(\mathbf{x}_{i'}^m(t_l) - \mathbf{x}_i^m(t_l)) \\ \vdots \end{bmatrix},$$

where $\boldsymbol{\varphi} = \{\varphi_{k_1,k_2}\}_{k_1,k_2=1}^K$, and

$$\mathbf{D}(\mathbf{X}) = \begin{bmatrix} \Xi_1(\mathbf{x}_1) & 0 & \cdots & 0 \\ 0 & \Xi_2(\mathbf{x}_2) & \cdots & 0 \\ \vdots & \vdots & \ddots & \vdots \\ 0 & 0 & \cdots & \Xi_N(\mathbf{x}_N) \end{bmatrix} \quad \text{and} \quad \mathbf{B}_t = \begin{bmatrix} \mathbf{B}_1(t) \\ \vdots \\ \mathbf{B}_N(t) \end{bmatrix} \in \mathbb{R}^{D\times D},$$

The observation data is given in a slightly different form, i.e., $\{\mathbf{X}_t, d\mathbf{X}_t\}_{t\in[0,T]}$ with $\mathbf{X}_0 \sim \mu$ is given, we find the minimizer from the following loss:

$$\begin{aligned} \mathcal{E}_{\mathcal{H}}^{\text{Sto}}(\varphi) = \mathbb{E}_{\mathbf{X}_0\sim\mu_0}\Big[\frac{1}{2T} \int_{t=0}^{T} &< \mathbf{f}_{\boldsymbol{\varphi}}, \mathbf{D}^{-2}(\mathbf{X}_t)\mathbf{f}_{\boldsymbol{\varphi}}(\mathbf{X}_t) > dt \\ &- \frac{1}{T} \int_{t=0}^{T} < \mathbf{f}_{\boldsymbol{\varphi}}(\mathbf{X}_t), \mathbf{D}^{-2}(\mathbf{X}_t) d\mathbf{X}_t > \Big]. \end{aligned}$$

In actual applications, we will only be given snapshots of the states, i.e., $\mathbf{X}_l^m l, m = 1^{L,M}$. We must approximate both $d\mathbf{X}_l^m$ and the time integral in the loss. An initial study of the algorithm and its convergence was investigated in [50], where the automatic learning of the noise σ and the drift term $\mathbf{f}_{\boldsymbol{\phi}}$ is combined.

Remark 4 Upon contemplation of a system comprising homogeneous agents, where the covariance matrix for each $\mathbf{B}_i$ adopts the structure $\Xi_i(\mathbf{x}) = \sigma \mathbf{I}_{d\times d}$ with $\sigma > 0$ as a constant shared across all $\mathbf{B}_i$, the loss function $\mathcal{E}_{\mathcal{H}}^{\text{Sto}}$ can be rendered in a simplified form:

$$\mathcal{E}_{\mathcal{H}}^{\text{Sto}}(\varphi) = \frac{1}{2\sigma^2 T} \mathbb{E}_{\mathbf{X}_0\sim\mu_0}\Big[\int_{t=0}^{T} (||\mathbf{f}_{\varphi}||^2 dt - 2 < \mathbf{f}_{\varphi}(\mathbf{X}_t), d\mathbf{X}_t >) \Big].$$

The learning scenario characterized by a constant noise level devoid of correlation with other components or agents, particularly in the context of homogeneous agents, has been thoroughly investigated and demonstrated to exhibit superior convergence properties as documented in [51]. Notably, the recently introduced loss function in [50] has expanded the scope of learning capabilities, enabling the accommodation of state-dependent and correlated noise across heterogeneous types of agents. Both of the aforementioned loss functions draw inspiration from the Girsanov theorem. Nevertheless, the novel loss function additionally accounts for the correlated noise scenario, as expounded in Theorem 7.4 within the framework presented in [52].

3.4 Riemannian Geometry Constraints

The state variable, i.e., $\mathbf{x}_i \in \mathbb{R}^d$, might be living on a low-dimensional manifold as d gets large. We consider the case when $\mathbf{x}_i \in \mathcal{M} \subset \mathbb{R}^d$, where $\mathcal{M}$ is a Riemannian manifold naturally embedded in $\mathbb{R}^d$. We further assume that we know enough information about the manifold, i.e., the pair $(\mathcal{M}, g)$, with g being the Riemannian metric, is given to us. We also consider that the updated first-order model is as follows:

$$\mathbf{v}_i = \sum_{i'=1, i' \neq i}^{N} \frac{1}{N_{k_i}} \phi_{k_i, k_{i'}}(||\mathbf{x}_{i'} - \mathbf{x}_i||)\mathbf{w}^{\mathbf{x}}(\mathbf{x}_i, \mathbf{x}_{i'}), \quad i = 1, \cdots, N. \tag{11}$$

Here $\mathbf{v}_i = \dot{\mathbf{x}}_i \in \mathcal{T}_{\mathbf{x}_i}\mathcal{M}$ (where $\mathcal{T}_{\mathbf{x}}\mathcal{M}$ is the tangent space of $\mathcal{M}$ at $\mathbf{x}$), and $\mathbf{w}^{\mathbf{x}}(\mathbf{x}_i, \mathbf{x}_{i'})$ gives the unit tangent direction on the geodesic from $\mathbf{x}_i$ to $\mathbf{x}_{i'}$ if $\mathbf{x}_{i'}$ is not in the cut locus set of $\mathbf{x}_i$; otherwise $\mathbf{w}^{\mathbf{x}}(\mathbf{x}_i, \mathbf{x}_{i'}) = \mathbf{0}$. We also assume that each ϕ_{k_1,k_2} is defined on $[0, R]$, where R is sufficiently small so that length-minimizing geodesics exist uniquely. Then, such a first-order model is a well-defined gradient flow model of a sum of pairwise potential energy. To respect the geometry, we update the loss function so that it has the information about the Riemannian metric:

$$\mathcal{E}^{\text{Rie}}_{L,M,\mathcal{H}}(\boldsymbol{\varphi}) = \frac{1}{LM} \sum_{l,m=1}^{L,M} ||\dot{\mathbf{X}}_l^m - \mathbf{f}_{\boldsymbol{\varphi}}(\mathbf{X}_l^m)||_g^2,$$

where the new norm, $||\cdot||_g$, is given as

$$||\dot{\mathbf{X}}_l^m - \mathbf{f}_{\boldsymbol{\varphi}}(\mathbf{X}_l^m)||_g^2 = \sum_{i=1}^{N} \frac{1}{N_{k_i}} ||\dot{\mathbf{x}}_{i,l}^m - \sum_{i'=1, i' \neq i}^{N} \frac{1}{N_{k_{i'}}} \varphi_{k_i, k_{i'}}(r_{i,i',l}^m)\mathbf{w}_{i,i',l}^m||^2_{\mathcal{T}_{\mathbf{x}_{i,l}^m}\mathcal{M}}.$$

Here $\mathbf{x}_{i,l}^m = \mathbf{x}_i^m(t_l)$, $r_{i,i',l}^m = |\mathbf{x}_{i'}^m(t_l) - \mathbf{x}_i^m(t_l)|$, and $\mathbf{w}_{i,i',l}^m = \mathbf{w}^{\mathbf{x}}(\mathbf{x}_{i,l}^m, \mathbf{x}_{i',l}^m)$. With this new norm together with an updated coercivity condition respecting the geometric

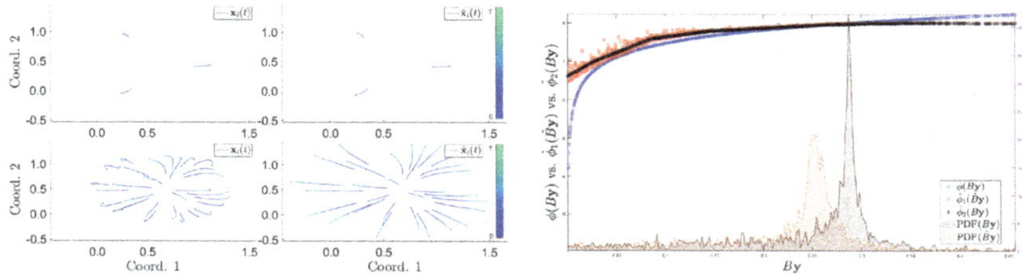

(a) (b)

Fig. 3 Power law dynamics introduced in [53], with $N = 3$ agents (learned from an updated algorithm); for other parameters see [31]. (**a**) $\mathbf{X}(t)$ vs $\hat{\mathbf{X}}(t)$. (**b**) ϕ vs $\hat{\phi}$ (with or without B)

structure of the data, we are able to preserve the convergence rate as if the data is in Euclidean space; see [29] for details (Fig. 3).

3.5 *Feature Map Learning*

Although there are many interacting-agent systems that demonstrate various complex emergent behaviors that are only involved with the interaction kernel functions depending on the pairwise distance as we described above, in real-life applications, the interactions may depend on some other unknown but a small number of variables (e.g., pairwise distances/angles of velocities, pairwise differences of phases). To consider this needed complexity, we consider the interacting-agent systems of N agents governed by the equations in the form (2), i.e.:

$$\dot{\mathbf{x}}_i(t) = \frac{1}{N} \sum_{i'=1}^{N} \Phi(\mathbf{x}_i(t), \mathbf{x}_{i'}(t))(\mathbf{x}_{i'}(t) - \mathbf{x}_i(t)) , \tag{12}$$

for $i = 1, \ldots, N$, $t \in [0, T]$, $\mathbf{x}_i \in \mathbb{R}^d$ is a state vector, and $\Phi : \mathbb{R}^d \times \mathbb{R}^d \to \mathbb{R}$ is the **interaction kernel**, governing how the state of agent i' influences the state of agent i. Note that while the state space of the system is dN-dimensional, the interaction kernel is a function of $2d$ dimensions. Moreover, we consider the case that the interaction kernel Φ is a function that depends on a smaller number of natural variables $\boldsymbol{\xi} \in \mathbb{R}^{d'}$, with $d' \ll 2d$, which are functions of pairs $(\mathbf{x}_i, \mathbf{x}_{i'}) \in \mathbb{R}^d \times \mathbb{R}^d$.

In other words, the interaction kernel Φ can be factorized (as composition of functions) as a **reduced interaction kernel** $\phi : \mathbb{R}^{d'} \to \mathbb{R}$, and **reduced variables** $\boldsymbol{\xi} : \mathbb{R}^d \times \mathbb{R}^d \to \mathbb{R}^{d'}$, i.e.:

$$\Phi(\mathbf{x}_i, \mathbf{x}_{i'}) = \phi(\boldsymbol{\xi}(\mathbf{x}_i, \mathbf{x}_{i'})). \tag{13}$$

When $\boldsymbol{\xi}(\mathbf{x}_i, \mathbf{x}_{i'}) := ||\mathbf{x}_i - \mathbf{x}_{i'}||$, $\Phi(\mathbf{x}_i, \mathbf{x}_{i'}) = \phi(\boldsymbol{\xi}(\mathbf{x}_i, \mathbf{x}_{i'})) = \phi(||\mathbf{x}_i - \mathbf{x}_{i'}||)$, with $\phi : \mathbb{R}_+ \to \mathbb{R}$, it becomes the simple model which is a special case of what we discussed above.

Given observations of trajectory data, i.e., $\{\mathbf{x}_i^{(m)}(t_l), \dot{\mathbf{x}}_i^{(m)}(t_l)\}_{i,l,m=1}^{N,L,M}$, with $0 = t_1 < \cdots < t_L = T$, initial conditions $\mathbf{x}_i^{(m)}(t_1) \sim \mu_0$ (some unknown probability distribution on the state space $\mathbb{R}^{dN}$) and $m = 1, \ldots, M$, we are interested in estimating the interaction kernel Φ, the only unknown quantity in (2). Note that this is again not a regression problem, but an inverse problem, since values of the function Φ are not observed, but only averages of such values, as per the right-hand side of (2), are.

We proceed by further assuming that $\boldsymbol{\xi}$ can be factorized as $\boldsymbol{\xi}(\mathbf{x}_i, \mathbf{x}_{i'}) := B\mathbf{z}(\mathbf{x}_i, \mathbf{x}_{i'})$, where $\mathbf{z} : \mathbb{R}^d \times \mathbb{R}^d \to \mathbb{R}^D$ is a *known* feature map from pairs of states to high-dimensional feature vectors, and the feature reduction map $B : \mathbb{R}^D \to \mathbb{R}^{d'}$ is an *unknown* linear map to be estimated. Since we will estimate the reduced interaction kernel ϕ in a nonparametric fashion, what really matters is the range of B, and so we may assume B to be orthogonal. Therefore, the factorization (13) may be rewritten in the form $\Phi(\mathbf{x}_i, \mathbf{x}_{i'}) = \phi(B\mathbf{z}(\mathbf{x}_i, \mathbf{x}_{i'}))$. To estimate ϕ and B, we proceed in two steps.

Step 1. Estimating the feature reduction map B, and the variables $\boldsymbol{\xi}$. While the choice of the feature map $\mathbf{z}$ is typically application-dependent, and can incorporate symmetries, or physical constraints on the system, a rather canonical choice is the map to polynomials in the states, and here we restrict ourselves to second-order polynomials, and therefore assume that

$$\mathbf{z}(\mathbf{x}_i, \mathbf{x}_{i'}) := \left[\mathbf{x}_i, \mathbf{x}_{i'}, ((\mathbf{x}_i)_j(\mathbf{x}_i)_{j'})_{j\le j'}, ((\mathbf{x}_{i'})_j(\mathbf{x}_{i'})_{j'})_{j\le j'}, ((\mathbf{x}_i)_j(\mathbf{x}_{i'})_{j'})_{j,j'=1,\ldots,d}\right]^T. \tag{14}$$

Here $\mathbf{z} \in \mathbb{R}^{D=2d^2+3d}$. Higher-order polynomials can be added for complicated interactions, and how to optimally choose such polynomial basis will be for future study.

The techniques of [54], similar to the previously existing multi-index regression works, are only applicable to the regression setting, and therefore not directly applicable. However, at this point, we note that if we allow the learning procedure to conduct "experiments" and have access to a simulator of (2) for different values of N, and in particular for $N = 2$, it can conduct observations of trajectories of two agents, and from those equations that reveal the values taken by the interaction kernels can be extracted, i.e., for $N = 2$:

$$\dot{\mathbf{x}}_1 = \frac{1}{2}\Phi(\mathbf{x}_1, \mathbf{x}_2)(\mathbf{x}_2 - \mathbf{x}_1),$$
$$\dot{\mathbf{x}}_2 = \frac{1}{2}\Phi(\mathbf{x}_2, \mathbf{x}_1)(\mathbf{x}_1 - \mathbf{x}_2),$$

we can then transform the equations as follows:

$$\frac{2\langle \dot{\mathbf{x}}_1, \mathbf{x}_2 - \mathbf{x}_1 \rangle}{||\mathbf{x}_2 - \mathbf{x}_1||^2} = \Phi(\mathbf{x}_1, \mathbf{x}_2), \quad \frac{2\langle \dot{\mathbf{x}}_2, \mathbf{x}_1 - \mathbf{x}_2 \rangle}{||\mathbf{x}_1 - \mathbf{x}_2||^2} = \Phi(\mathbf{x}_2, \mathbf{x}_1).$$

Consider $\mathbf{z}_{1,2} := \mathbf{z}(\mathbf{x}_1, \mathbf{x}_2) \in \mathbb{R}^D$ defined in (14), since we assume that $\Phi(\mathbf{x}_1, \mathbf{x}_2) = \phi(B\mathbf{z}_{1,2})$, we obtain a regression problem:

$$\phi(B\mathbf{z}_{1,2}) = \psi_{1,2} := 2\frac{\langle \dot{\mathbf{x}}_1, \mathbf{x}_2 - \mathbf{x}_1 \rangle}{||\mathbf{x}_2 - \mathbf{x}_1||^2} = \Phi(\mathbf{x}_1, \mathbf{x}_2).$$

Similar definitions are used for $\mathbf{z}_{2,1}$ and $\psi_{2,1}$.

Therefore, it takes us back to a regression setting, and we can apply the Multiplicatively Perturbed Least Squares (MPLS) approach of [54] to obtain an estimate $\hat{B}$ of the feature reduction map B. MPLS decomposes the regression function into linear and nonlinear components: $\Phi(\mathbf{x}_i, \mathbf{x}_{i'}) = \phi(B\mathbf{z}_{i,i'}) = \langle \beta, \mathbf{z}_{i,i'} \rangle + g(A\mathbf{z}_{i,i'})$, where $(i, i') = \{(1, 2), (2, 1)\}$ and g is orthogonal to linear polynomials. The intrinsic domain of Φ, the row space of B, is hence spanned by β and the rows of A, which are estimated by an ordinary linear approximation and, respectively, by the top right singular vectors of a matrix of "slope perturbations." Let us re-index the observations as $\{\mathbf{z}_q\}_{q=1}^{Q} := \{\mathbf{z}_{1,2}^{(m)}(t_l), \mathbf{z}_{2,1}^{(m)}(t_l)\}_{l,m=1}^{L,M}$, with the corresponding $\{\psi_q\}_{q=1}^{Q} := \{\psi_{1,2}^{(m)}(t_l), \psi_{2,1}^{(m)}(t_l)\}_{l,m=1}^{L,M}$, where $Q = 2LM$. Then, we can estimate $B \in \mathbb{R}^{d' \times D}$ via the MPLS algorithm as follows.

Algorithm [MPLS] Inputs: training data $\{\mathbf{z}_q, \psi_q\}_{q=1}^{Q}$, partitioned into subsets $\mathcal{S}$ and $\mathcal{S}'$, each of size, $|\mathcal{S}|$ and $|\mathcal{S}'|$, at least $\lfloor \frac{Q}{2} \rfloor$; parameters K and λ, with $K \gtrsim d' \log d'$ and $\lambda \approx \frac{1}{D}$ (choosing K and λ is discussed in [54]).

1. Compute an ordinary least squares linear approximation $\hat{\beta}$ to $\mathcal{S}'$:

$$\hat{\beta} := \arg\min_{\beta' \in \mathbb{R}^D} \frac{2}{|\mathcal{S}'|} \sum_{(\mathbf{z}', \psi') \in \mathcal{S}'} \left(\psi' - \langle \boldsymbol{\beta}', \mathbf{z}' \rangle\right)^2.$$

2. Let $\mathcal{R} := \left\{(\mathbf{z} - \langle \hat{\beta}, \mathbf{z} \rangle \|\hat{\beta}\|^{-2}\hat{\beta}, \psi - \langle \hat{\beta}, \mathbf{z} \rangle) \middle| (\mathbf{z}, \psi) \in \mathcal{S}\right\}$ be the residual data of this approximation on $\mathcal{S}$, paired with $\mathbf{z}$s projected away from $\hat{\beta}$.
3. Pick $\mathbf{u}_1, \ldots, \mathbf{u}_K$ in $\mathbb{R}^D$ (e.g., a random subset of $\{\mathbf{z}_q\}_{q=1}^{Q}$). For each $\mathbf{u}_i$, center the residuals to their weighted mean: with $w(\tilde{\mathbf{z}}, \mathbf{u}_i) = \exp(-\lambda ||\tilde{\mathbf{z}} - \mathbf{u}_i||^2)$:

$$\widetilde{\mathcal{R}} := \left\{ \left(\tilde{\mathbf{z}}, r - \frac{\sum_{(\tilde{\mathbf{z}},r) \in \mathcal{R}} w(\tilde{\mathbf{z}}; \mathbf{u}_i) r}{\sum_{(\tilde{\mathbf{z}},r) \in \mathcal{R}} w(\tilde{\mathbf{z}}; \mathbf{u}_i)} \right) \middle| (\tilde{\mathbf{z}}, r) \in \mathcal{R} \right\},$$

and compute the slope perturbation via least squares:

$$\hat{\mathbf{p}}_i := \arg\min_{\mathbf{p}\in\mathbb{R}^D} \frac{2}{|\mathcal{S}|} \sum_{(\tilde{\mathbf{z}},\tilde{r})\in\widetilde{\mathcal{R}}} (w_k(\tilde{\mathbf{z}}, \mathbf{u}_i)\tilde{r} - \langle \mathbf{p}, \tilde{\mathbf{z}}\rangle)^2$$

4. Let $\hat{P} \in \mathbb{R}^{K\times D}$ have rows $\hat{\mathbf{p}}_i$'s, $i = 1, \ldots, K$, and compute the rank-d' singular value decomposition of $\hat{P} \approx U_{d'}\Sigma_{d'}V_{d'}^T$. Return $\hat{A} := V_{d'}^T \in \mathbb{R}^{d'\times D}$ and $\hat{\beta}$.

While further study on how to optimally add the higher-order terms in (14) is ongoing, the MPLS algorithm we applied is not cursed by the dimension D, making it possible to increase the dimension of (the range of) the feature map $\mathbf{z}$ with relatively small additional sampling requirements.

Step 2. Estimating the reduced interaction kernel ϕ. Once an estimate $\hat{B}$ for the feature reduction map B has been constructed, we proceed to estimate ϕ. Always in the case $N = 2$, we proceed by projecting the pairs of states, using the estimated feature reduction map $\hat{B}$, to $\mathbb{R}^{d'}$, and use a nonparametric regression technique on that subspace, aimed at minimizing, over a suitable set of functions $\psi \in \mathcal{H}$, the error functional:

$$\mathcal{E}_{L,M,\mathcal{H}}(\varphi) := \frac{1}{LM} \sum_{l,m=1}^{L,M} ||\dot{\mathbf{X}}_l^m - \mathbf{f}_\varphi(\mathbf{X}_l^m)||_{\mathcal{S}}^2,$$

where

$$\mathbf{f}_\varphi(\mathbf{X}_l^m) = \begin{bmatrix} \vdots \\ \frac{1}{N}\sum_{i'=1,i'\neq i}^{N} \varphi(\hat{B}\mathbf{z}_{i,i',l}^m)(\mathbf{x}_{i'}^m(t_l) - \mathbf{x}_i^m(t_l)) \\ \vdots \end{bmatrix},$$

where $\mathbf{z}_{i,i',l}^m = \mathbf{z}(\mathbf{x}_i^m(t_l), \mathbf{x}_{i'}^m(t_l))$. We will choose $\mathcal{H}$ to be a convex and compact (in the L^∞ norm) subset of a subspace of functions of the estimated variables $\hat{B}\mathbf{y}$, e.g., spanned by splines with knots on a grid, or piecewise polynomials. The dimension of $\mathcal{H}$ will be chosen as a suitably increasing function of the number of training trajectories M, following the ideas of [27] and [32] (in the case of multiple reduced variables); see [31] for detailed discussion.

3.6 Coupled Systems

We explore two types of coupled systems to introduce more intricate sets of states. Initially, we consider the state variable $\mathbf{y}_i$ as $\mathbf{y}_i = [\mathbf{x}_i\ \zeta_i]$, where $\mathbf{x}_i \in \mathbb{R}^d$ and $\zeta_i \in \mathbb{R}$ (we can also take $\zeta_i \in \mathbb{R}^{d'}$ for any $d' \geq 1$), and thus, $\mathbf{y}_i \in \mathbb{R}^{d+1}$. Typically, $\mathbf{x}_i$

describes the position, while ζ_i denotes the phase, excitation, or opinion (as seen in the swarmalator model, for instance [36]). The interaction arises not only from each $\mathbf{y}_i$ affecting the agents but also from the intra-agent interaction of $\mathbf{x}_i$ and ζ_i, resulting in a set of more complex patterns. The evolution of $\mathbf{y}_i$ is governed by the following system of ODEs:

$$\begin{cases} \dot{\mathbf{x}}_i &= F^{\mathbf{x}}(\mathbf{y}_i) + \sum_{i'=1, i' \neq i}^{N} \frac{1}{N_{k_i}} \phi^E_{k_i, k_{i'}}(\boldsymbol{\xi}^E(\mathbf{y}_i, \mathbf{y}_{i'}))(\mathbf{x}_{i'} - \mathbf{x}_i) \\ \dot{\zeta}_i &= F^{\zeta}(\mathbf{y}_i) + \sum_{i'=1, i' \neq i}^{N} \frac{1}{N_{k_i}} \phi^{\zeta}_{k_i, k_{i'}}(\boldsymbol{\xi}^{\zeta}(\mathbf{y}_i, \mathbf{y}_{i'}))(\zeta_{i'} - \zeta_i) \end{cases}, \quad i = 1, \cdots, N. \tag{15}$$

Here, $F^{\mathbf{x}} : \mathbb{R}^{d+1} \rightarrow \mathbb{R}^d$ represents a force governing how the interaction of the ith agent and its surrounding environment affects the change of $\mathbf{x}_i$. Similarly, $F^{\zeta} : \mathbb{R}^{d+1} \rightarrow \mathbb{R}$ acts on ζ_i. Additionally, $\boldsymbol{\xi}^E : \mathbb{R}^{2d+2} \rightarrow \mathbb{R}^{d^E}$ is an energy-based reduced variable, where $1 \leq d^E \ll 2d + 2$. The function $\phi^E : \mathbb{R}^{d^E} \rightarrow \mathbb{R}$ is the energy-based reduced interaction kernel. Similarly, $\boldsymbol{\xi}^{\zeta} : \mathbb{R}^{2d+2} \rightarrow \mathbb{R}^{d^{\zeta}}$ is a ζ-based reduced variable, and ϕ^{ζ} is its corresponding reduced interaction kernel. Given the observations $\mathbf{y}_i^m(t_l), \dot{\mathbf{y}}_i^m(t_l)i, l, m = 1^{N,L,M}$, we aim to find the set of reduced interaction kernels[4] by minimizing the following loss functionals. First, for $\boldsymbol{\varphi}^E = \{\varphi^E_{k_1,k_2}\}_{k_1,k_2=1}^K$, we minimize:

$$\mathcal{E}^{\mathbf{x}}_{L,M,\mathcal{H}^E}(\boldsymbol{\varphi}^E) = \frac{1}{LM} \sum_{l,m=1}^{L,M} ||\dot{\mathbf{Y}}_l^m - \mathbf{f}_{\boldsymbol{\varphi}^E}(\mathbf{Y}_l^m)||^2_{\mathcal{S}'}$$

for $\boldsymbol{\varphi}^E = \{\varphi^E_{k_1,k_2} \in \mathcal{H}^E_{k_1,k_2}\}_{k_1,k_2=1}^K$ and $\mathcal{H}^E = \oplus_{k_1,k_2=1}^K \mathcal{H}^E_{k_1,k_2}$,

$$\mathbf{Y}_l^m = \begin{bmatrix} \vdots \\ \mathbf{x}_i^m(t_l) \\ \xi_i^m(t_l) \\ \vdots \end{bmatrix}$$

and

$$\mathbf{f}_{\boldsymbol{\varphi}^E}(\mathbf{X}_l^m) = \begin{bmatrix} \vdots \\ F^{\mathbf{x}}(\mathbf{y}_{i,l}^m) + \sum_{i'=1, i' \neq i}^{N} \frac{1}{N_{k_i}} \varphi^E_{k_i, k_{i'}}(\boldsymbol{\xi}^E(\mathbf{y}_{i,l}^m, \mathbf{y}_{i',l}^m))(\mathbf{x}_{i',l}^m - \mathbf{x}_{i,l}^m) \\ \vdots \end{bmatrix}$$

[4] Here we assume that the reduced variables $\boldsymbol{\xi}^E$ and $\boldsymbol{\xi}^{\zeta}$ are known to us.

Here $\mathbf{x}_{i,l}^m = \mathbf{x}_i^m(t_l)$ and $\mathbf{y}_{i,l}^m = \mathbf{y}_i^m(t_l)$. Similarly for learning $\boldsymbol{\varphi}^\zeta = \{\varphi_{k_1,k_2}^\zeta\}_{k_1,k_2=1}^K$:

$$\mathcal{E}_{L,M,\mathcal{H}^\zeta}^\zeta(\boldsymbol{\varphi}^\zeta) = \frac{1}{LM}\sum_{l,m=1}^{L,M} ||\dot{\mathbf{Y}}_l^m - \mathbf{f}_{\boldsymbol{\varphi}^\zeta}(\mathbf{Y}_l^m)||_{\mathcal{S}'}^2$$

for $\boldsymbol{\varphi}^\zeta = \{\varphi_{k_1,k_2}^\zeta \in \mathcal{H}_{k_1,k_2}^\zeta\}_{k_1,k_2=1}^K$ and $\mathcal{H}^E = \oplus_{k_1,k_2=1}^K \mathcal{H}_{k_1,k_2}^\zeta$ and

$$\mathbf{f}_{\boldsymbol{\varphi}^\zeta}(\mathbf{Y}_l^m) = \begin{bmatrix} \vdots \\ F^\zeta(\mathbf{y}_{i,l}^m) + \sum_{i'=1,i'\neq i}^N \frac{1}{N_{k_i}}\varphi_{k_i,k_{i'}}^\zeta(\boldsymbol{\xi}^\zeta(\mathbf{y}_{i,l}^m, \mathbf{y}_{i',l}^m))(\zeta_{i',l}^m - \zeta_{i,l}^m) \\ \vdots \end{bmatrix}$$

Here $\zeta_{i,l}^m = \zeta_i^m(t_l)$.

Next, we consider the state variable $\mathbf{y}_i$ as $\mathbf{y}_i = \begin{bmatrix}\mathbf{x}_i & \mathbf{v}_i & \zeta_i\end{bmatrix}$, where $\mathbf{x}_i, \mathbf{v}_i \in \mathbb{R}^d$, and $\zeta_i \in \mathbb{R}$. Specifically, we require that $\mathbf{v}_i = \dot{\mathbf{x}}_i$. The change of $\mathbf{y}_i$ is governed by the following second-order ODE system:

$$\begin{cases} \dot{\mathbf{x}}_i &= \mathbf{v}_i, \\ m_i\dot{\mathbf{v}}_i &= F^{\mathbf{v}}(\mathbf{y}_i) + \sum_{i'=1,i'\neq i}^N \frac{1}{N_{k_i}}\Big[\phi_{k_i,k_{i'}}^E(\boldsymbol{\xi}^E(\mathbf{y}_i, \mathbf{y}_{i'}))(\mathbf{x}_{i'} - \mathbf{x}_i) \\ &\quad +\phi_{k_i,k_{i'}}^A(\boldsymbol{\xi}^A(\mathbf{y}_i, \mathbf{y}_{i'}))(\mathbf{x}_{i'} - \mathbf{x}_i)\Big] \\ \dot{\zeta}_i &= F^\zeta(\mathbf{y}_i) + \sum_{i'=1,i'\neq i}^N \frac{1}{N_{k_i}}\phi_{k_i,k_{i'}}^\zeta(\boldsymbol{\xi}^\zeta(\mathbf{y}_i, \mathbf{y}_{i'}))(\zeta_{i'} - \zeta_i) \end{cases}, \quad i = 1, \cdots, N. \tag{16}$$

Similarly, $F^{\mathbf{v}} : \mathbb{R}^{2d+1} \to \mathbb{R}^d$ is a force that governs how the interaction of the ith agent and its surrounding environment affects the change of $\mathbf{v}_i$. Furthermore, $F^\zeta : \mathbb{R}^{2d+1} \to \mathbb{R}$ acts on ζ_i. Additionally, $\boldsymbol{\xi}^E : \mathbb{R}^{4d+2} \to \mathbb{R}^{d^E}$ represents an energy-based reduced variable (where $1 \leq d^E \ll 4d+2$), and $\phi^E : \mathbb{R}^{d^E} \to \mathbb{R}$ is the corresponding energy-based reduced interaction kernel. Similarly, $\boldsymbol{\xi}^A : \mathbb{R}^{4d+2} \to \mathbb{R}^{d^A}$ represents an alignment-based reduced variable (where $1 \leq d^A \ll 4d+2$), and $\phi^A : \mathbb{R}^{d^A} \to \mathbb{R}$ is the corresponding alignment-based reduced interaction kernel. Finally, $\boldsymbol{\xi}^\zeta : \mathbb{R}^{4d+2} \to \mathbb{R}^{d^\zeta}$ represents a ζ-based reduced variable, and ϕ^ζ is its corresponding reduced interaction kernel.

In order to learn $\boldsymbol{\phi}^E = \{\phi_{k_1,k_2}^E\}_{k_1,k_2=1}^K$ and $\boldsymbol{\phi}^A = \{\phi_{k_1,k_2}^A\}_{k_1,k_2=1}^K$, we use a slightly updated loss functional:

$$\mathcal{E}_{L,M,\mathcal{H}^E,\mathcal{H}^A}^{E+A}(\boldsymbol{\varphi}^E, \boldsymbol{\varphi}^A) = \frac{1}{LM}\sum_{l,m=1}^{L,M} ||\dot{\mathbf{Y}}_l^m - \mathbf{f}_{\boldsymbol{\varphi}^E,\boldsymbol{\varphi}^A}(\mathbf{Y}_l^m)||_{\mathcal{S}'}^2$$

where

$$\mathbf{f}_{\varphi^E,\varphi^A}(\mathbf{Y}_l^m) = \begin{bmatrix} \vdots \\ F^{\mathbf{v}}(\mathbf{y}_{i,l}^m) + \sum_{i'=1,i'\neq i}^{N} \frac{1}{N_{k_i}} \Big[\varphi_{k_i,k_{i'}}^E(\boldsymbol{\xi}^E(\mathbf{y}_{i,l}^m, \mathbf{y}_{i',l}^m))(\mathbf{x}_{i',l}^m - \mathbf{x}_{i,l}^m) \\ +\varphi_{k_i,k_{i'}}^A(\boldsymbol{\xi}^A(\mathbf{y}_{i,l}^m, \mathbf{y}_{i',l}^m))(\mathbf{v}_{i',l}^m - \mathbf{v}_{i,l}^m) \Big] \\ \vdots \end{bmatrix}$$

where $\mathbf{v}_{i,l}^m = \mathbf{v}_i^m(t_l)$. We can use a similar loss as in $\mathcal{E}_{L,M,\mathcal{H}}^{\text{Mul}}$ for $\mathbf{f}_{\varphi^E,\varphi^A}(\mathbf{Y}_l^m)$. See [32] where feature maps, i.e., $\boldsymbol{\xi}^E(\mathbf{y}_{i,l}^m, \mathbf{y}_{i',l}^m)$ and $\boldsymbol{\xi}^A(\mathbf{y}_{i,l}^m, \mathbf{y}_{i',l}^m)$, are known.

3.7 *Learning with Gaussian Priors*

Gaussian process regression (GPR) serves as a nonparametric Bayesian machine learning technique designed for supervised learning, equipped with an inherent framework for quantifying uncertainty. Consequently, GPR has found application in the study of ordinary differential equations (ODEs), stochastic differential equations (SDEs), and partial differential equations (PDEs) [55–65], resulting in more accurate and robust models for dynamical systems. Given the unique characteristics of dynamical data, it necessitates novel concepts and substantial efforts tailored to specific types of dynamical systems and data regimes. In our work, we model latent interaction kernels as Gaussian processes, embedding them with the underlying structure of our governing equations, including translation and rotational invariance. This distinguishes our approach from most other works, which model state variables as Gaussian processes.

Despite the challenges involved, recent mathematical advancements have led to the development of a general physical model based on Newton's second law, such as the methods we discussed in Sect. 3.1.1. This model has been demonstrated to capture a wide range of collective behaviors accurately. Specifically, the model describes a system of N agents that interact according to a system of ODEs, where for each agent $i = 1, \cdots, N$:

$$m_i\ddot{\mathbf{x}}_i = F_i(\mathbf{x}_i, \dot{\mathbf{x}}_i, \boldsymbol{\alpha}_i) + \sum_{i'=1}^{N} \frac{1}{N}\Big[\phi^E(|\mathbf{x}_{i'} - \mathbf{x}_i|)(\mathbf{x}_{i'} - \mathbf{x}_i) + \phi^A(|\mathbf{x}_{i'} - \mathbf{x}_i|)(\dot{\mathbf{x}}_{i'} - \dot{\mathbf{x}}_i)\Big], \tag{17}$$

where $m_i \geq 0$ is the mass of the agent i; $\ddot{\mathbf{x}}_i \in \mathbb{R}^d$ is the acceleration; $\dot{\mathbf{x}}_i \in \mathbb{R}^d$ is the velocity; $\mathbf{x}_i \in \mathbb{R}^d$ is the position of agent i; the first term F_i is a parametric function of position and velocities, modeling self-propulsion and frictions of agent

i with the environment, and the scalar parameters $\boldsymbol{\alpha}_i$ describing their strength; $|\mathbf{x}_j - \mathbf{x}_i|$ is the Euclidean distance; and the 1D functions $\phi^E, \phi^A : \mathbb{R}^+ \to \mathbb{R}$ are called the *energy- and alignment-based radial interaction kernels*, respectively. The ϕ^E term describes the alignment of positions based on the difference of positions; the ϕ^A term describes the alignment of velocities based on the difference of velocities. Our primary objective is to infer the interaction kernels $\boldsymbol{\phi} = \{\phi^E, \phi^A\}$ as well as the unknown scalar parameters $\boldsymbol{\alpha}$ and potentially $\boldsymbol{m}$ from the observed trajectory data. Subsequently, we utilize the learned governing equations to make predictions regarding future events or simulate new datasets.

To learn the model given by (17), we initiate the process by modeling the interaction kernel functions ϕ^E and ϕ^A with the priors as two independent Gaussian processes:

$$\phi^E \sim \mathcal{GP}(0, K_{\theta^E}(r, r')), \qquad \phi^A \sim \mathcal{GP}(0, K_{\theta^A}(r, r')), \tag{18}$$

where K_{θ^E}, K_{θ^A} are covariance functions with hyperparameters $\boldsymbol{\theta} = (\theta^E, \theta^A)$. $\boldsymbol{\theta}$ can either be chosen by the modeler or tuned via a data-driven procedure discussed later.

Then, given the noisy observational data $\mathbb{Y} = [\mathbf{Y}^{(1,1)}, \ldots, \mathbf{Y}^{(M,L)}]^T \in \mathbb{R}^{dNML}$, and $\mathbb{Z} = [\mathbf{Z}_{\sigma^2}^{(1,1)}, \ldots, \mathbf{Z}_{\sigma^2}^{(M,L)}]^T \in \mathbb{R}^{dNML}$ where

$$\mathbf{Y}^{(m,l)} = \mathbf{Y}^{(m)}(t_l) := \begin{bmatrix} \mathbf{X}(t) \\ \dot{\mathbf{X}}(t) \end{bmatrix} \in \mathbb{R}^{2dN}, \tag{19}$$

and

$$\boldsymbol{m}\mathbf{Z}_{\sigma^2}^{(m,l)} = \boldsymbol{m}\ddot{\mathbf{X}}_{\sigma^2}^{(m,l)} := F_{\boldsymbol{\alpha}}(\mathbf{Y}^{(m,l)}) + \mathbf{f}_{\boldsymbol{\phi}}(\mathbf{Y}^{(m,l)}) + \epsilon^{(m,l)}, \tag{20}$$

with $\mathbf{f}_{\boldsymbol{\phi}}(\mathbf{Y}(t)) = \mathbf{f}_{\phi^E, \phi^A}(\mathbf{Y}(t))$ represents the sum of energy and alignment-based interactions as in (17), and $\epsilon^{(m,l)} \sim \mathcal{N}(0, \sigma^2 I_{dN})$ are i.i.d noise, based on the properties of Gaussian processes. With the priors of ϕ^E, ϕ^A, we can train the hyperparameters $\boldsymbol{m}, \boldsymbol{\alpha}, \boldsymbol{\theta}$ and σ by maximizing the probability of the observational data, which is equivalent to minimizing the negative log marginal likelihood (NLML) (see Chapter 4 in [66])

$$\begin{aligned} &- \log p(\boldsymbol{m}\mathbb{Z} | \mathbb{Y}, \boldsymbol{\alpha}, \boldsymbol{\theta}, \sigma^2) \\ &= \frac{1}{2}(\boldsymbol{m}\mathbb{Z} - F_{\boldsymbol{\alpha}}(\mathbb{Y}))^T (K_{\mathbf{f}_{\boldsymbol{\phi}}}(\mathbb{Y}, \mathbb{Y}; \boldsymbol{\theta}) + \sigma^2 I_{dNML})^{-1} (\boldsymbol{m}\mathbb{Z} - F_{\boldsymbol{\alpha}}(\mathbb{Y})) \\ &\quad + \frac{1}{2} \log |K_{\mathbf{f}_{\boldsymbol{\phi}}}(\mathbb{Y}, \mathbb{Y}; \boldsymbol{\theta}) + \sigma^2 I_{dNML}| + \frac{dNML}{2} \log 2\pi. \end{aligned} \tag{21}$$

and the posterior/predictive distribution for the interaction kernels $\phi^{\text{type}}(r^*)$, type $=$ E or A, can be obtained by

$$p(\phi^{\text{type}}(r^*)|\mathbb{Y}, \mathbb{Z}, r^*) \sim \mathcal{N}(\bar{\phi}^{\text{type}}, var(\bar{\phi}^{\text{type}})), \tag{22}$$

where

$$\bar{\phi}^{\text{type}} = K_{\phi^{\text{type}},\mathbf{f}_{\boldsymbol{\phi}}}(r^*, \mathbb{Y})(K_{\mathbf{f}_{\boldsymbol{\phi}}}(\mathbb{Y}, \mathbb{Y}) + \sigma^2 I_{dNML})^{-1}(\boldsymbol{m}\mathbb{Z} - F_{\boldsymbol{\alpha}}(\mathbb{Y})), \tag{23}$$

$$\begin{aligned} &var(\bar{\phi}^{\text{type}}) \\ &= K_{\theta^{\text{type}}}(r^*, r^*) - K_{\phi^{\text{type}},\mathbf{f}_{\boldsymbol{\phi}}}(r^*, \mathbb{Y})(K_{\mathbf{f}_{\boldsymbol{\phi}}}(\mathbb{Y}, \mathbb{Y}) + \sigma^2 I_{dNML})^{-1} K_{\mathbf{f}_{\boldsymbol{\phi}},\phi^{\text{type}}}(\mathbb{Y}, r^*). \end{aligned} \tag{24}$$

and $F_{\boldsymbol{\alpha}}(\mathbb{Y}) = \text{Vec}(\{F_{\boldsymbol{\alpha}}(\mathbf{Y}^{(m,l)})\}_{m=1,l=1}^{M,L}) \in \mathbb{R}^{dNML}$, $K_{\mathbf{f}_{\boldsymbol{\phi}}}(\mathbb{Y}, \mathbb{Y}; \theta) \in \mathbb{R}^{dNML \times dNML}$ is the covariance matrix between $\mathbf{f}_{\boldsymbol{\phi}}(\mathbb{Y})$ and $\mathbf{f}_{\boldsymbol{\phi}}(\mathbb{Y})$, which can be compute elementwisely based on the covariance functions K_{θ^E}, K_{θ^A}, $K_{\mathbf{f}_{\boldsymbol{\phi}},\phi^{\text{type}}}(\mathbb{Y}, r^*) = K_{\phi^{\text{type}},\mathbf{f}_{\boldsymbol{\phi}}}(r^*, \mathbb{Y})^T$ denotes the covariance matrix between $\mathbf{f}_{\boldsymbol{\phi}}(\mathbb{Y})$ and $\phi^{\text{type}}(r^*)$.

Note here the marginal likelihood does not simply favor the models that fit the training data best, but induces an automatic trade-off between data fit and model complexity. We can use the posterior mean estimators of $\boldsymbol{\phi}$ in trajectory prediction by performing numerical simulations of the equations:

$$\boldsymbol{m}\hat{\mathbf{Z}}(t) = F_{\hat{\boldsymbol{\alpha}}}(\mathbf{Y}(t)) + \hat{\mathbf{f}}_{\bar{\boldsymbol{\phi}}}(\mathbf{Y}(t)). \tag{25}$$

and the posterior variance $var(\bar{\phi}^{\text{type}})$ can be used as a good indicator for the uncertainty of the estimation $\bar{\phi}^{\text{type}}$ based on our Bayesian approach; see Fig. 4 and also other examples in [34, 35].

In classical regression setting [66], there is an interesting link between GP regression with the kernel ridge regression (KRR), where the posterior mean can be viewed as a KRR estimator to solve a regularized least squares empirical risk functional. In our setting, we have noisy functional observations of the interaction kernels, i.e., the $\{r_{\mathbb{X}}, r_{\mathbb{V}}, \mathbb{Z}\}$ instead of the pairs $\{r_{\mathbb{X}}, \phi^E(r_{\mathbb{X}}), \phi^A(r_{\mathbb{X}})\}$, where $r_{\mathbb{X}}, r_{\mathbb{V}}, \in \mathbb{R}^{MLN^2}$ are the sets contains all the pairwise distances in $\mathbb{X}$, and $\mathbb{V}$, so we face an inverse problem here, instead of a classical regression problem. Thanks to the linearity of the inverse problem, we can still derive a Representer theorem [68] that helps understand the role of the hyperparameters.

Theorem 2 (Theorem 3.2 in [35]) *Given the training data* $\{\mathbb{Y}, \mathbb{Z}\}$*, if the priors* $\phi^E \sim \mathcal{GP}(0, \tilde{K}^E)$*,* $\phi^A \sim \mathcal{GP}(0, \tilde{K}^A)$ *with* $\tilde{K}^E = \frac{\sigma^2 K^E}{MNL\lambda^E}$*,* $\tilde{K}^A = \frac{\sigma^2 K^A}{MNL\lambda^A}$ *for some* $\lambda^E, \lambda^A > 0$*, then the posterior mean* $\bar{\boldsymbol{\phi}} = (\bar{\phi}^E, \bar{\phi}^A)$ *in (23) coincides with the minimizer,* $\boldsymbol{\phi}^{\lambda,M}_{\mathcal{H}_{K^E} \times \mathcal{H}_{K^A}}$*, of the regularized empirical risk functional* $\mathcal{E}^{\lambda,M}(\cdot)$ *on*

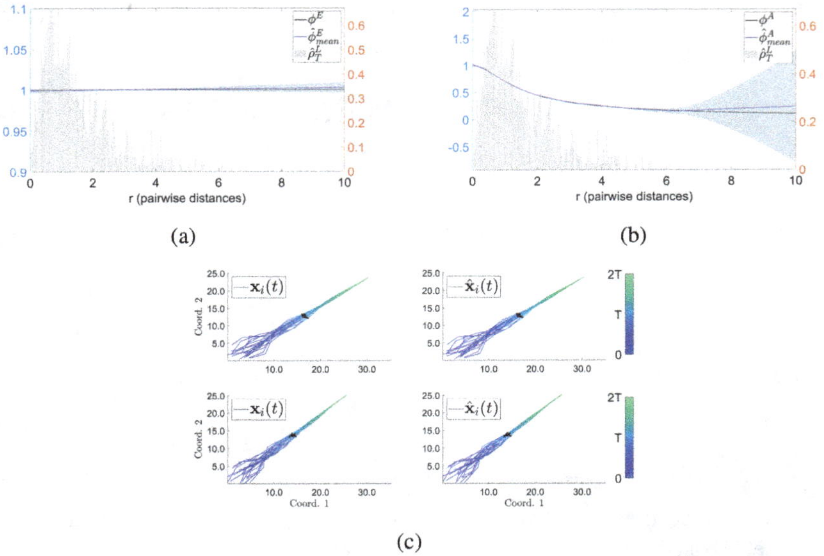

Fig. 4 Flocking with external potential (FwEP) model [67] with $\{N, L, M, \sigma\} = \{20, 6, 3, 0.01\}$, and $\phi^E(r) = 1$, $\phi^A(r) = \frac{1}{(1+r^2)^{1/2}}$, for other parameters see [32]. The light blue regions are two-standard-deviation bands around the means which indicate the uncertainty of the estimators. (**a**) ϕ^E vs $\hat{\phi}^E$. (**b**) ϕ^A vs $\hat{\phi}^A$. (**c**) $\mathbf{X}(t)$ vs $\hat{\mathbf{X}}(t)$

$\mathcal{H}_{K^E} \times \mathcal{H}_{K^A}$ where $\mathcal{E}^{\lambda,M}(\cdot)$ is defined by

$$\mathcal{E}^{\lambda,M}(\boldsymbol{\varphi}) := \frac{1}{LM} \sum_{l=1,m=1}^{L,M} \|\mathbf{f}_{\boldsymbol{\varphi}}(\mathbf{Y}^{(m,l)}) - \mathbf{Z}_{\sigma^2}^{(m,l)}\|^2 + \lambda^E \|\varphi^E\|_{\mathcal{H}_{K^E}}^2 + \lambda^A \|\varphi^A\|_{\mathcal{H}_{K^A}}^2. \tag{26}$$

From the theorem, it is clear to see how hyperparameters affect the prediction of interaction kernels: θ^E, θ^A, and σ jointly affect the choice of Mercer kernels and regularization constant.

In order to ensure the asymptotic identifiability of the true interaction kernels as the number of observational data snapshots goes to infinity, we study the well-posedness under a statistical inverse problem setting, and provide the coercivity condition in this case:

Definition 2 (Definition 3.3 in [35]) We say that the system (17) satisfies the coercivity condition if $\forall \boldsymbol{\varphi} \in \mathcal{H}_{K^E} \times \mathcal{H}_{K^A}$:

$$\|A\boldsymbol{\varphi}\|_{L^2(\rho_{\mathbf{Y}})}^2 = \|\mathbf{f}_{\boldsymbol{\varphi}}\|_{L^2(\rho_{\mathbf{Y}})}^2 \geq c_{\mathcal{H}_{K^E}} \|\varphi^E\|_{L^2(\tilde{\rho}_r^E)}^2 + c_{\mathcal{H}_{K^A}} \|\varphi^A\|_{L^2(\tilde{\rho}_r^A)}^2 \tag{27}$$

for some constants $c_{\mathcal{H}_{K^E}}, c_{\mathcal{H}_{K^A}} > 0$.

One can prove the well-posedness on a suitable subspace determined by the source conditions on ϕ^E, ϕ^A with some mild assumptions if the coercivity condition (27) holds, and both kernels can be recovered with a statistically optimal rate in M under the corresponding RKHS norm:

$$\|\boldsymbol{\phi}^{\lambda,M}_{\mathcal{H}_{K^E}\times\mathcal{H}_{K^A}} - \boldsymbol{\phi}\|_{\mathcal{H}_{K^E}\times\mathcal{H}_{K^A}} \lesssim M^{\frac{-\gamma}{2\gamma+2}}, \tag{28}$$

where $0 < \gamma \leq \frac{1}{2}$; see Section 3.2 in [35] for detailed discussion. A novel operator-theoretical framework is established in [34] for the single-kernel systems (2), which proves the reconstruction error converges at an upper rate in M under Hölder-type source conditions on ϕ (Theorem 25 in [34]). One can extend this result to the double-kernel case following the same theoretical framework, and we leave it for future investigation. This result generalized the analysis of kernel regression methods [66] and linear inverse problems to interacting particle systems, and we believe one can obtain more refined rates and bounds using our framework as the bridge in the future.

Compared with the previous works that focused on learning interaction kernels, this method using GPs has the following advantages:

(1) it can handle more difficult yet more practical scenarios, i.e., joint inference of scalar parameters $\boldsymbol{\alpha}$ and ϕ, as both are often unknown in practical scenarios. Therefore, our method can learn the governing equations (17).
(2) It provides uncertainty quantification on estimators. In the ideal data regime, we provide a rigorous analysis and show how it depends on the system parameters. This uncertainty measures the reliability of our estimators; in particular, it can be used to measure the mismatch between our proposed models with the real-world systems.
(3) It has a powerful training procedure to select a data-driven prior, and this overcomes the drawback of the previous least squares algorithms: there is no criterion to select the optimal choice of function spaces (in terms of both basis and dimensions) for learning so as to minimize the generalization error. See more examples and discussions about the comparisons in [34].

4 Conclusion

In summary, our paper has presented a thorough exploration of methodologies for deducing the governing structure of collective dynamics from observational data. A detailed examination of the learning framework for first-order models underscores the significance of formulating a variational inverse problem approach based on the intrinsic low-dimensional properties of the dynamical right-hand side. Our methods exhibit efficacy in extending the fundamental approach to encompass multispecies systems, stochastic noise, dynamics constrained on Riemannian manifolds, missing features, coupled systems, and learning with Gaussian priors. Despite the escalating complexity of model equations, once cast into the standard $\dot{\mathbf{Y}} = \mathbf{f}(\mathbf{Y})$ form, the formulation of the loss function—the crux of the learning process—becomes evident. We have provided references to our original papers for readers seeking a more in-depth understanding, and additionally, we have conducted comparisons with three other methods, namely, SINDy, NeuralODE, and random feature learning.

Numerous avenues warrant consideration for future research. In instances where a single-time observation of the steady state is the sole available data, we have initiated the application of the reproducing kernel Hilbert space approach introduced in [69, 70] to provide reasonable estimations of ϕ, up to certain scaling factors. Regarding dynamics with geometric structures, we focus on developing learning methods capable of inferring both the dynamics and the associated Riemannian manifolds. While noise in observation data, encompassing both $\mathbf{Y}$ and $\dot{\mathbf{Y}}$, has been addressed in [27], the scenario of having noisy $\mathbf{Y}$ without information on $\dot{\mathbf{Y}}$ remains an open question. Additionally, exploring the learning of multispecies dynamics without prior knowledge of type information, as introduced in [23], provides an intriguing starting point. As the number of agents in the system (N) grows large, i.e., $N \to \infty$, computational challenges arise in handling extensive data. Utilizing the corresponding PDEs stemming from the mean-field limit as a guide for learning, as proposed in [22, 69–71], offers promising insights. Partial observation data may render our current learning approach inadequate in capturing the underlying structure; however, we propose a synergistic approach by combining our existing learning methodology with supplementary information derived from the mean-field distribution. The utilization of $\frac{1}{N}$ averaging in the model equations facilitates the derivation of corresponding mean-field PDEs. However, from the perspective of agent-based modeling, such averaging may not align with the actual modeling context. A potential avenue for future exploration involves addressing topological averaging; specifically, considering the right-hand side as $\sum_{j=1}^{N} \frac{\phi(||\mathbf{x}_j - \mathbf{x}_i||)}{\sum_{k=1}^{N} \phi(||\mathbf{x}_k - \mathbf{x}_i||)}$ instead of $\frac{1}{N}\sum_{j=1}^{N} \phi(||\mathbf{x}_j - \mathbf{x}_i||)$ could be a promising direction for further investigation.

Acknowledgments JF and MZ made equal contributions to both the research and the composition of the manuscript. They extend their gratitude to Mauro Maggioni from Johns Hopkins University for the invaluable discussions pertaining to this review. Partial support for MZ is acknowledged from NSF-AoF-2225507 and the startup fund provided by Illinois Tech.

References

1. S. M. Stigler. *The History of Statistics: The Measurement of Uncertainty Before* 1900. Harvard University Press, Cambridge, MA, 1st edition, 1986.
2. U. Krause. A discrete nonlinear and non-autonomous model of consensus formation. *Communications in difference equations*, pages 227–236, 2000.

3. Felipe Cucker and Steve Smale. On the mathematical foundations of learning. *Bulletin of the American mathematical society*, 39(1):1–49, 2002.
4. T. Vicsek and A. Zafeiris. Collective motion. *Physics Reports*, 517:71–140, 2012.
5. T. Vicsek, A. Czirók, E. Ben-Jacob, I. Cohen, and O. Shochet. Novel Type of Phase Transition in a System of Self-Driven Particles. *Physical Review Letters*, 75:1226–1229, August 1995.
6. Yao li Chuang, Maria R. D'Orsogna, Daniel Marthaler, Andrea L. Bertozzi, and Lincoln S. Chayes. State transitions and the continuum limit for a 2D interacting, self-propelled particle system. *Physica D: Nonlinear Phenomena*, 232(1):33–47, 2007.
7. Chad M. Topaz, Maria R. D'Orsogna, Leah Edelstein-Keshet, and Andrew J. Bernoff. Locust dynamics: Behavioral phase change and swarming. *PLoS Comput Biol.*, 8(8):e1002642, 2012.
8. S. H. Strogatz. From Kuramoto to Crawford: exploring the onset of synchronization in populations of coupled oscillators. *Physica D*, (143):1–20, 2000.
9. Kevin P. O'Keeffe, Hyunsuk Hong, and Steven H. Strogatz. Oscillators that sync and swarm. *Nature Communications*, 8(1):1–12, 2017.
10. Nicola Bellomo, Pierre Degond, and Eitan Tadmor, editors. *Active Particles, Volume 1*. Springer International Publishing AG, Switerland, 2017.
11. Eitan Tadmor. On the mathematics of swarming: emergent behavior in alignment dynamics. *Notices of the AMS*, 68(4):493–503, 2021.
12. Eitan Tadmor. Long time and large crowd dynamics of fully discrete cucker-smale alignment models. *Pure and Applied Functional Analysis*, 8(2):603–626, 2023.
13. Raphael A Kasonga. Maximum likelihood theory for large interacting systems. *SIAM Journal on Applied Mathematics*, 50(3):865–875, 1990.
14. Jaya Prakash Narayan Bishwal et al. Estimation in interacting diffusions: Continuous and discrete sampling. *Applied Mathematics*, 2(9):1154–1158, 2011.
15. Susana N Gomes, Andrew M Stuart, and Marie-Therese Wolfram. Parameter estimation for macroscopic pedestrian dynamics models from microscopic data. *SIAM Journal on Applied Mathematics*, 79(4):1475–1500, 2019.
16. Xiaohui Chen. Maximum likelihood estimation of potential energy in interacting particle systems from single-trajectory data. *Electronic Communications in Probability*, 26:1–13, 2021.
17. Louis Sharrock, Nikolas Kantas, Panos Parpas, and Grigorios A Pavliotis. Parameter estimation for the mckean-vlasov stochastic differential equation. *arXiv preprint arXiv:2106.13751*, 2021.
18. Valentine Genon-Catalot and Catherine Larédo. Inference for ergodic mckean-vlasov stochastic differential equations with polynomial interactions. *hal-03866218v2*, 2022.
19. Laetitia Della Maestra and Marc Hoffmann. The lan property for mckean-vlasov models in a mean-field regime. *arXiv preprint arXiv:2205.05932*, 2022.
20. Rentian Yao, Xiaohui Chen, and Yun Yang. Mean-field nonparametric estimation of interacting particle systems. *arXiv preprint arXiv:2205.07937*, 2022.
21. Steven L. Brunton, Joshua L. Proctor, and J. Nathan Kutz. Discovering governing equations from data by sparse identification of nonlinear dynamical systems. *Proceedings of the National Academy of Sciences*, 113(15):3932–3937, 2016.
22. Daniel A Messenger and David M Bortz. Learning mean-field equations from particle data using wsindy. *Physica D: Nonlinear Phenomena*, 439:133406, 2022.
23. Daniel A Messenger, Graycen E Wheeler, Xuedong Liu, and David M Bortz. Learning anisotropic interaction rules from individual trajectories in a heterogeneous cellular population. *Journal of the Royal Society Interface*, 19(195):20220412, 2022.
24. M. Raissi, P. Perdikaris, and G.E. Karniadakis. Physics-informed neural networks: A deep learning framework for solving forward and inverse problems involving nonlinear partial differential equations. *Journal of Computational Physics*, 378:686–707, 2019.
25. Ricky T. Q. Chen, Yulia Rubanova, Jesse Bettencourt, and David K Duvenaud. Neural ordinary differential equations. In S. Bengio, H. Wallach, H. Larochelle, K. Grauman, N. Cesa-Bianchi, and R. Garnett, editors, *Advances in Neural Information Processing Systems*, volume 31. Curran Associates, Inc., 2018.

26. Scott Camazine, Jean-Louis Deneubourg, Nigel R. Franks, James Sneyd, Guy Theraula, and Eric Bonabeau. *Self-Organization in Biological Systems*. Princeton University Press, NJ, 1st edition, 2003.
27. Fei Lu, Ming Zhong, Sui Tang, and Mauro Maggioni. Nonparametric inference of interaction laws in systems of agents from trajectory data. *Proceedings of the National Academy of Sciences*, 116(29):14424–14433, 2019.
28. Ming Zhong, Jason Miller, and Mauro Maggioni. Data-driven discovery of emergent behaviors in collective dynamics. *Physica D: Nonlinear Phenomena*, 411:132542, 2020.
29. Mauro Maggioni, Jason J Miller, Hongda Qiu, and Ming Zhong. Learning interaction kernels for agent systems on riemannian manifolds. In Marina Meila and Tong Zhang, editors, *Proceedings of the 38th International Conference on Machine Learning*, volume 139 of *Proceedings of Machine Learning Research*, pages 7290–7300. PMLR, 18–24 Jul 2021.
30. Fei Lu, Mauro Maggioni, and Sui Tang. Learning interaction kernels in heterogeneous systems of agents from multiple trajectories. *The Journal of Machine Learning Research*, 22(1):1518–1584, 2021.
31. Jinchao Feng, Mauro Maggioni, Patrick Martin, and Ming Zhong. Learning interaction variables and kernels from observations of agent-based systems. *IFAC-PapersOnLine*, 55(30):162–167, 2022. 25th International Symposium on Mathematical Theory of Networks and Systems MTNS 2022.
32. Jason Miller, Sui Tang, Ming Zhong, and Mauro Maggioni. Learning theory for inferring interaction kernels in second-order interacting agent systems. *Sampling Theory, Signal Processing, and Data Analysis*, 21(1):21, 2023.
33. Ming Zhong, Jason Miller, and Mauro Maggioni. Machine learning for discovering effective interaction kernels between celestial bodies from ephemerides, 2021.
34. Jinchao Feng, Charles Kulick, Yunxiang Ren, and Sui Tang. Learning particle swarming models from data with Gaussian processes. *Mathematics of Computation*, 2023.
35. Jinchao Feng, Charles Kulick, and Sui Tang. Data-driven model selections of second-order particle dynamics via integrating gaussian processes with low-dimensional interacting structures. *arXiv:2311.00902*, 2023.
36. Baoli Hao, Ming Zhong, and Kevin O'Keeffe. Attractive and repulsive interactions in the one-dimensional swarmalator model. *Physical Review E*, 108(6):064214, dec 2023.
37. R. Shu and E. Tadmor. Flocking hydrodynamics with external potentials. *Archive for Rational Mechanics and Analysis*, 238:347–381, 2020.
38. R. Shu and E. Tadmor. Anticipation breeds alignment. *Archive for Rational Mechanics and Analysis*, 240:203–241, 2021.
39. Audrey Filella, Francois Nadal, Clement Sire, Eva Kanso, and Christophe Eloy. Model of collective fish behavior with hydrodynamic interactions. *Phys. Rev. Lett.*, 120:198101, 2018.
40. Trenton Gerew and Ming Zhong. Concurrent emergence of clustering, flocking and synchronization in systems of interacting agents, 2023.
41. James M Greene, Eitan Tadmor, and Ming Zhong. The emergence of lines of hierarchy in collective motion of biological systems. *Physical Biology*, 20(5):055001, jun 2023.
42. Mattia Bongini, Massimo Fornasier, Markus Hansen, and Mauro Maggioni. Inferring interaction rules from observations of evolutive systems I: The variational approach. *Mathematical Models and Methods in Applied Sciences*, 27(05):909–951, 2017.
43. S. Mostch and E. Tadmor. Heterophilious Dynamics Enhances Consensus. *SIAM Rev.*, 56(4):577–621, 2014.
44. F. Cucker and S. Smale. On the mathematical foundations of learning. *Bull. Amer. Math. Soc*, 39(1):1–49, 2002.
45. Yuxuan Liu, Scott G McCalla, and Hayden Schaeffer. Random feature models for learning interacting dynamical systems. *Proceedings of the Royal Society A*, 479(2275):20220835, 2023.
46. Daniel E. Shea, Steven L. Brunton, and J. Nathan Kutz. Sindy-bvp: Sparse identification of nonlinear dynamics for boundary value problems. *ArXiv*, abs/2005.10756, 2020.

47. Kadierdan Kaheman, J. Nathan Kutz, and Steven L. Brunton. Sindy-pi: a robust algorithm for parallel implicit sparse identification of nonlinear dynamics. *Proceedings. Mathematical, Physical, and Engineering Sciences*, 476, 2020.
48. Daniel A. Messenger and David M. Bortz. Weak sindy for partial differential equations. *Journal of Computational Physics*, 443:110525, 2021.
49. Y. Chen and T. Kolokolnikov. A minimal model of predator-swarm interactions. *J. R. Soc. Interface*, 11:20131208, 2013.
50. Ziheng Guo, Igor Cialenco, and Ming Zhong. Learning stochastic dynamics from data, 2024.
51. Fei Lu, Maruo Maggioni, and Sui Tang. Learning interaction kernels in stochastic systems of interacting particles from multiple trajectories. *Foundations of Computational Mathematics*, 22:1013–1067, 2022.
52. Simo Särkkä" and Arno Solin. *Applied Stochastic Differential Equations*. Institute of Mathematical Statistics Textbooks. Cambridge University Press, 2019.
53. T. Kolokolnikov, H. Sun, D. Uminsky, and A. Bertozzi. A theory of complex patterns arising from 2d particle interactions. *Phys Rev E, Rapid Communications*, 84:015203(R), 2011.
54. M.P. Martin. *Multiplicatively Perturbed Least Squares for Dimension Reduction*. PhD thesis, Johns Hopkins University, 2021.
55. Markus Heinonen, Cagatay Yildiz, Henrik Mannerström, Jukka Intosalmi, and Harri Lähdesmäki. Learning unknown ode models with Gaussian processes. In *International Conference on Machine Learning*, pages 1959–1968. PMLR, 2018.
56. Cedric Archambeau, Dan Cornford, Manfred Opper, and John Shawe-Taylor. Gaussian process approximations of stochastic differential equations. In *Gaussian Processes in Practice*, pages 1–16. PMLR, 2007.
57. Cagatay Yildiz, Markus Heinonen, Jukka Intosalmi, Henrik Mannerstrom, and Harri Lahdesmaki. Learning stochastic differential equations with Gaussian processes without gradient matching. In *2018 IEEE 28th International Workshop on Machine Learning for Signal Processing (MLSP)*, pages 1–6. IEEE, 2018.
58. Zheng Zhao, Filip Tronarp, Roland Hostettler, and Simo Särkkä. State-space Gaussian process for drift estimation in stochastic differential equations. In *ICASSP 2020-2020 IEEE International Conference on Acoustics, Speech and Signal Processing (ICASSP)*, pages 5295–5299. IEEE, 2020.
59. Maziar Raissi, Paris Perdikaris, and George Em Karniadakis. Machine learning of linear differential equations using Gaussian processes. *Journal of Computational Physics*, 348:683–693, 2017.
60. Jiuhai Chen, Lulu Kang, and Guang Lin. Gaussian process assisted active learning of physical laws. *Technometrics*, pages 1–14, 2020.
61. Hongqiao Wang and Xiang Zhou. Explicit estimation of derivatives from data and differential equations by Gaussian process regression. *International Journal for Uncertainty Quantification*, 11(4), 2021.
62. Yifan Chen, Bamdad Hosseini, Houman Owhadi, and Andrew M Stuart. Solving and learning nonlinear pdes with gaussian processes. *Journal of Computational Physics*, 447:110668, 2021.
63. Seungjoon Lee, Mahdi Kooshkbaghi, Konstantinos Spiliotis, Constantinos I Siettos, and Ioannis G Kevrekidis. Coarse-scale PDEs from fine-scale observations via machine learning. *Chaos: An Interdisciplinary Journal of Nonlinear Science*, 30(1):013141, 2020.
64. J-L Akian, Luc Bonnet, Houman Owhadi, and Éric Savin. Learning "best" kernels from data in Gaussian process regression. With application to aerodynamics. *Journal of Computational Physics*, 470:111595, 2022.
65. Matthieu Darcy, Boumediene Hamzi, Jouni Susiluoto, Amy Braverman, and Houman Owhadi. Learning dynamical systems from data: a simple cross-validation perspective, part ii: nonparametric kernel flows. *preprint*, 2021.
66. Christopher KI Williams and Carl Edward Rasmussen. *Gaussian processes for machine learning*, volume 2. MIT press Cambridge, MA, 2006.
67. Ruiwen Shu and Eitan Tadmor. Flocking hydrodynamics with external potentials. *Archive for Rational Mechanics and Analysis*, 238:347–381, 2020.

68. Houman Owhadi and Clint Scovel. *Operator-adapted wavelets, fast solvers, and numerical homogenization: from a game theoretic approach to numerical approximation and algorithm design*, volume 35. Cambridge University Press, 2019.
69. Quanjun Lang and Fei Lu. Identifiability of interaction kernels in mean-field equations of interacting particles. *Foundations of Data Science*, 5(4):480–502, 2023.
70. Quanjun Lang and Fei Lu. Learning interaction kernels in mean-field equations of first-order systems of interacting particles. *SIAM Journal on Scientific Computing*, 44(1):A260–A285, 2022.
71. Louis Sharrock, Nikolas Kantas, Panos Parpas, and Grigorios A. Pavliotis. Online parameter estimation for the mckean–vlasov stochastic differential equation. *Stochastic Processes and their Applications*, 162:481–546, 2023.

Provably Accelerating Ill-Conditioned Low-Rank Estimation via Scaled Gradient Descent, Even with Overparameterization

Cong Ma, Xingyu Xu, Tian Tong, and Yuejie Chi

Abstract Many problems encountered in data science can be formulated as estimating a low-rank object (e.g., matrices and tensors) from incomplete, and possibly corrupted, linear measurements. Through the lens of matrix and tensor factorization, one of the most popular approaches is to employ simple iterative algorithms such as gradient descent (GD) to recover the low-rank factors directly, which allow for small memory and computation footprints. However, the convergence rate of GD depends linearly, and sometimes even quadratically, on the condition number of the low-rank object, and therefore, GD slows down dramatically when the problem is ill-conditioned. This chapter reviews an algorithmic approach, dubbed scaled gradient descent (ScaledGD), that provably converges linearly at a constant rate independent of the condition number of the low-rank object while maintaining the low per-iteration cost of gradient descent for a variety of tasks including sensing, robust principal component analysis, and completion. With a small variation, ScaledGD continues to admit fast global convergence to a near-optimal solution, again almost independent of the condition number, from a small random initialization when the rank is overspecified. In total, ScaledGD highlights the power of appropriate preconditioning in accelerating nonconvex statistical estimation, where the iteration-varying preconditioners promote desirable invariance properties of the trajectory with respect to the symmetry in low-rank factorization without hurting generalization.

C. Ma
University of Chicago, Chicago, IL, USA
e-mail: congm@uchicago.edu

X. Xu · Y. Chi (✉)
Carnegie Mellon University, Pittsburgh, PA, USA
e-mail: xingyuxu@andrew.cmu.edu; yuejiechi@cmu.edu

T. Tong
Amazon, Seattle, WA, USA
e-mail: tongtn@amazon.com

S. Foucart, S. Wojtowytsch (eds.), *Explorations in the Mathematics of Data Science*,
Applied and Numerical Harmonic Analysis,
https://doi.org/10.1007/978-3-031-66497-7_7

1 Introduction

Low-rank matrix and tensor estimation is a critical task in fields such as machine learning, signal processing, imaging science, and many others. The central task can be regarded as recovering a d-dimensional object $\mathcal{X}_\star \in \mathbb{R}^{n_1 \times \cdots \times n_d}$ from its highly incomplete observation $\boldsymbol{y} \in \mathbb{R}^m$ given by

$$\boldsymbol{y} \approx \mathcal{A}(\mathcal{X}_\star).$$

Here, $\mathcal{A} : \mathbb{R}^{n_1 \times \cdots \times n_d} \mapsto \mathbb{R}^m$ represents a certain linear map modeling the data collection process. Importantly, the number m of observations is often much smaller than the ambient dimension $\prod_{i=1}^{d} n_i$ of the data object due to resource or physical constraints, necessitating the need of exploiting low-rank structures to allow for meaningful recovery. It is natural to minimize the least-squares loss function:

$$\underset{\mathcal{X} \in \mathbb{R}^{n_1 \times \cdots \times n_d}}{\text{minimize}} \quad f(\mathcal{X}) := \frac{1}{2} \|\mathcal{A}(\mathcal{X}) - \boldsymbol{y}\|_2^2 \tag{1}$$

subject to some rank constraint. However, naively imposing the rank constraint is computationally intractable, and moreover, as the size of the object increases, the costs involved in optimizing over the full space (i.e., $\mathbb{R}^{n_1 \times \cdots \times n_d}$) are prohibitive in terms of both memory and computation.

To cope with these challenges, one popular approach is to represent the object of interest via its low-rank factors, which take a more economical form, and then optimize over the factors instead. Although this leads to a nonconvex optimization problem over the factors, recent breakthroughs have shown that simple iterative algorithms such as vanilla gradient descent (GD), when properly initialized (e.g., via the spectral method), can provably converge to the true low-rank factors under mild statistical assumptions; see [19] for an overview. This enables us to tap into the scalability of gradient descent in solving large-scale problems due to its amenability to computing advances such as parallelism [4].

Nonetheless, upon closer examination, the computational cost of vanilla gradient descent is still cumbersome, especially for ill-conditioned instances. Although the per-iteration cost is small, the iteration complexity of vanilla gradient descent scales linearly with respect to the condition number of the low-rank matrix [59], which degenerates even worse for higher-order tensors [28]. In fact, the issue of ill-conditioning is quite ubiquitous in real-world data modeling with many contributing factors. On one end, extracting fine-grained and weak information often manifests to estimating ill-conditioned object of interest, when the goals are to separate close-located sources of intelligence, to identify a weak mode nearby a strong one, to predict individualized responses for similar objects, and so on. While the impact of condition numbers on the computational efficacy cannot be ignored in practice, it unfortunately has not been properly addressed in recent algorithmic advances, which often assume the problem is well-conditioned. These together raise an important question:

Is it possible to design a first-order algorithm with a comparable per-iteration cost as gradient descent, but converges much faster at a rate that is independent of the condition number in a provable manner for a wide variety of low-rank matrix and tensor estimation tasks?

In this chapter, we answer this question affirmatively by reviewing a recently proposed algorithmic approach dubbed scaled gradient descent (`ScaledGD`), which is instantiated for an array of low-rank matrix and tensor estimation tasks. Our treatment is largely built upon, and in some cases reproduced from, a sequence of our recent papers [21, 55, 58, 66], which aims to provide a thorough and coherent presentation of the methodology and its performance.

1.1 An Overview of ScaledGD

ScaledGD for Low-Rank Matrix Estimation By parametrizing the matrix object $\boldsymbol{X} = \boldsymbol{L}\boldsymbol{R}^\top$ via two low-rank factors $\boldsymbol{L} \in \mathbb{R}^{n_1 \times r}$ and $\boldsymbol{R} \in \mathbb{R}^{n_2 \times r}$ in (1), where r is the rank of the true low-rank object $\boldsymbol{X}_\star$, we arrive at the objective function:

$$\underset{\boldsymbol{L}\in\mathbb{R}^{n_1\times r},\boldsymbol{R}\in\mathbb{R}^{n_2\times r}}{\text{minimize}} \quad \mathcal{L}(\boldsymbol{L}, \boldsymbol{R}) := f(\boldsymbol{L}\boldsymbol{R}^\top). \tag{2}$$

Given an initialization $(\boldsymbol{L}_0, \boldsymbol{R}_0)$, `ScaledGD` proceeds as follows:

$$\begin{aligned} \boldsymbol{L}_{t+1} &= \boldsymbol{L}_t - \eta \nabla_{\boldsymbol{L}} \mathcal{L}(\boldsymbol{L}_t, \boldsymbol{R}_t)(\boldsymbol{R}_t^\top \boldsymbol{R}_t)^{-1}, \\ \boldsymbol{R}_{t+1} &= \boldsymbol{R}_t - \eta \nabla_{\boldsymbol{R}} \mathcal{L}(\boldsymbol{L}_t, \boldsymbol{R}_t)(\boldsymbol{L}_t^\top \boldsymbol{L}_t)^{-1}, \end{aligned} \tag{3}$$

where $\eta > 0$ is the step size and $\nabla_{\boldsymbol{L}}\mathcal{L}(\boldsymbol{L}_t, \boldsymbol{R}_t)$ (resp. $\nabla_{\boldsymbol{R}}\mathcal{L}(\boldsymbol{L}_t, \boldsymbol{R}_t)$) is the gradient of the loss function $\mathcal{L}$ with respect to the factor $\boldsymbol{L}_t$ (resp. $\boldsymbol{R}_t$) at the tth iteration. Comparing to vanilla gradient descent, the search directions of the low-rank factors $\boldsymbol{L}_t, \boldsymbol{R}_t$ in (3) are *scaled* by $(\boldsymbol{R}_t^\top \boldsymbol{R}_t)^{-1}$ and $(\boldsymbol{L}_t^\top \boldsymbol{L}_t)^{-1}$, respectively. Intuitively, the scaling serves as a preconditioner as in quasi-Newton-type algorithms, with the hope of improving the quality of the search direction to allow larger step sizes. Since computing the Hessian is extremely expensive, it is necessary to design preconditioners that are both theoretically sound and practically cheap to compute. Such requirements are met by `ScaledGD`, where the preconditioners are computed by inverting two $r \times r$ matrices, whose size is much smaller than the dimension of matrix factors. Theoretically, we confirm that `ScaledGD` achieves linear convergence at a rate *independent of* the condition number of the matrix when initialized properly, e.g., using the standard spectral method, for several canonical problems: low-rank matrix sensing, robust PCA, and matrix completion.

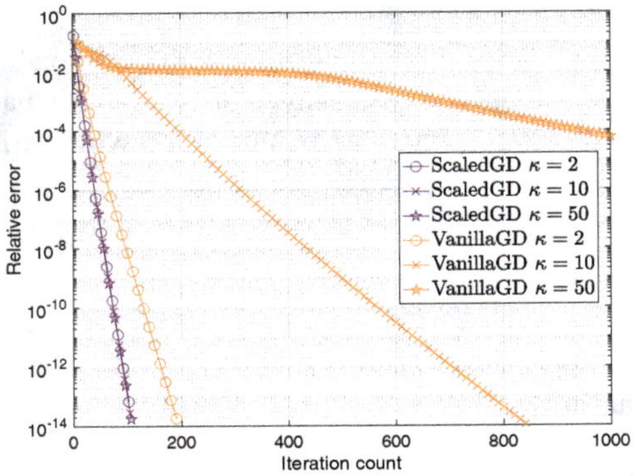

Fig. 1 Performance of ScaledGD and vanilla GD for completing a 1000×1000 incoherent matrix of rank 10 with different condition numbers $\kappa = 2, 10, 50$, where each entry is observed independently with probability 0.2. Here, both methods are initialized via the spectral method. It can be seen that ScaledGD converges much faster than vanilla GD even for moderately large condition numbers. Figure reproduced from [55]

Performance of ScaledGD for Low-Rank Matrix Completion

Figure 1 illustrates the relative error of completing a 1000×1000 incoherent matrix (cf. Definition 2) of rank 10 with varying condition numbers from 20% of its entries, using either ScaledGD or vanilla GD with spectral initialization. Even for moderately ill-conditioned matrices, the convergence rate of vanilla GD slows down dramatically, while it is evident that ScaledGD converges at a rate independent of the condition number and therefore is much more efficient.

ScaledGD for Low-Rank Tensor Estimation Turning to the tensor case, we focus on one of the most widely adopted low-rank structures for tensors under the *Tucker* decomposition [60], by assuming the true tensor $\mathcal{X}_\star$ to be low multilinear rank, or simply low rank, when its multilinear rank $\boldsymbol{r} = (r_1, r_2, r_3)$. By parameterizing the order-3 tensor object[1] as $\mathcal{X} = (\boldsymbol{U}, \boldsymbol{V}, \boldsymbol{W}) \cdot \mathcal{S}$, where $\boldsymbol{F} := (\boldsymbol{U}, \boldsymbol{V}, \boldsymbol{W}, \mathcal{S})$ consists of the factors $\boldsymbol{U} \in \mathbb{R}^{n_1 \times r_1}$, $\boldsymbol{V} \in \mathbb{R}^{n_2 \times r_2}$, $\boldsymbol{W} \in \mathbb{R}^{n_3 \times r_3}$, and $\mathcal{S} \in \mathbb{R}^{r_1 \times r_2 \times r_3}$, we aim to optimize the objective function:

$$\underset{\boldsymbol{F}}{\text{minimize}} \quad \mathcal{L}(\boldsymbol{F}) := \frac{1}{2} \left\| \mathcal{A}\left((\boldsymbol{U}, \boldsymbol{V}, \boldsymbol{W}) \cdot \mathcal{S}\right) - \boldsymbol{y} \right\|_2^2 . \tag{4}$$

[1] For ease of presentation, we focus on 3-way tensors; our algorithm and theory can be generalized to higher-order tensors in a straightforward manner.

Given an initialization $\boldsymbol{F}_0 = (\boldsymbol{U}_0, \boldsymbol{V}_0, \boldsymbol{W}_0, \mathcal{S}_0)$, `ScaledGD` proceeds as follows:

$$\begin{aligned}
\boldsymbol{U}_{t+1} &= \boldsymbol{U}_t - \eta \nabla_{\boldsymbol{U}} \mathcal{L}(\boldsymbol{F}_t)\big(\breve{\boldsymbol{U}}_t^\top \breve{\boldsymbol{U}}_t\big)^{-1}, \\
\boldsymbol{V}_{t+1} &= \boldsymbol{V}_t - \eta \nabla_{\boldsymbol{V}} \mathcal{L}(\boldsymbol{F}_t)\big(\breve{\boldsymbol{V}}_t^\top \breve{\boldsymbol{V}}_t\big)^{-1}, \\
\boldsymbol{W}_{t+1} &= \boldsymbol{W}_t - \eta \nabla_{\boldsymbol{W}} \mathcal{L}(\boldsymbol{F}_t)\big(\breve{\boldsymbol{W}}_t^\top \breve{\boldsymbol{W}}_t\big)^{-1}, \\
\mathcal{S}_{t+1} &= \mathcal{S}_t - \eta \left((\boldsymbol{U}_t^\top \boldsymbol{U}_t)^{-1}, (\boldsymbol{V}_t^\top \boldsymbol{V}_t)^{-1}, (\boldsymbol{W}_t^\top \boldsymbol{W}_t)^{-1}\right) \boldsymbol{\cdot} \nabla_{\mathcal{S}} \mathcal{L}(\boldsymbol{F}_t),
\end{aligned} \tag{5}$$

where $\nabla_{\boldsymbol{U}} \mathcal{L}(\boldsymbol{F})$, $\nabla_{\boldsymbol{V}} \mathcal{L}(\boldsymbol{F})$, $\nabla_{\boldsymbol{W}} \mathcal{L}(\boldsymbol{F})$, and $\nabla_{\mathcal{S}} \mathcal{L}(\boldsymbol{F})$ are the partial derivatives of $\mathcal{L}(\boldsymbol{F})$ with respect to the corresponding variables, and

$$\begin{aligned}
\breve{\boldsymbol{U}}_t &:= \mathcal{M}_1\left((\boldsymbol{I}_{r_1}, \boldsymbol{V}_t, \boldsymbol{W}_t) \boldsymbol{\cdot} \mathcal{S}_t\right)^\top = (\boldsymbol{W}_t \otimes \boldsymbol{V}_t)\mathcal{M}_1(\mathcal{S}_t)^\top, \\
\breve{\boldsymbol{V}}_t &:= \mathcal{M}_2\left((\boldsymbol{U}_t, \boldsymbol{I}_{r_2}, \boldsymbol{W}_t) \boldsymbol{\cdot} \mathcal{S}_t\right)^\top = (\boldsymbol{W}_t \otimes \boldsymbol{U}_t)\mathcal{M}_2(\mathcal{S}_t)^\top, \\
\breve{\boldsymbol{W}}_t &:= \mathcal{M}_3\left((\boldsymbol{U}_t, \boldsymbol{V}_t, \boldsymbol{I}_{r_3}) \boldsymbol{\cdot} \mathcal{S}_t\right)^\top = (\boldsymbol{V}_t \otimes \boldsymbol{U}_t)\mathcal{M}_3(\mathcal{S}_t)^\top.
\end{aligned} \tag{6}$$

Here, $\mathcal{M}_k(\mathcal{S})$ is the matricization of the tensor $\mathcal{S}$ along the kth mode ($k = 1, 2, 3$), and $\otimes$ denotes the Kronecker product. We investigate the theoretical properties of `ScaledGD` for tensor regression, tensor robust PCA, and tensor completion, which are notably more challenging than the matrix counterpart. It is demonstrated that `ScaledGD`—when initialized properly using appropriate spectral methods—again achieves linear convergence at a rate *independent* of the condition number of the ground truth tensor with near-optimal sample complexities.

ScaledGD for Low-Rank Tensor Completion

Figure 2 illustrates the number of iterations needed to achieve a relative error $\|\mathcal{X} - \mathcal{X}_\star\|_{\mathsf{F}} \le 10^{-3}\|\mathcal{X}_\star\|_{\mathsf{F}}$ for `ScaledGD` and regularized GD [28] under different condition numbers for tensor completion under the Bernoulli sampling model. Clearly, the iteration complexity of GD deteriorates at a superlinear rate with respect to the condition number κ, while `ScaledGD` enjoys an iteration complexity that is independent of κ as predicted by our theory. Indeed, with a seemingly small modification, `ScaledGD` takes merely 17 iterations to achieve the desired accuracy over the entire range of κ, while GD takes thousands of iterations even with a moderate condition number!

To highlight, `ScaledGD` possesses many desirable properties appealing to practitioners:

- *Low per-iteration cost:* as a preconditioned GD or quasi-Newton algorithm, `ScaledGD` updates the factors along the descent direction of a scaled gradient,

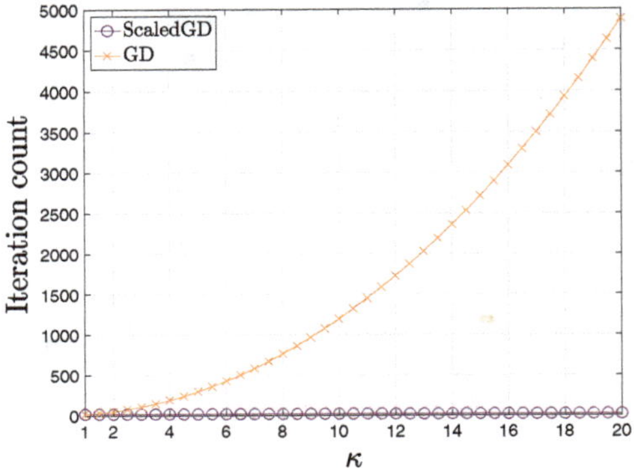

Fig. 2 The iteration complexities of `ScaledGD` and regularized GD to achieve $\|\mathcal{X} - \mathcal{X}_\star\|_\mathsf{F} \le 10^{-3}\|\mathcal{X}_\star\|_\mathsf{F}$ with respect to different condition numbers for low-rank tensor completion with $n_1 = n_2 = n_3 = 100$, $r_1 = r_2 = r_3 = 5$, and the probability of observation $p = 0.1$. Figure reproduced from [58]

where the preconditioners can be viewed as the inverse of the diagonal blocks of the Hessian for the population loss (i.e., matrix factorization and tensor factorization). As the sizes of the preconditioners are proportional to the rank rather than the ambient dimension, the matrix inverses are cheap to compute with a minimal overhead and the overall per-iteration cost is still low and linear in the time it takes to read the input data.

- *Equivariance to parameterization:* one crucial property of `ScaledGD` is that if we reparameterize the factors by some invertible transformation, the entire trajectory will go through the same reparameterization, leading to an *invariant* sequence of low-rank updates regardless of the parameterization being adopted.
- *Implicit balancing:* `ScaledGD` optimizes the natural loss function in an *unconstrained* manner without requiring additional regularizations or orthogonalizations used in prior literature [24, 28, 35], and the factors stay balanced in an automatic manner as if they are implicitly regularized [41].

In total, the fast convergence rate of `ScaledGD`, together with its low computational and memory costs by operating in the factor space, makes it a highly scalable and desirable method for low-rank estimation tasks.

1.2 Related Works

Our work contributes to the growing literature of design and analysis of provable nonconvex optimization procedures for high-dimensional signal estimation; see e.g., [15, 19, 31] for recent overviews. A growing number of problems have been demonstrated to possess benign geometry that is amenable for optimization [43] either globally or locally under appropriate statistical models. On one end, it is shown that there are no spurious local minima in the optimization landscape of matrix sensing and completion [3, 26, 27, 46], phase retrieval [20, 53], dictionary learning [52], kernel PCA [12], and linear neural networks [1, 36]. Such landscape analysis facilitates the adoption of generic saddle-point escaping algorithms [25,

34, 44] to ensure global convergence. However, the resulting iteration complexity is typically high. On the other end, local refinements with carefully designed initializations often admit fast convergence, for example, in phase retrieval [7, 42], matrix sensing [33, 62, 70], matrix completion [13, 16, 17, 42, 54, 71], blind deconvolution [37, 42, 50], quadratic sampling [38], and robust PCA [18, 45, 67], to name a few.

Existing approaches for asymmetric low-rank matrix estimation often require additional regularization terms to balance the two factors, either in the form of $\frac{1}{2}\|\boldsymbol{L}^\top\boldsymbol{L} - \boldsymbol{R}^\top\boldsymbol{R}\|_{\mathsf{F}}^2$ [46, 59] or $\frac{1}{2}\|\boldsymbol{L}\|_{\mathsf{F}}^2 + \frac{1}{2}\|\boldsymbol{R}\|_{\mathsf{F}}^2$ [17, 18, 72], which ease the theoretical analysis but are often unnecessary for the practical success, as long as the initialization is balanced. Some recent work studies the unregularized gradient descent for low-rank matrix factorization and sensing including [11, 22, 41]. However, the iteration complexity of all these approaches scales at least linearly with respect to the condition number κ of the low-rank matrix, e.g., $O(\kappa\log(1/\epsilon))$, to reach ϵ-accuracy; therefore, they converge slowly when the underlying matrix becomes ill-conditioned. In contrast, `ScaledGD` enjoys a local convergence rate of $O(\log(1/\epsilon))$, therefore incurring a much smaller computational footprint when κ is large. Last but not least, alternating minimization [29, 33] (which alternatively updates $\boldsymbol{L}_t$ and $\boldsymbol{R}_t$) or singular value projection [32, 45] (which operates in the matrix space) also converges at the rate $O(\log(1/\epsilon))$, but the per-iteration cost is much higher than `ScaledGD`. Another notable algorithm is the Riemannian gradient descent algorithm in [62], which also converges at the rate $O(\log(1/\epsilon))$ under the same sample complexity for low-rank matrix sensing, but requires a higher memory complexity since it operates in the matrix space rather than the factor space.

Turning to the tensor case, unfolding-based approaches typically result in suboptimal sample complexities since they do not fully exploit the tensor structure. Yuan and Zhang [68] studied directly minimizing the nuclear norm of the tensor, which regrettably is not computationally tractable. Xia and Yuan [63] proposed a Grassmannian gradient descent algorithm over the factors other than the core tensor for exact tensor completion, whose iteration complexity is not characterized. The statistical rates of tensor completion, together with a spectral method, were investigated in [64, 69], and uncertainty quantifications were recently dealt with in [65]. In addition, for low-rank tensor regression, [47] proposed a general convex optimization approach based on decomposable regularizers, and [48] developed an iterative hard thresholding algorithm. A concurrent work [40] proposed a Riemannian Gauss-Newton algorithm and obtained an impressive quadratic convergence rate for tensor regression. Compared with `ScaledGD`, this algorithm runs in the tensor space, and the update rule is more sophisticated with higher per-iteration cost by solving a least-squares problem and performing a truncated HOSVD every iteration. Another recent work [6] studies the Riemannian gradient descent algorithm which also achieves an iteration complexity free of condition number; however, the initialization scheme was not studied therein. Riemannian gradient descent is also applied to low-rank tensor completion with Tucker decomposition in [61].

1.3 Chapter Organization and Notation

The rest of this chapter is organized as follows. Sections 2 and 3 describe `ScaledGD` and details its application to sensing, robust PCA, and completion with theoretical guarantees in terms of both statistical and computational complexities for the matrix and tensor case, respectively. Section 4 discusses a variant of `ScaledGD` when the rank is not specified exactly. Section 5 illustrates the empirical performance of `ScaledGD` on real data, with a particular focus on the issues of rank selection. Finally, we conclude in Sect. 6.

Before continuing, we introduce several notations used throughout the chapter. First of all, we use boldfaced symbols for vectors and matrices (e.g., $\boldsymbol{x}$ and $\boldsymbol{X}$) and boldface calligraphic letters (e.g., $\mathcal{X}$) to denote tensors. For a vector $\boldsymbol{v}$, we use $\|\boldsymbol{v}\|_0$ to denote its ℓ_0 counting norm and $\|\boldsymbol{v}\|_2$ to denote the ℓ_2 norm. For any matrix $\boldsymbol{A}$, we use $\sigma_i(\boldsymbol{A})$ to denote its ith largest singular value and $\sigma_{\max}(\boldsymbol{A})$ (resp. $\sigma_{\min}(\boldsymbol{A})$) to denote its largest (resp. smallest) nonzero singular value. Let $\mathcal{P}_{\mathsf{diag}}(\boldsymbol{A})$ denote the projection that keeps only the diagonal entries of $\boldsymbol{A}$ and $\mathcal{P}_{\mathsf{off-diag}}(\boldsymbol{A}) = \boldsymbol{A} - \mathcal{P}_{\mathsf{diag}}(\boldsymbol{A})$, for a square matrix $\boldsymbol{A}$. Let $\boldsymbol{A}_{i,\cdot}$ and $\boldsymbol{A}_{\cdot,j}$ denote its ith row and jth column, respectively. In addition, $\|\boldsymbol{A}\|_{\mathsf{F}}$, $\|\boldsymbol{A}\|_{1,\infty}$, $\|\boldsymbol{A}\|_{2,\infty}$, and $\|\boldsymbol{A}\|_\infty$ stand for the Frobenius norm, the $\ell_{1,\infty}$ norm (i.e., the largest ℓ_1 norm of the rows), the $\ell_{2,\infty}$ norm (i.e., the largest ℓ_2 norm of the rows), and the entrywise ℓ_∞ norm (the largest magnitude of all entries) of a matrix $\boldsymbol{A}$, respectively. The set of invertible matrices in $\mathbb{R}^{r\times r}$ is denoted by $\mathrm{GL}(r)$. The $r \times r$ identity matrix is denoted by $\boldsymbol{I}_r$.

The mode-1 matricization $\mathcal{M}_1(\mathcal{X}) \in \mathbb{R}^{n_1\times(n_2 n_3)}$ of a tensor $\mathcal{X} \in \mathbb{R}^{n_1\times n_2\times n_3}$ is given by $[\mathcal{M}_1(\mathcal{X})]\big(i_1, i_2 + (i_3 - 1)n_2\big) = \mathcal{X}(i_1, i_2, i_3)$, for $1 \le i_k \le n_k$, $k = 1, 2, 3$; $\mathcal{M}_2(\mathcal{X})$ and $\mathcal{M}_3(\mathcal{X})$ can be defined in a similar manner. The inner product between two tensors is defined as

$$\langle \mathcal{X}_1, \mathcal{X}_2\rangle = \sum_{i_1,i_2,i_3} \mathcal{X}_1(i_1, i_2, i_3)\mathcal{X}_2(i_1, i_2, i_3),$$

and the Frobenius norm of a tensor is defined as $\|\mathcal{X}\|_{\mathsf{F}} = \sqrt{\langle \mathcal{X}, \mathcal{X}\rangle}$. Define the ℓ_∞ norm of $\mathcal{X}$ as $\|\mathcal{X}\|_\infty = \max_{i_1,i_2,i_3} |\mathcal{X}(i_1, i_2, i_3)|$. With slight abuse of terminology, denote

$$\sigma_{\max}(\mathcal{X}) = \max_{k=1,2,3} \sigma_{\max}(\mathcal{M}_k(\mathcal{X})) \quad \text{and} \quad \sigma_{\min}(\mathcal{X}) = \min_{k=1,2,3} \sigma_{\min}(\mathcal{M}_k(\mathcal{X}))$$

as the maximum and minimum nonzero singular values of $\mathcal{X}$, respectively.

For a general tensor $\mathcal{X}$, define $\mathcal{H}_{\boldsymbol{r}}(\mathcal{X})$ as the top-$\boldsymbol{r}$ higher-order SVD (HOSVD) of $\mathcal{X}$ with $\boldsymbol{r} = (r_1, r_2, r_3)$, given by

$$\mathcal{H}_{\boldsymbol{r}}(\mathcal{X}) = (\boldsymbol{U}, \boldsymbol{V}, \boldsymbol{W}) \boldsymbol{\cdot} \mathcal{S}, \tag{7}$$

where U is the top-r_1 left singular vectors of $\mathcal{M}_1(\mathcal{X})$, V is the top-r_2 left singular vectors of $\mathcal{M}_2(\mathcal{X})$, W is the top-r_3 left singular vectors of $\mathcal{M}_3(\mathcal{X})$, and $\mathcal{S} = (U^\top, V^\top, W^\top) \cdot \mathcal{X}$ is the core tensor.

Let $a \vee b = \max\{a, b\}$ and $a \wedge b = \min\{a, b\}$. Throughout, $f(n) \lesssim g(n)$ or $f(n) = O(g(n))$ means $|f(n)|/|g(n)| \leq C$ for some constant $C > 0$ when n is sufficiently large; $f(n) \gtrsim g(n)$ means $|f(n)|/|g(n)| \geq C$ for some constant $C > 0$ when n is sufficiently large. Last but not least, we use the terminology "with overwhelming probability" to denote the event happens with probability at least $1 - c_1 n^{-c_2}$, where $c_1, c_2 > 0$ are some universal constants, whose values may vary from line to line.

2 ScaledGD for Low-Rank Matrix Estimation

This section is devoted to introducing `ScaledGD` and establishing its statistical and computational guarantees for various low-rank matrix estimation problems; the majority of the discussed results are sampled from [55]. Table 1 summarizes the performance guarantees of `ScaledGD` in terms of both statistical and computational complexities with comparisons to prior algorithms using GD.

2.1 Assumptions

Denote by $U_\star \Sigma_\star V_\star^\top$ the compact singular value decomposition (SVD) of the rank-r matrix $X_\star \in \mathbb{R}^{n_1 \times n_2}$, i.e.,

$$X_\star = U_\star \Sigma_\star V_\star^\top.$$

Here, $U_\star \in \mathbb{R}^{n_1 \times r}$ and $V_\star \in \mathbb{R}^{n_2 \times r}$ are composed of r left and right singular vectors, respectively, and $\Sigma_\star \in \mathbb{R}^{r \times r}$ is a diagonal matrix consisting of r singular values of $X_\star$ organized in a nonincreasing order, i.e., $\sigma_1(X_\star) \geq \cdots \geq \sigma_r(X_\star) > 0$. Define the ground truth low-rank factors as

$$L_\star := U_\star \Sigma_\star^{1/2}, \qquad \text{and} \qquad R_\star := V_\star \Sigma_\star^{1/2}, \tag{8}$$

so that $X_\star = L_\star R_\star^\top$. Correspondingly, denote the stacked factor matrix as

$$F_\star := \begin{bmatrix} L_\star \\ R_\star \end{bmatrix} \in \mathbb{R}^{(n_1+n_2)\times r}. \tag{9}$$

Key Parameters The condition number of $X_\star$ is defined as follows.

Table 1 Comparisons of ScaledGD with GD when tailored to various problems (with spectral initialization) [59, 67, 71], where they have comparable per-iteration costs. Here, we say that the output $\boldsymbol{X}$ of an algorithm reaches ϵ-accuracy, if it satisfies $\|\boldsymbol{X} - \boldsymbol{X}_\star\|_\mathsf{F} \le \epsilon\sigma_r(\boldsymbol{X}_\star)$. Here, $n := n_1 \vee n_2 = \max\{n_1, n_2\}$, and κ and μ are the condition number and incoherence parameter of $\boldsymbol{X}_\star$, respectively

	Matrix sensing		Matrix robust PCA		Matrix completion	
Algorithms	Sample complexity	Iteration complexity	Corruption fraction	Iteration complexity	Sample complexity	Iteration complexity
GD	$nr^2\kappa^2$	$\kappa \log \frac{1}{\epsilon}$	$\frac{1}{\mu r^{3/2}\kappa^{3/2} \vee \mu r \kappa^2}$	$\kappa \log \frac{1}{\epsilon}$	$(\mu \vee \log n)\mu n r^2 \kappa^2$	$\kappa \log \frac{1}{\epsilon}$
ScaledGD	$nr^2\kappa^2$	$\log \frac{1}{\epsilon}$	$\frac{1}{\mu r^{3/2}\kappa}$	$\log \frac{1}{\epsilon}$	$(\mu\kappa^2 \vee \log n)\mu n r^2 \kappa^2$	$\log \frac{1}{\epsilon}$

Definition 1 (Matrix Condition Number) Define

$$\kappa := \frac{\sigma_1(\boldsymbol{X}_\star)}{\sigma_r(\boldsymbol{X}_\star)} \tag{10}$$

as the condition number of $\boldsymbol{X}_\star$.

We next introduce the incoherence condition, which is known to be crucial for reliable estimation of the low-rank matrix $\boldsymbol{X}_\star$ in matrix completion and robust PCA [14].

Definition 2 (Matrix Incoherence) A rank-r matrix $\boldsymbol{X}_\star \in \mathbb{R}^{n_1 \times n_2}$ with compact SVD as $\boldsymbol{X}_\star = \boldsymbol{U}_\star \boldsymbol{\Sigma}_\star \boldsymbol{V}_\star^\top$ is said to be μ-incoherent if

$$\|\boldsymbol{U}_\star\|_{2,\infty} \le \sqrt{\frac{\mu}{n_1}}\|\boldsymbol{U}_\star\|_{\mathsf{F}} = \sqrt{\frac{\mu r}{n_1}}, \quad \text{and} \quad \|\boldsymbol{V}_\star\|_{2,\infty} \le \sqrt{\frac{\mu}{n_2}}\|\boldsymbol{V}_\star\|_{\mathsf{F}} = \sqrt{\frac{\mu r}{n_2}}.$$

2.2 *Matrix sensing*

Observation Model Assume that we have collected a set of linear measurements about a rank-r matrix $\boldsymbol{X}_\star \in \mathbb{R}^{n_1 \times n_2}$, given as

$$\boldsymbol{y} = \mathcal{A}(\boldsymbol{X}_\star) \in \mathbb{R}^m, \tag{11}$$

where $\mathcal{A}(\boldsymbol{X}) = \{\langle \boldsymbol{A}_i, \boldsymbol{X}\rangle\}_{i=1}^m : \mathbb{R}^{n_1 \times n_2} \mapsto \mathbb{R}^m$ is the linear map modeling the measurement process. The goal of low-rank matrix sensing is to recover $\boldsymbol{X}_\star$ from $\boldsymbol{y}$ when the number of measurements $m \ll n_1 n_2$, which has wide applications in medical imaging, signal processing, and data compression [9].

Algorithm Development Writing $\boldsymbol{X} \in \mathbb{R}^{n_1 \times n_2}$ into a factored form $\boldsymbol{X} = \boldsymbol{L}\boldsymbol{R}^\top$, we consider the following optimization problem:

$$\underset{\boldsymbol{F} \in \mathbb{R}^{(n_1+n_2)\times r}}{\text{minimize}} \quad \mathcal{L}(\boldsymbol{F}) = \frac{1}{2}\left\|\mathcal{A}(\boldsymbol{L}\boldsymbol{R}^\top) - \boldsymbol{y}\right\|_2^2. \tag{12}$$

Here as before, $\boldsymbol{F}$ denotes the stacked factor matrix $[\boldsymbol{L}^\top, \boldsymbol{R}^\top]^\top$. We suggest running `ScaledGD` (3) with the spectral initialization to solve (12), which performs the top-r SVD on $\mathcal{A}^*(\boldsymbol{y})$, where $\mathcal{A}^*(\cdot)$ is the adjoint operator of $\mathcal{A}(\cdot)$. The full algorithm is stated in Algorithm 1. The low-rank matrix can be estimated as $\boldsymbol{X}_T = \boldsymbol{L}_T \boldsymbol{R}_T^\top$ after running T iterations of `ScaledGD`.

Theoretical Guarantee To understand the performance of `ScaledGD` for low-rank matrix sensing, we adopt a standard assumption on the sensing operator $\mathcal{A}(\cdot)$, namely, the restricted isometry property (RIP).

Algorithm 1 ScaledGD for low-rank matrix sensing with spectral initialization

Spectral initialization: Let $U_0 \Sigma_0 V_0^\top$ be the top-r SVD of $\mathcal{A}^*(y)$, and set

$$L_0 = U_0 \Sigma_0^{1/2}, \quad \text{and} \quad R_0 = V_0 \Sigma_0^{1/2}. \tag{13}$$

Scaled gradient updates: for $t = 0, 1, 2, \dots, T-1$ **do**

$$\begin{aligned} L_{t+1} &= L_t - \eta \mathcal{A}^*(\mathcal{A}(L_t R_t^\top) - y) R_t (R_t^\top R_t)^{-1}, \\ R_{t+1} &= R_t - \eta \mathcal{A}^*(\mathcal{A}(L_t R_t^\top) - y)^\top L_t (L_t^\top L_t)^{-1}. \end{aligned} \tag{14}$$

Definition 3 (Matrix RIP [49]) The linear map $\mathcal{A}(\cdot)$ is said to obey the rank-r RIP with a constant $\delta_r \in [0, 1)$, if for all matrices $M \in \mathbb{R}^{n_1 \times n_2}$ of rank at most r, one has

$$(1 - \delta_r)\|M\|_\mathsf{F}^2 \le \|\mathcal{A}(M)\|_2^2 \le (1 + \delta_r)\|M\|_\mathsf{F}^2.$$

It is well known that many measurement ensembles satisfy the RIP [9, 49]. For example, if the entries of A_is are composed of i.i.d. Gaussian entries $\mathcal{N}(0, 1/m)$, then the RIP is satisfied for a constant δ_r as long as m is on the order of $(n_1+n_2)r/\delta_r^2$. With the RIP condition in place, the following theorem demonstrates that ScaledGD converges linearly at a constant rate as long as the sensing operator $\mathcal{A}(\cdot)$ has a sufficiently small RIP constant.

Theorem 1 ([55]) *Suppose that $\mathcal{A}(\cdot)$ obeys the $2r$-RIP with $\delta_{2r} \le 0.02/(\sqrt{r}\kappa)$. If the step size obeys $0 < \eta \le 2/3$, then for all $t \ge 0$, the iterates of the* ScaledGD *method in Algorithm 1 satisfy*

$$\left\| L_t R_t^\top - X_\star \right\|_\mathsf{F} \le (1 - 0.6\eta)^t 0.15 \sigma_r(X_\star).$$

Theorem 1 establishes that the reconstruction error $\|L_t R_t^\top - X_\star\|_\mathsf{F}$ contracts linearly at a constant rate, as long as the sample size satisfies $m = O(nr^2\kappa^2)$ with Gaussian random measurements [49], where we recall that $n = n_1 \vee n_2$. To reach ϵ-accuracy, i.e., $\|L_t R_t^\top - X_\star\|_\mathsf{F} \le \epsilon \sigma_r(X_\star)$, ScaledGD takes at most $T = O(\log(1/\epsilon))$ iterations, which is *independent* of the condition number κ of $X_\star$. In comparison, GD with spectral initialization in [59] converges in $O(\kappa \log(1/\epsilon))$ iterations as long as $m = O(nr^2\kappa^2)$. Therefore, ScaledGD converges at a much faster rate than GD at the same sample complexity while maintaining a similar per-iteration cost (cf. Table 1).

2.3 Matrix Robust Principal Component Analysis

Observation Model Assume that we have observed the data matrix

$$Y = X_\star + S_\star,$$

which is a superposition of a rank-r matrix $X_\star$, modeling the clean data, and a sparse matrix $S_\star$, modeling the corruption or outliers. The goal of robust PCA [8, 10] is to separate the two matrices $X_\star$ and $S_\star$ from their mixture Y.

Following [10, 45, 67], we consider a deterministic sparsity model for $S_\star$, in which $S_\star$ contains at most α-fraction of nonzero entries per row and column for some $\alpha \in [0, 1)$, i.e., $S_\star \in \mathcal{S}_\alpha$, where we denote

$$\mathcal{S}_\alpha := \{S \in \mathbb{R}^{n_1 \times n_2} : \|S_{i,\cdot}\|_0 \leq \alpha n_2 \text{ for all } i, \text{ and } \|S_{\cdot,j}\|_0 \leq \alpha n_1 \text{ for all } j\}. \tag{15}$$

Algorithm Development Writing $X \in \mathbb{R}^{n_1 \times n_2}$ into the factored form $X = LR^\top$, we consider the following optimization problem:

$$\underset{F \in \mathbb{R}^{(n_1+n_2)\times r}, S \in \mathcal{S}_\alpha}{\text{minimize}} \quad \mathcal{L}(F, S) = \frac{1}{2} \left\| LR^\top + S - Y \right\|_\mathsf{F}^2. \tag{16}$$

It is thus natural to alternatively update $F = [L^\top, R^\top]^\top$ and S, where F is updated via the proposed `ScaledGD` algorithm, and S is updated by hard thresholding, which trims the small entries of the residual matrix $Y - LR^\top$. More specifically, for some truncation level $0 \leq \bar{\alpha} \leq 1$, we define the sparsification operator that only keeps $\bar{\alpha}$ fraction of largest entries in each row and column:

$$(\mathcal{T}_{\bar{\alpha}}[A])_{i,j} = \begin{cases} A_{i,j}, & \text{if } |A|_{i,j} \geq |A|_{i,(\bar{\alpha} n_2)}, \text{ and } |A|_{i,j} \geq |A|_{(\bar{\alpha} n_1),j} \\ 0, & \text{otherwise} \end{cases}, \tag{17}$$

where $|A|_{i,(k)}$ (resp. $|A|_{(k),j}$) denote the kth largest element in magnitude in the ith row (resp. jth column). The `ScaledGD` algorithm with the spectral initialization for solving robust PCA is formally stated in Algorithm 2. Note that, comparing with [67], we do not require a balancing term $\|L^\top L - R^\top R\|_\mathsf{F}^2$ in the loss function (16), nor the projection of the low-rank factors onto the $\ell_{2,\infty}$ ball in each iteration.

Theoretical Guarantee The following theorem establishes the performance guarantee of `ScaledGD` as long as the fraction α of corruptions is sufficiently small.

Theorem 2 ([55]) *Suppose that $X_\star$ is μ-incoherent and that the corruption fraction α obeys $\alpha \leq c/(\mu r^{3/2}\kappa)$ for some sufficiently small constant $c > 0$. If the step size obeys $0.1 \leq \eta \leq 2/3$, then for all $t \geq 0$, the iterates of `ScaledGD` in*

Algorithm 2 `ScaledGD` for robust PCA with spectral initialization

Spectral initialization: Let $U_0\Sigma_0 V_0^\top$ be the top-r SVD of $Y - \mathcal{T}_\alpha[Y]$, and set

$$L_0 = U_0\Sigma_0^{1/2}, \quad \text{and} \quad R_0 = V_0\Sigma_0^{1/2}. \tag{18}$$

Scaled gradient updates: for $t = 0, 1, 2, \ldots, T-1$ **do**

$$\begin{aligned} S_t &= \mathcal{T}_{2\alpha}[Y - L_t R_t^\top], \\ L_{t+1} &= L_t - \eta(L_t R_t^\top + S_t - Y)R_t(R_t^\top R_t)^{-1}, \\ R_{t+1} &= R_t - \eta(L_t R_t^\top + S_t - Y)^\top L_t(L_t^\top L_t)^{-1}. \end{aligned} \tag{19}$$

Algorithm 2 satisfy

$$\left\| L_t R_t^\top - X_\star \right\|_{\mathsf{F}} \leq (1 - 0.6\eta)^t 0.03\sigma_r(X_\star).$$

Theorem 2 establishes that the reconstruction error $\|L_t R_t^\top - X_\star\|_{\mathsf{F}}$ contracts linearly at a constant rate, as long as the fraction of corruptions satisfies $\alpha \lesssim 1/(\mu r^{3/2}\kappa)$. To reach ϵ-accuracy, i.e., $\|L_t R_t^\top - X_\star\|_{\mathsf{F}} \leq \epsilon\sigma_r(X_\star)$, `ScaledGD` takes at most $T = O(\log(1/\epsilon))$ iterations, which is *independent* of κ. In comparison, projected gradient descent with spectral initialization in [67] converges in $O(\kappa \log(1/\epsilon))$ iterations as long as $\alpha \lesssim 1/(\mu r^{3/2}\kappa^{3/2} \vee \mu r\kappa^2)$. Therefore, `ScaledGD` converges at a much faster rate than GD while maintaining a comparable per-iteration cost (cf. Table 1). In addition, our theory unveils that `ScaledGD` automatically maintains the incoherence and balancedness of the low-rank factors without imposing explicit regularizations.

2.4 Matrix Completion

Observation Model Assume that we have observed a subset Ω of entries of $X_\star$ given as $\mathcal{P}_\Omega(X_\star)$, where $\mathcal{P}_\Omega : \mathbb{R}^{n_1\times n_2} \mapsto \mathbb{R}^{n_1\times n_2}$ is a projection such that

$$(\mathcal{P}_\Omega(X))_{i,j} = \begin{cases} X_{i,j}, & \text{if } (i,j) \in \Omega, \\ 0, & \text{otherwise} \end{cases}. \tag{20}$$

Here, Ω is generated according to the Bernoulli model in the sense that each $(i, j) \in \Omega$ independent with probability $p \in (0, 1]$. The goal of matrix completion is to recover the matrix $X_\star$ from its partial observation $\mathcal{P}_\Omega(X_\star)$.

Algorithm Development Again, writing $X \in \mathbb{R}^{n_1 \times n_2}$ into the factored form $X = LR^\top$, we consider the following optimization problem:

$$\underset{F \in \mathbb{R}^{(n_1+n_2)\times r}}{\text{minimize}} \quad \mathcal{L}(F) := \frac{1}{2p} \left\| \mathcal{P}_\Omega(LR^\top - X_\star) \right\|_\mathsf{F}^2. \tag{21}$$

Similarly to robust PCA, the underlying low-rank matrix $X_\star$ needs to be incoherent (cf. Definition 2) to avoid ill-posedness. One typical strategy to ensure the incoherence condition is to perform projection after the gradient update, by projecting the iterates to maintain small $\ell_{2,\infty}$ norms of the factor matrices. However, the standard projection operator [16] is not covariant with respect to invertible transforms and, consequently, needs to be modified when using scaled gradient updates. To that end, we introduce the following new projection operator: for every $\widetilde{F} \in \mathbb{R}^{(n_1+n_2)\times r} = [\widetilde{L}^\top, \widetilde{R}^\top]^\top$,

$$\begin{aligned} \mathcal{P}_B(\widetilde{F}) = \underset{F \in \mathbb{R}^{(n_1+n_2)\times r}}{\operatorname{argmin}} \quad & \|(L - \widetilde{L})(\widetilde{R}^\top \widetilde{R})^{1/2}\|_\mathsf{F}^2 + \|(R - \widetilde{R})(\widetilde{L}^\top \widetilde{L})^{1/2}\|_\mathsf{F}^2 \\ \text{s.t.} \quad & \sqrt{n_1} \left\| L(\widetilde{R}^\top \widetilde{R})^{1/2} \right\|_{2,\infty} \vee \sqrt{n_2} \left\| R(\widetilde{L}^\top \widetilde{L})^{1/2} \right\|_{2,\infty} \le B, \end{aligned} \tag{22}$$

which finds a factored matrix that is closest to $\widetilde{F}$ and stays incoherent in a weighted sense. Luckily, the solution to the above scaled projection admits a simple closed-form solution, given by

$$\begin{aligned} \mathcal{P}_B(\widetilde{F}) := \begin{bmatrix} L \\ R \end{bmatrix}, \quad \text{where } L_{i,\cdot} &:= \left(1 \wedge \frac{B}{\sqrt{n_1} \|\widetilde{L}_{i,\cdot} \widetilde{R}^\top\|_2}\right) \widetilde{L}_{i,\cdot}, \; 1 \le i \le n_1, \\ R_{j,\cdot} &:= \left(1 \wedge \frac{B}{\sqrt{n_2} \|\widetilde{R}_{j,\cdot} \widetilde{L}^\top\|_2}\right) \widetilde{R}_{j,\cdot}, \; 1 \le j \le n_2. \end{aligned} \tag{23}$$

With the new projection operator in place, we propose the following `ScaledGD` method with spectral initialization for solving matrix completion, formally stated in Algorithm 3.

Theoretical Guarantee The following theorem establishes the performance guarantee of `ScaledPGD` as long as the number of observations is sufficiently large.

Theorem 3 ([55]) *Suppose that* $X_\star$ *is* μ*-incoherent and that* p *satisfies* $p \ge C(\mu\kappa^2 \vee \log(n_1 \vee n_2))\mu r^2 \kappa^2/(n_1 \wedge n_2)$ *for some sufficiently large constant* C*. Set the projection radius as* $B = C_B\sqrt{\mu r}\sigma_1(X_\star)$ *for some constant* $C_B \ge 1.02$*. If the step size obeys* $0 < \eta \le 2/3$*, then with probability at least* $1 - c_1(n_1 \vee n_2)^{-c_2}$,

Algorithm 3 `ScaledGD` for matrix completion with spectral initialization

Spectral initialization: Let $U_0\Sigma_0 V_0^\top$ be the top-r SVD of $\frac{1}{p}\mathcal{P}_\Omega(X_\star)$, and set

$$\begin{bmatrix} L_0 \\ R_0 \end{bmatrix} = \mathcal{P}_B\left(\begin{bmatrix} U_0\Sigma_0^{1/2} \\ V_0\Sigma_0^{1/2} \end{bmatrix}\right). \tag{24}$$

Scaled projected gradient updates: for $t = 0, 1, 2, \ldots, T-1$ **do**

$$\begin{bmatrix} L_{t+1} \\ R_{t+1} \end{bmatrix} = \mathcal{P}_B\left(\begin{bmatrix} L_t - \frac{\eta}{p}\mathcal{P}_\Omega(L_t R_t^\top - X_\star) R_t (R_t^\top R_t)^{-1} \\ R_t - \frac{\eta}{p}\mathcal{P}_\Omega(L_t R_t^\top - X_\star)^\top L_t (L_t^\top L_t)^{-1} \end{bmatrix}\right). \tag{25}$$

for all $t \geq 0$, *the iterates of* `ScaledGD` *in* (25) *satisfy*

$$\left\| L_t R_t^\top - X_\star \right\|_{\mathsf{F}} \leq (1 - 0.6\eta)^t 0.03\sigma_r(X_\star).$$

Here $c_1, c_2 > 0$ *are two universal constants.*

Theorem 3 establishes that the reconstruction error $\|L_t R_t^\top - X_\star\|_{\mathsf{F}}$ contracts linearly at a constant rate, as long as the probability of observation satisfies $p \gtrsim (\mu\kappa^2 \vee \log(n_1 \vee n_2))\mu r^2\kappa^2/(n_1 \wedge n_2)$. To reach ϵ-accuracy, i.e., $\|L_t R_t^\top - X_\star\|_{\mathsf{F}} \leq \epsilon\sigma_r(X_\star)$, `ScaledPGD` takes at most $T = O(\log(1/\epsilon))$ iterations, which is *independent* of κ. In comparison, projected gradient descent [71] with spectral initialization converges in $O(\kappa\log(1/\epsilon))$ iterations as long as $p \gtrsim (\mu \vee \log(n_1 \vee n_2))\mu r^2\kappa^2/(n_1 \wedge n_2)$. Therefore, `ScaledGD` achieves much faster convergence than its unscaled counterpart, at a slightly higher sample complexity, which we believe can be further improved by finer analysis (cf. Table 1).

2.5 A Glimpse of the Analysis

At the heart of our analysis is a proper metric to measure the progress of the `ScaledGD` iterates $F_t := [L_t^\top, R_t^\top]^\top$. Obviously, the factored representation is not unique in that for any invertible matrix $Q \in \mathrm{GL}(r)$, one has $LR^\top = (LQ)(RQ^{-\top})^\top$. Therefore, the reconstruction error metric needs to take into account this identifiability issue. More importantly, we need a diagonal scaling in the distance error metric to properly account for the effect of preconditioning. To provide intuition, note that the update rule (3) can be viewed as finding the best local quadratic approximation of $\mathcal{L}(\cdot)$ in the following sense:

$$\begin{aligned}&(L_{t+1}, R_{t+1}) \\ &= \operatorname*{argmin}_{L,R}\ \mathcal{L}(L_t, R_t) + \langle \nabla_L \mathcal{L}(L_t, R_t), L - L_t\rangle + \langle \nabla_R \mathcal{L}(L_t, R_t), R - R_t\rangle\end{aligned}$$

$$+\frac{1}{2\eta}\left(\left\|(\boldsymbol{L}-\boldsymbol{L}_t)(\boldsymbol{R}_t^\top\boldsymbol{R}_t)^{1/2}\right\|_{\mathsf{F}}^2+\left\|(\boldsymbol{R}-\boldsymbol{R}_t)(\boldsymbol{L}_t^\top\boldsymbol{L}_t)^{1/2}\right\|_{\mathsf{F}}^2\right),$$

where it is different from the common interpretation of gradient descent in the way the quadratic approximation is taken by a scaled norm. When $\boldsymbol{L}_t \approx \boldsymbol{L}_\star$ and $\boldsymbol{R}_t \approx \boldsymbol{R}_\star$ are approaching the ground truth, the additional scaling factors can be approximated by $\boldsymbol{L}_t^\top\boldsymbol{L}_t \approx \boldsymbol{\Sigma}_\star$ and $\boldsymbol{R}_t^\top\boldsymbol{R}_t \approx \boldsymbol{\Sigma}_\star$, leading to the following error metric:

$$\mathrm{dist}^2(\boldsymbol{F},\boldsymbol{F}_\star) := \inf_{\boldsymbol{Q}\in\mathrm{GL}(r)} \left\|(\boldsymbol{L}\boldsymbol{Q}-\boldsymbol{L}_\star)\boldsymbol{\Sigma}_\star^{1/2}\right\|_{\mathsf{F}}^2+\left\|(\boldsymbol{R}\boldsymbol{Q}^{-\top}-\boldsymbol{R}_\star)\boldsymbol{\Sigma}_\star^{1/2}\right\|_{\mathsf{F}}^2. \tag{26}$$

The design and analysis of this new distance metric are of crucial importance in obtaining the improved rate of `ScaledGD`. In comparison, the previously studied distance metrics (proposed mainly for GD) either do not include the diagonal scaling [41, 59] or only consider the ambiguity class up to orthonormal transforms [59], which fail to unveil the benefit of `ScaledGD`.

3 ScaledGD for Low-Rank Tensor Estimation

This section is devoted to introducing `ScaledGD` and establishing its statistical and computational guarantees for various low-rank tensor estimation problems; the majority of the results are sampled from [21, 58].

3.1 Assumptions

Suppose the ground truth tensor $\mathcal{X}_\star = [\mathcal{X}_\star(i_1,i_2,i_3)] \in \mathbb{R}^{n_1\times n_2\times n_3}$ admits the following Tucker decomposition:

$$\mathcal{X}_\star(i_1,i_2,i_3)=\sum_{j_1=1}^{r_1}\sum_{j_2=1}^{r_2}\sum_{j_3=1}^{r_3}\boldsymbol{U}_\star(i_1,j_1)\boldsymbol{V}_\star(i_2,j_2)\boldsymbol{W}_\star(i_3,j_3)\mathcal{G}_\star(j_1,j_2,j_3) \tag{27}$$

for $1\le i_k\le n_k$, or more compactly,

$$\mathcal{X}_\star=(\boldsymbol{U}_\star,\boldsymbol{V}_\star,\boldsymbol{W}_\star)\cdot\mathcal{G}_\star, \tag{28}$$

where $\mathcal{G}_\star = [\mathcal{G}_\star(j_1,j_2,j_3)] \in \mathbb{R}^{r_1\times r_2\times r_3}$ is the core tensor of multilinear rank $\boldsymbol{r}=(r_1,r_2,r_3)$, and $\boldsymbol{U}_\star=[\boldsymbol{U}_\star(i_1,j_1)]\in\mathbb{R}^{n_1\times r_1}$, $\boldsymbol{V}_\star=[\boldsymbol{V}_\star(i_2,j_2)]\in\mathbb{R}^{n_2\times r_2}$, $\boldsymbol{W}_\star=[\boldsymbol{W}_\star(i_3,j_3)]\in\mathbb{R}^{n_3\times r_3}$ are the factor matrices of each mode. Letting $\mathcal{M}_k(\mathcal{X}_\star)$

be the mode-k matricization of $\mathcal{X}_\star$, we have

$$\mathcal{M}_1(\mathcal{X}_\star) = \boldsymbol{U}_\star \mathcal{M}_1(\mathcal{G}_\star)(\boldsymbol{W}_\star \otimes \boldsymbol{V}_\star)^\top, \tag{29a}$$

$$\mathcal{M}_2(\mathcal{X}_\star) = \boldsymbol{V}_\star \mathcal{M}_2(\mathcal{G}_\star)(\boldsymbol{W}_\star \otimes \boldsymbol{U}_\star)^\top, \tag{29b}$$

$$\mathcal{M}_3(\mathcal{X}_\star) = \boldsymbol{W}_\star \mathcal{M}_3(\mathcal{G}_\star)(\boldsymbol{V}_\star \otimes \boldsymbol{U}_\star)^\top. \tag{29c}$$

It is straightforward to see that the Tucker decomposition is not uniquely specified: for any invertible matrices $\boldsymbol{Q}_k \in \mathbb{R}^{r_k \times r_k}$, $k = 1, 2, 3$, one has

$$(\boldsymbol{U}_\star, \boldsymbol{V}_\star, \boldsymbol{W}_\star) \cdot \mathcal{G}_\star = (\boldsymbol{U}_\star \boldsymbol{Q}_1, \boldsymbol{V}_\star \boldsymbol{Q}_2, \boldsymbol{W}_\star \boldsymbol{Q}_3) \cdot ((\boldsymbol{Q}_1^{-1}, \boldsymbol{Q}_2^{-1}, \boldsymbol{Q}_3^{-1}) \cdot \mathcal{G}_\star).$$

We shall fix the ground truth factors such that $\boldsymbol{U}_\star$, $\boldsymbol{V}_\star$ and $\boldsymbol{W}_\star$ are orthonormal matrices consisting of left singular vectors in each mode. Furthermore, the core tensor $\mathcal{S}_\star$ is related to the singular values in each mode as

$$\mathcal{M}_k(\mathcal{G}_\star)\mathcal{M}_k(\mathcal{G}_\star)^\top = \boldsymbol{\Sigma}_{\star,k}^2, \qquad k = 1, 2, 3, \tag{30}$$

where $\boldsymbol{\Sigma}_{\star,k} := \mathrm{diag}[\sigma_1(\mathcal{M}_k(\mathcal{X}_\star)), \ldots, \sigma_{r_k}(\mathcal{M}_k(\mathcal{X}_\star))]$ is a diagonal matrix where the diagonal elements are composed of the nonzero singular values of $\mathcal{M}_k(\mathcal{X}_\star)$ and $r_k = \mathrm{rank}(\mathcal{M}_k(\mathcal{X}_\star))$ for $k = 1, 2, 3$.

Key Parameters Of particular interest is a sort of condition number of $\mathcal{X}_\star$, which plays an important role in governing the computational efficiency of first-order algorithms.

Definition 4 (Tensor Condition Number) The condition number of $\mathcal{X}_\star$ is defined as

$$\kappa := \frac{\sigma_{\max}(\mathcal{X}_\star)}{\sigma_{\min}(\mathcal{X}_\star)} = \frac{\max_{k=1,2,3} \sigma_1(\mathcal{M}_k(\mathcal{X}_\star))}{\min_{k=1,2,3} \sigma_{r_k}(\mathcal{M}_k(\mathcal{X}_\star))}. \tag{31}$$

Another parameter is the incoherence parameter, which plays an important role in governing the well-posedness of low-rank tensor RPCA and completion.

Definition 5 (Tensor Incoherence) The incoherence parameter of $\mathcal{X}_\star$ is defined as

$$\mu := \max\left\{\frac{n_1}{r_1}\|\boldsymbol{U}_\star\|_{2,\infty}^2, \frac{n_2}{r_2}\|\boldsymbol{V}_\star\|_{2,\infty}^2, \frac{n_3}{r_3}\|\boldsymbol{W}_\star\|_{2,\infty}^2\right\}. \tag{32}$$

Roughly speaking, a small incoherence parameter ensures that the energy of the tensor is evenly distributed across its entries, so that a small random subset of its elements still reveals substantial information about the latent structure of the entire tensor.

3.2 Tensor Sensing

Observation Model We first consider tensor sensing—also known as tensor regression—with Gaussian design. Assume that we have access to a set of observations given as

$$y_i = \langle \mathcal{A}_i, \mathcal{X}_\star \rangle, \quad i = 1, \dots, m, \qquad \text{or concisely,} \qquad \boldsymbol{y} = \mathcal{A}(\mathcal{X}_\star), \tag{33}$$

where $\mathcal{A}_i \in \mathbb{R}^{n_1 \times n_2 \times n_3}$ is the ith measurement tensor composed of i.i.d. Gaussian entries drawn from $\mathcal{N}(0, 1/m)$, and $\mathcal{A}(\mathcal{X}) = \{\langle \mathcal{A}_i, \mathcal{X} \rangle\}_{i=1}^m$ is a linear map from $\mathbb{R}^{n_1 \times n_2 \times n_3}$ to $\mathbb{R}^m$, whose adjoint operator is given by $\mathcal{A}^*(\boldsymbol{y}) = \sum_{i=1}^m y_i \mathcal{A}_i$. The goal is to recover $\mathcal{X}_\star$ from $\boldsymbol{y}$, by leveraging the low-rank structure of $\mathcal{X}_\star$.

Algorithm Development It is natural to minimize the following loss function:

$$\underset{\boldsymbol{F}=(\boldsymbol{U},\boldsymbol{V},\boldsymbol{W},\mathcal{G})}{\text{minimize}} \quad \mathcal{L}(\boldsymbol{F}) := \frac{1}{2} \left\| \mathcal{A}((\boldsymbol{U}, \boldsymbol{V}, \boldsymbol{W}) \cdot \mathcal{G}) - \boldsymbol{y} \right\|_2^2. \tag{34}$$

The proposed `ScaledGD` algorithm to minimize (34) is described in Algorithm 4, where the algorithm is initialized by applying HOSVD to $\mathcal{A}^*(\boldsymbol{y})$, followed by scaled gradient updates given in (35).

Theoretical Guarantee Encouragingly, we can guarantee that `ScaledGD` provably recovers the ground truth tensor as long as the sample size is sufficiently large, which is given in the following theorem.

Algorithm 4 `ScaledGD` for low-rank tensor sensing

Input parameters: step size η, multilinear rank $\boldsymbol{r} = (r_1, r_2, r_3)$.

Spectral initialization: Let $(\boldsymbol{U}_0, \boldsymbol{V}_0, \boldsymbol{W}_0, \mathcal{G}_0)$ be the factors in the top-$\boldsymbol{r}$ HOSVD of $\mathcal{A}^*(\boldsymbol{y})$ (cf. (7)).

Scaled gradient updates: for $t = 0, 1, 2, \dots, T-1$

$$\begin{aligned}
\boldsymbol{U}_{t+1} &= \boldsymbol{U}_t - \eta \mathcal{M}_1\left(\mathcal{A}^*(\mathcal{A}((\boldsymbol{U}_t, \boldsymbol{V}_t, \boldsymbol{W}_t) \cdot \mathcal{G}_t) - \boldsymbol{y})\right) \breve{\boldsymbol{U}}_t^\top \left(\breve{\boldsymbol{U}}_t^\top \breve{\boldsymbol{U}}_t\right)^{-1}, \\
\boldsymbol{V}_{t+1} &= \boldsymbol{V}_t - \eta \mathcal{M}_2\left(\mathcal{A}^*(\mathcal{A}((\boldsymbol{U}_t, \boldsymbol{V}_t, \boldsymbol{W}_t) \cdot \mathcal{G}_t) - \boldsymbol{y})\right) \breve{\boldsymbol{V}}_t^\top \left(\breve{\boldsymbol{V}}_t^\top \breve{\boldsymbol{V}}_t\right)^{-1}, \\
\boldsymbol{W}_{t+1} &= \boldsymbol{W}_t - \eta \mathcal{M}_3\left(\mathcal{A}^*(\mathcal{A}((\boldsymbol{U}_t, \boldsymbol{V}_t, \boldsymbol{W}_t) \cdot \mathcal{G}_t) - \boldsymbol{y})\right) \breve{\boldsymbol{W}}_t^\top \left(\breve{\boldsymbol{W}}_t^\top \breve{\boldsymbol{W}}_t\right)^{-1}, \\
\mathcal{G}_{t+1} &= \mathcal{G}_t - \eta \left((\boldsymbol{U}_t^\top \boldsymbol{U}_t)^{-1} \boldsymbol{U}_t^\top, (\boldsymbol{V}_t^\top \boldsymbol{V}_t)^{-1} \boldsymbol{V}_t^\top, (\boldsymbol{W}_t^\top \boldsymbol{W}_t)^{-1} \boldsymbol{W}_t^\top\right) \\
&\qquad \cdot \mathcal{A}^*(\mathcal{A}((\boldsymbol{U}_t, \boldsymbol{V}_t, \boldsymbol{W}_t) \cdot \mathcal{G}_t) - \boldsymbol{y}),
\end{aligned} \tag{35}$$

where $\breve{\boldsymbol{U}}_t$, $\breve{\boldsymbol{V}}_t$, and $\breve{\boldsymbol{W}}_t$ are defined in (6).

Theorem 4 ([58]) *Let* $n = \max_{k=1,2,3} n_k$ *and* $r = \max_{k=1,2,3} r_k$. *With Gaussian design, suppose that* m *satisfies*

$$m \gtrsim \epsilon_0^{-1}\sqrt{n_1 n_2 n_3} r^{3/2}\kappa^2 + \epsilon_0^{-2}(nr^2\kappa^4 \log n + r^4\kappa^2)$$

for some small constant $\epsilon_0 > 0$. *If the step size obeys* $0 < \eta \le 2/5$, *then with probability at least* $1 - c_1 n^{-c_2}$ *for universal constants* $c_1, c_2 > 0$, *for all* $t \ge 0$, *the iterates of Algorithm 4 satisfy*

$$\|(\boldsymbol{U}_t, \boldsymbol{V}_t, \boldsymbol{W}_t) \cdot \mathcal{G}_t - \mathcal{X}_\star\|_{\mathsf{F}} \le 3\epsilon_0(1 - 0.6\eta)^t \sigma_{\min}(\mathcal{X}_\star).$$

Theorem 4 ensures that the reconstruction error $\|(\boldsymbol{U}_t, \boldsymbol{V}_t, \boldsymbol{W}_t) \cdot \mathcal{G}_t - \mathcal{X}_\star\|_{\mathsf{F}}$ contracts linearly at a constant rate independent of the condition number of $\mathcal{X}_\star$; to find an ε-accurate estimate, i.e., $\|(\boldsymbol{U}_t, \boldsymbol{V}_t, \boldsymbol{W}_t) \cdot \mathcal{G}_t - \mathcal{X}_\star\|_{\mathsf{F}} \le \varepsilon\sigma_{\min}(\mathcal{X}_\star)$, `ScaledGD` needs at most $O(\log(1/\varepsilon))$ iterations, as long as the sample complexity satisfies

$$m \gtrsim n^{3/2} r^{3/2} \kappa^2,$$

where again we keep only terms with dominating orders of n. Compared with regularized GD [28], `ScaledGD` achieves a low computation complexity with robustness to ill-conditioning, improving its iteration complexity by a factor of κ^2, and does not require any explicit regularization.

3.3 Tensor Robust Principal Component Analysis

Observation Model Suppose that we collect a set of corrupted observations of $\mathcal{X}_\star$ as

$$\mathcal{Y} = \mathcal{X}_\star + \mathcal{S}_\star, \tag{36}$$

where $\mathcal{S}_\star$ is the corruption tensor. The problem of tensor RPCA seeks to separate $\mathcal{X}_\star$ and $\mathcal{S}_\star$ from their sum $\mathcal{Y}$ as efficiently and accurately as possible. Similar to the matrix case, we consider a deterministic sparsity model following the matrix case [10, 45, 67], where $\mathcal{S}_\star$ contains at most a small fraction of nonzero entries per fiber. Formally, the corruption tensor $\mathcal{S}_\star$ is said to be α-fraction sparse, i.e., $\mathcal{S}_\star \in \mathcal{S}_\alpha$, where

$$\mathcal{S}_\alpha := \Big\{ \mathcal{S} \in \mathbb{R}^{n_1 \times n_2 \times n_3} : \ \|\mathcal{S}_{i_1,i_2,:}\|_0 \le \alpha n_3, \ \|\mathcal{S}_{i_1,:,i_3}\|_0 \le \alpha n_2, \\ \|\mathcal{S}_{:,i_2,i_3}\|_0 \le \alpha n_1, \ \text{for all } 1 \le i_k \le n_k, \quad k = 1, 2, 3 \Big\}. \tag{37}$$

ScaledGD Algorithm It is natural to optimize the following objective function:

$$\underset{F,\mathcal{S}}{\text{minimize}}\ \mathcal{L}(F,\mathcal{S}) := \frac{1}{2}\left\|\left(U,V,W\right)\cdot\mathcal{G}+\mathcal{S}-\mathcal{Y}\right\|_{\mathsf{F}}^2, \tag{38}$$

where $F = (U, V, W, \mathcal{G})$ and $\mathcal{S}$ are the optimization variables for the tensor factors and the corruption tensor, respectively. Our algorithm alternates between corruption removal and factor refinements, as detailed in Algorithm 5. To remove the corruption, we use the following soft-shrinkage operator that trims the magnitudes of the entries by the amount of some carefully preset threshold.

Definition 6 (Soft-Shrinkage Operator) For an order-3 tensor $\mathcal{X}$, the soft-shrinkage operator $\mathcal{T}_{\zeta}^{\mathsf{soft}}[\cdot] : \mathbb{R}^{n_1\times n_2\times n_3} \mapsto \mathbb{R}^{n_1\times n_2\times n_3}$ with threshold $\zeta > 0$ is defined as

$$\left[\mathcal{T}_{\zeta}^{\mathsf{soft}}[\mathcal{X}]\right]_{i_1,i_2,i_3} := \operatorname{sgn}\left([\mathcal{X}]_{i_1,i_2,i_3}\right)\cdot\max\left(0, \left|[\mathcal{X}]_{i_1,i_2,i_3}\right|-\zeta\right).$$

The soft-shrinkage operator sets entries with magnitudes smaller than ζ to 0, while uniformly shrinking the magnitudes of the other entries by ζ. At the beginning of each iteration, the corruption tensor is updated via applying the soft-thresholding operator to the current residual $\mathcal{Y}-\left(U_t, V_t, W_t\right)\cdot\mathcal{G}_t$ using some properly selected threshold ζ_t, followed by updating the tensor factors F_t via scaled gradient descent with respect to $\mathcal{L}(F_t, \mathcal{S}_{t+1})$ in (38). To complete the algorithm description, we still need to specify how to initialize the algorithm. This is again achieved by the spectral method, which computes the rank-$\boldsymbol{r}$ HOSVD of the observation after applying the soft-shrinkage operator

---

Algorithm 5 `ScaledGD` for tensor robust principal component analysis

---

Input: observed tensor $\mathcal{Y}$, multilinear rank $\boldsymbol{r}$, learning rate η, and threshold schedule $\{\zeta_t\}_{t=0}^{T}$.
Spectral initialization: Set $\mathcal{S}_0 = \mathcal{T}_{\zeta_0}^{\mathsf{soft}}[\mathcal{Y}]$ and $\left(U_0, V_0, W_0, \mathcal{G}_0\right)$ as the factors in the top-$\boldsymbol{r}$ HOSVD of $\mathcal{Y}-\mathcal{S}_0$ (cf. (7)).
Scaled gradient updates: **for** $t = 0, 1, 2, \ldots, T-1$ **do**

$$\mathcal{S}_{t+1} = \mathcal{T}_{\zeta_{t+1}}^{\mathsf{soft}}\left[\mathcal{Y}-\left(U_t, V_t, W_t\right)\cdot\mathcal{G}_t\right], \tag{39a}$$

$$U_{t+1} = U_t - \eta\left(U_t\breve{U}_t^{\top} + \mathcal{M}_1(\mathcal{S}_{t+1}) - \mathcal{M}_1(\mathcal{Y})\right)\breve{U}_t(\breve{U}_t^{\top}\breve{U}_t)^{-1}, \tag{39b}$$

$$V_{t+1} = V_t - \eta\left(V_t\breve{V}_t^{\top} + \mathcal{M}_2(\mathcal{S}_{t+1}) - \mathcal{M}_2(\mathcal{Y})\right)\breve{V}_t(\breve{V}_t^{\top}\breve{V}_t)^{-1}, \tag{39c}$$

$$W_{t+1} = W_t - \eta\left(W_t\breve{W}_t^{\top} + \mathcal{M}_1(\mathcal{S}_{t+1}) - \mathcal{M}_1(\mathcal{Y})\right)\breve{W}_t(\breve{W}_t^{\top}\breve{W}_t)^{-1}, \tag{39d}$$

$$\begin{aligned}\mathcal{G}_{t+1} = \mathcal{G}_t - \eta&\left((U_t^{\top}U_t)^{-1}U_t^{\top}, (V_t^{\top}V_t)^{-1}V_t^{\top}, (W_t^{\top}W_t)^{-1}W_t^{\top}\right)\\ &\cdot\left(\left(U_t, V_t, W_t\right)\cdot\mathcal{G}_t + \mathcal{S}_{t+1} - \mathcal{Y}\right),\end{aligned} \tag{39e}$$

where $\breve{U}_t$, $\breve{V}_t$, and $\breve{W}_t$ are defined in (6).

Theoretical Guarantee Fortunately, the `ScaledGD` algorithm provably recovers the ground truth tensor—as long as the fraction of corruptions is not too large—with proper choices of the tuning parameters, as captured in following theorem.

Theorem 5 ([21]) *Let $\mathcal{Y} = \mathcal{X}_\star + \mathcal{S}_\star \in \mathbb{R}^{n_1 \times n_2 \times n_3}$, where $\mathcal{X}_\star$ is μ-incoherent with multilinear rank $\boldsymbol{r} = (r_1, r_2, r_3)$, and $\mathcal{S}_\star$ is α-sparse. Suppose that the threshold values $\{\zeta_k\}_{k=0}^{\infty}$ obey that $\|\mathcal{X}_\star\|_\infty \le \zeta_0 \le 2\,\|\mathcal{X}_\star\|_\infty$ and $\zeta_{t+1} = \rho\zeta_t$, $t \ge 1$, for some properly tuned $\zeta_1 := 8\sqrt{\frac{\mu^3 r_1 r_2 r_3}{n_1 n_2 n_3}}\sigma_{\min}(\mathcal{X}_\star)$ and $\frac{1}{7} \le \eta \le \frac{1}{4}$, where $\rho = 1 - 0.45\eta$. Then, the iterates $\mathcal{X}_t = \big(\boldsymbol{U}_t, \boldsymbol{V}_t, \boldsymbol{W}_t\big) \cdot \mathcal{G}_t$ satisfy*

$$\|\mathcal{X}_t - \mathcal{X}_\star\|_{\mathsf{F}} \le 0.03\rho^t \sigma_{\min}(\mathcal{X}_\star), \tag{40a}$$

$$\|\mathcal{X}_t - \mathcal{X}_\star\|_\infty \le 8\rho^t \sqrt{\frac{\mu^3 r_1 r_2 r_3}{n_1 n_2 n_3}}\sigma_{\min}(\mathcal{X}_\star) \tag{40b}$$

for all $t \ge 0$, as long as the level of corruptions obeys $\alpha \le \frac{c_0}{\mu^2 r_1 r_2 r_3 \kappa}$ for some sufficiently small $c_0 > 0$.

Theorem 5 implies that upon appropriate choices of the parameters, if the level of corruptions α is small enough, i.e., not exceeding the order of $\frac{1}{\mu^2 r_1 r_2 r_3 \kappa}$, we can ensure that `ScaledGD` converges at a linear rate independent of the condition number and exactly recovers the ground truth tensor $\mathcal{X}_\star$ even when the gross corruptions are arbitrary and adversarial. Furthermore, when $\mu = O(1)$ and $r = O(1)$, the entrywise error bound (40b)—which is smaller than the Frobenius error (40a) by a factor of $\sqrt{\frac{1}{n_1 n_2 n_3}}$—suggests the errors are distributed in an even manner across the entries for incoherent and low-rank tensors.

3.4 Tensor Completion

Observation Model Assume that we have observed a subset of entries in $\mathcal{X}_\star$, given as $\mathcal{Y} = \mathcal{P}_\Omega(\mathcal{X}_\star)$, where $\mathcal{P}_\Omega : \mathbb{R}^{n_1 \times n_2 \times n_3} \mapsto \mathbb{R}^{n_1 \times n_2 \times n_3}$ is a projection such that

$$[\mathcal{P}_\Omega(\mathcal{X}_\star)](i_1, i_2, i_3) = \begin{cases} \mathcal{X}_\star(i_1, i_2, i_3), & \text{if } (i_1, i_2, i_3) \in \Omega, \\ 0, & \text{otherwise.} \end{cases} \tag{41}$$

Here, Ω is generated according to the Bernoulli observation model in the sense that

$$(i_1, i_2, i_3) \in \Omega \quad \text{independently with probability } p \in (0, 1]. \tag{42}$$

The goal is to recover the tensor $\mathcal{X}_\star$ from its partial observation $\mathcal{P}_\Omega(\mathcal{X}_\star)$, which can be achieved by minimizing the loss function

$$\min_{F=(U,V,W,\mathcal{S})} \mathcal{L}(F) := \frac{1}{2p} \left\| \mathcal{P}_\Omega\big((U, V, W) \cdot \mathcal{S}\big) - \mathcal{Y} \right\|_{\mathsf{F}}^2. \tag{43}$$

Algorithm Development To guarantee faithful recovery from partial observations, the underlying low-rank tensor $\mathcal{X}_\star$ needs to be incoherent (cf. Definition 5) to avoid ill-posedness. One typical strategy, as employed in the matrix setting, to ensure the incoherence condition is to trim the rows of the factors after the scaled gradient update. We introduce the scaled projection as follows:

$$(U, V, W, \mathcal{S}) = \mathcal{P}_B(U_+, V_+, W_+, \mathcal{S}_+), \tag{44}$$

where $B > 0$ is the projection radius, and

$$\begin{aligned}
U(i_1, :) &= \left(1 \wedge \frac{B}{\sqrt{n_1}\|U_+(i_1, :)\breve{U}_+^\top\|_2}\right) U_+(i_1, :), \qquad 1 \le i_1 \le n_1; \\
V(i_2, :) &= \left(1 \wedge \frac{B}{\sqrt{n_2}\|V_+(i_2, :)\breve{V}_+^\top\|_2}\right) V_+(i_2, :), \qquad 1 \le i_2 \le n_2; \\
W(i_3, :) &= \left(1 \wedge \frac{B}{\sqrt{n_3}\|W_+(i_3, :)\breve{W}_+^\top\|_2}\right) W_+(i_3, :), \qquad 1 \le i_3 \le n_3; \\
\mathcal{S} &= \mathcal{S}_+.
\end{aligned}$$

Here, we recall $\breve{U}_+$, $\breve{V}_+$, $\breve{W}_+$ are analogously defined in (6) using $(U_+, V_+, W_+, \mathcal{S}_+)$. As can be seen, each row of U_+ (resp. V_+ and W_+) is scaled by a scalar based on the row ℓ_2 norms of $U_+\breve{U}_+^\top$ (resp. $V_+\breve{V}_+^\top$ and $W_+\breve{W}_+^\top$), which is the mode-1 (resp. mode-2 and mode-3) matricization of the tensor $(U_+, V_+, W_+) \cdot \mathcal{S}_+$. It is a straightforward observation that the projection can be computed efficiently.

With the scaled projection $\mathcal{P}_B(\cdot)$ defined in hand, we are in a position to describe the details of the proposed `ScaledGD` algorithm, summarized in Algorithm 6. It consists of two stages: spectral initialization followed by iterative refinements using the scaled projected gradient updates in (45). For the spectral initialization, we take advantage of the subspace estimators proposed in [5, 64] for highly unbalanced matrices. Specifically, we estimate the subspace spanned by $U_\star$ by that spanned by top-r_1 eigenvectors U_+ of the diagonally deleted Gram matrix of $p^{-1}\mathcal{M}_1(\mathcal{Y})$, denoted as

$$\mathcal{P}_{\mathsf{off\text{-}diag}}(p^{-2}\mathcal{M}_1(\mathcal{Y})\mathcal{M}_1(\mathcal{Y})^\top),$$

Algorithm 6 ScaledGD for low-rank tensor completion

Input parameters: step size η, multilinear rank $\boldsymbol{r} = (r_1, r_2, r_3)$, probability of observation p, projection radius B.
Spectral initialization: Let $\boldsymbol{U}_+$ be the top-r_1 eigenvectors of $\mathcal{P}_{\text{off-diag}}(p^{-2}\mathcal{M}_1(\mathcal{Y})\mathcal{M}_1(\mathcal{Y})^\top)$, and similarly for $\boldsymbol{V}_+$, $\boldsymbol{W}_+$, and $\mathcal{S}_+ = p^{-1}(\boldsymbol{U}_+^\top, \boldsymbol{V}_+^\top, \boldsymbol{W}_+^\top) \cdot \mathcal{Y}$. Set $(\boldsymbol{U}_0, \boldsymbol{V}_0, \boldsymbol{W}_0, \mathcal{S}_0) = \mathcal{P}_B(\boldsymbol{U}_+, \boldsymbol{V}_+, \boldsymbol{W}_+, \mathcal{S}_+)$.
Scaled gradient updates: for $t = 0, 1, 2, \dots, T-1$, **do**

$$\begin{aligned}
\boldsymbol{U}_{t+} &= \boldsymbol{U}_t - \frac{\eta}{p}\mathcal{M}_1\left(\mathcal{P}_\Omega\big((\boldsymbol{U}_t, \boldsymbol{V}_t, \boldsymbol{W}_t)\cdot\mathcal{S}_t\big) - \mathcal{Y}\right)\breve{\boldsymbol{U}}_t(\breve{\boldsymbol{U}}_t^\top\breve{\boldsymbol{U}}_t)^{-1},\\
\boldsymbol{V}_{t+} &= \boldsymbol{V}_t - \frac{\eta}{p}\mathcal{M}_2\left(\mathcal{P}_\Omega\big((\boldsymbol{U}_t, \boldsymbol{V}_t, \boldsymbol{W}_t)\cdot\mathcal{S}_t\big) - \mathcal{Y}\right)\breve{\boldsymbol{V}}_t(\breve{\boldsymbol{V}}_t^\top\breve{\boldsymbol{V}}_t)^{-1},\\
\boldsymbol{W}_{t+} &= \boldsymbol{W}_t - \frac{\eta}{p}\mathcal{M}_3\left(\mathcal{P}_\Omega\big((\boldsymbol{U}_t, \boldsymbol{V}_t, \boldsymbol{W}_t)\cdot\mathcal{S}_t\big) - \mathcal{Y}\right)\breve{\boldsymbol{W}}_t(\breve{\boldsymbol{W}}_t^\top\breve{\boldsymbol{W}}_t)^{-1},\\
\mathcal{S}_{t+} &= \mathcal{S}_t - \frac{\eta}{p}\left((\boldsymbol{U}_t^\top\boldsymbol{U}_t)^{-1}\boldsymbol{U}_t^\top, (\boldsymbol{V}_t^\top\boldsymbol{V}_t)^{-1}\boldsymbol{V}_t^\top, (\boldsymbol{W}_t^\top\boldsymbol{W}_t)^{-1}\boldsymbol{W}_t^\top\right)\\
&\qquad \cdot\left(\mathcal{P}_\Omega\big((\boldsymbol{U}_t, \boldsymbol{V}_t, \boldsymbol{W}_t)\cdot\mathcal{S}_t\big) - \mathcal{Y}\right),
\end{aligned} \tag{45}$$

where $\breve{\boldsymbol{U}}_t$, $\breve{\boldsymbol{V}}_t$, and $\breve{\boldsymbol{W}}_t$ are defined in (6). Set

$$(\boldsymbol{U}_{t+1}, \boldsymbol{V}_{t+1}, \boldsymbol{W}_{t+1}, \mathcal{S}_{t+1}) = \mathcal{P}_B(\boldsymbol{U}_{t+}, \boldsymbol{V}_{t+}, \boldsymbol{W}_{t+}, \mathcal{S}_{t+}).$$

---

and the other two factors $\boldsymbol{V}_+$ and $\boldsymbol{W}_+$ are estimated similarly. The core tensor is then estimated as

$$\mathcal{S}_+ = p^{-1}(\boldsymbol{U}_+^\top, \boldsymbol{V}_+^\top, \boldsymbol{W}_+^\top)\cdot\mathcal{Y}.$$

To ensure the initialization is incoherent, we pass it through the scaled projection operator to obtain the final initial estimate:

$$(\boldsymbol{U}_0, \boldsymbol{V}_0, \boldsymbol{W}_0, \mathcal{S}_0) = \mathcal{P}_B\big(\boldsymbol{U}_+, \boldsymbol{V}_+, \boldsymbol{W}_+, \mathcal{S}_+\big).$$

Theoretical Guarantee The following theorem establishes the performance guarantee of ScaledGD for tensor completion, as soon as the sample size is sufficiently large.

Theorem 6 ([58]) *Let $n = \max_{k=1,2,3} n_k$ and $r = \max_{k=1,2,3} r_k$. Suppose that $\mathcal{X}_\star$ is μ-incoherent, $n_k \gtrsim \epsilon_0^{-1}\mu r_k^{3/2}\kappa^2$ for $k = 1, 2, 3$, and that p satisfies*

$$pn_1n_2n_3 \gtrsim \epsilon_0^{-1}\sqrt{n_1n_2n_3}\mu^{3/2}r^{5/2}\kappa^3\log^3 n + \epsilon_0^{-2}n\mu^3r^4\kappa^6\log^5 n$$

for some small constant $\epsilon_0 > 0$. Set the projection radius as $B = C_B\sqrt{\mu r}\sigma_{\max}(\mathcal{X}_\star)$ for some constant $C_B \geq (1+\epsilon_0)^3$. If the step size obeys $0 < \eta \leq 2/5$, then with probability at least $1 - c_1 n^{-c_2}$ for universal constants $c_1, c_2 > 0$, for all $t \geq 0$, the

Table 2 Comparisons of ScaledGD with existing algorithms for tensor completion when the tensor is incoherent and low rank under the Tucker decomposition. Here, we say that the output $\mathcal{X}$ of an algorithm reaches ε-accuracy, if it satisfies $\|\mathcal{X} - \mathcal{X}_\star\|_\mathsf{F} \le \varepsilon\sigma_{\min}(\mathcal{X}_\star)$. Here, κ and $\sigma_{\min}(\mathcal{X}_\star)$ are the condition number and the minimum singular value of $\mathcal{X}_\star$ (defined in Sect. 3.1). For simplicity, we let $n = \max_{k=1,2,3} n_k$ and $r = \max_{k=1,2,3} r_k$ and assume $r \vee \kappa \ll n^\delta$ for some small constant δ to keep only terms with dominating orders of n

Algorithms	Sample complexity	Iteration complexity	Parameter space
Unfolding + nuclear norm min. [30]	$n^2 r \log^2 n$	polynomial	tensor
Tensor nuclear norm min. [68]	$n^{3/2} r^{1/2} \log^{3/2} n$	NP-hard	tensor
Grassmannian GD [63]	$n^{3/2} r^{7/2} \kappa^4 \log^{7/2} n$	N/A	factor
`ScaledGD`	$n^{3/2} r^{5/2} \kappa^3 \log^3 n$	$\log \frac{1}{\varepsilon}$	factor

iterates of Algorithm 6 satisfy

$$\|(\boldsymbol{U}_t, \boldsymbol{V}_t, \boldsymbol{W}_t) \cdot \mathcal{S}_t - \mathcal{X}_\star\|_\mathsf{F} \le 3\epsilon_0(1 - 0.6\eta)^t \sigma_{\min}(\mathcal{X}_\star).$$

Theorem 6 ensures that to find an ε-accurate estimate, i.e., $\|(\boldsymbol{U}_t, \boldsymbol{V}_t, \boldsymbol{W}_t) \cdot \mathcal{S}_t - \mathcal{X}_\star\|_\mathsf{F} \le \varepsilon\sigma_{\min}(\mathcal{X}_\star)$, `ScaledGD` takes at most $O(\log(1/\varepsilon))$ iterations, which is independent of the condition number of $\mathcal{X}_\star$, as long as the sample complexity is large enough. Assuming that $\mu = O(1)$ and $r \vee \kappa \ll n^\delta$ for some small constant δ to keep only terms with dominating orders of n, the sample complexity simplifies to

$$pn_1n_2n_3 \gtrsim n^{3/2} r^{5/2} \kappa^3 \log^3 n,$$

which is near optimal in view of the conjecture that no polynomial-time algorithm will be successful if the sample complexity is less than the order of $n^{3/2}$ for tensor completion [2]. Compared with existing algorithms collected in Table 2, `ScaledGD` is the *first* algorithm that simultaneously achieves a near-optimal sample complexity and a near-linear run time complexity in a provable manner.

4 Preconditioning Meets Overparameterization

In this section we treat a more complicated scenario, where the correct rank r of the ground truth $\boldsymbol{X}_\star$ is not known a priori . In this case, a practical solution is to *overparameterize*, i.e., to choose some $r' > r$, and proceed as if r' is the correct rank. It turns out that `ScaledGD` needs some simple modification to work robustly in this setting. The modified algorithm, which we call `ScaledGD(λ)`, will be introduced in the rest of this section, along with theoretical analysis on its global convergence.

Motivation We begin with inspecting the behavior of `ScaledGD` in the overparameterized setting. Assume we already find some $\boldsymbol{L}_t \in \mathbb{R}^{n_1 \times r'}$, $\boldsymbol{R}_t \in \mathbb{R}^{n_2 \times r'}$ that are close to the ground truth: the first r columns of $\boldsymbol{L}_t$ are close to $\boldsymbol{L}_\star$, while the rest $r' - r$ columns are close to zero, mutatis mutandis for $\boldsymbol{R}_t$.

Recall that in the update Eq. (3) of `ScaledGD`:

$$\boldsymbol{L}_{t+1} = \boldsymbol{L}_t - \eta \nabla_{\boldsymbol{L}} \mathcal{L}(\boldsymbol{L}_t, \boldsymbol{R}_t)(\boldsymbol{R}_t^\top \boldsymbol{R}_t)^{-1},$$
$$\boldsymbol{R}_{t+1} = \boldsymbol{R}_t - \eta \nabla_{\boldsymbol{R}} \mathcal{L}(\boldsymbol{L}_t, \boldsymbol{R}_t)(\boldsymbol{L}_t^\top \boldsymbol{L}_t)^{-1},$$

the preconditioners are chosen to be the inverse of the $r' \times r'$ matrices $\boldsymbol{L}_t^\top \boldsymbol{L}_t$, $\boldsymbol{R}_t^\top \boldsymbol{R}_t$. However, since the last $r' - r$ columns for $\boldsymbol{L}_t$ (respectively, $\boldsymbol{R}_t$) are close to zero, it is clear that $\boldsymbol{L}_t^\top \boldsymbol{L}_t$ (respectively, $\boldsymbol{R}_t^\top \boldsymbol{R}_t$) are approximately of rank at most r. When $r' > r$, this means $\boldsymbol{L}_t^\top \boldsymbol{L}_t$ and $\boldsymbol{R}_t^\top \boldsymbol{R}_t$ are close to being degenerate since their approximate ranks (no larger than r) are smaller than their dimensions r', thus taking inverse of them are numerically unstable.

`ScaledGD(λ)`: `ScaledGD` with Overparameterization One of the simplest remedies to such instability is to *regularize* the preconditioner. Before taking inverse of $\boldsymbol{L}_t^\top \boldsymbol{L}_t$ and $\boldsymbol{R}_t^\top \boldsymbol{R}_t$, we add a regularizer $\lambda \boldsymbol{I}$ to avoid degeneracy, where $\lambda > 0$ is a regularization parameter; the preconditioner thus becomes $(\boldsymbol{L}_t^\top \boldsymbol{L}_t + \lambda \boldsymbol{I})^{-1}$ and $(\boldsymbol{R}_t^\top \boldsymbol{R}_t + \lambda \boldsymbol{I})^{-1}$. The new update rule is specified by

$$\boldsymbol{L}_{t+1} = \boldsymbol{L}_t - \eta \nabla_{\boldsymbol{L}} \mathcal{L}(\boldsymbol{L}_t, \boldsymbol{R}_t)(\boldsymbol{R}_t^\top \boldsymbol{R}_t + \lambda \boldsymbol{I})^{-1},$$
$$\boldsymbol{R}_{t+1} = \boldsymbol{R}_t - \eta \nabla_{\boldsymbol{R}} \mathcal{L}(\boldsymbol{L}_t, \boldsymbol{R}_t)(\boldsymbol{L}_t^\top \boldsymbol{L}_t + \lambda \boldsymbol{I})^{-1}. \tag{46}$$

This regularized version of `ScaledGD` is called `ScaledGD(λ)`. It turns out that the simple regularization trick works out well: `ScaledGD(λ)` not only (almost) inherits the κ-free convergence rate, but also has the advantage of being robust to overparameterization and enjoying provable global convergence to any prescribed accuracy level, when initialized by a small random initialization. This is established in [66] formally for the matrix sensing setting studied in Sect. 2.4, where the two factors $\boldsymbol{L}_\star$, $\boldsymbol{R}_\star$ in $\boldsymbol{X}_\star = \boldsymbol{L}_\star \boldsymbol{R}_\star^\top$ are equal, i.e., $\boldsymbol{X}_\star = \boldsymbol{L}_\star \boldsymbol{L}_\star^\top$. Under such situation, `ScaledGD(λ)` instantiates to

$$\boldsymbol{L}_{t+1} = \boldsymbol{L}_t - \eta \mathcal{A}^*(\mathcal{A}(\boldsymbol{L}_t \boldsymbol{L}_t^\top) - \boldsymbol{y})\boldsymbol{L}_t(\boldsymbol{L}_t^\top \boldsymbol{L}_t + \lambda \boldsymbol{I})^{-1}. \tag{47}$$

Similar to [39, 51], we set the initialization $\boldsymbol{L}_0$ to be a small random matrix, i.e.,

$$\boldsymbol{L}_0 = \alpha \boldsymbol{G}, \tag{48}$$

where $\boldsymbol{G} \in \mathbb{R}^{n \times r'}$ is some matrix considered to be normalized and $\alpha > 0$ controls the magnitude of the initialization. To simplify exposition, we take $\boldsymbol{G}$ to be a standard random Gaussian matrix, that is, $\boldsymbol{G}$ is a random matrix with i.i.d. entries distributed as $\mathcal{N}(0, 1/n)$.

Theoretical Guarantee We have the performance guarantee of `ScaledGD(λ)` under the standard RIP assumption as follows.

Theorem 7 ([66]) *Suppose that $\mathcal{A}(\cdot)$ satisfies the rank-$(r+1)$ RIP with $\delta_{r+1} =: \delta$. Furthermore, there exist a sufficiently small constant $c_\delta > 0$ and a sufficiently large constant $C_\delta > 0$ such that*

$$\delta \le c_\delta r^{-1/2} \kappa^{-C_\delta}. \tag{49}$$

Assume there exist some universal constants $c_\eta, c_\lambda, C_\alpha > 0$ such that (η, λ, α) in `ScaledGD(λ)` *satisfy the following conditions:*

$$\text{(Learning rate)} \qquad \eta \le c_\eta, \tag{50a}$$

$$\text{(Damping parameter)} \qquad \frac{1}{100} c_\lambda \sigma_{\min}^2(\boldsymbol{L}_\star) \le \lambda \le c_\lambda \sigma_{\min}^2(\boldsymbol{L}_\star), \tag{50b}$$

$$\text{(Initialization size)} \qquad \log \frac{\|\boldsymbol{L}_\star\|}{\alpha} \ge \frac{C_\alpha}{\eta} \log(2\kappa) \cdot \log(2\kappa n). \tag{50c}$$

With high probability (with respect to the realization of the random initialization $\boldsymbol{G}$), there exists a universal constant $C_{\min} > 0$ such that for some $T \le T_{\min} := \frac{C_{\min}}{\eta} \log \frac{\|\boldsymbol{L}_\star\|}{\alpha}$, the iterates $\boldsymbol{L}_t$ of (47) obey

$$\|\boldsymbol{L}_T \boldsymbol{L}_T^\top - \boldsymbol{X}_\star\|_{\mathsf{F}} \le \alpha^{1/3} \|\boldsymbol{L}_\star\|^{5/3}.$$

In particular, for any prescribed accuracy target $\epsilon \in (0, 1)$, by choosing a sufficiently small α fulfilling both (50c) *and $\alpha \le \epsilon^3 \|\boldsymbol{L}_\star\|$, we have $\|\boldsymbol{L}_T \boldsymbol{L}_T^\top - \boldsymbol{X}_\star\|_{\mathsf{F}} \le \epsilon \|\boldsymbol{X}_\star\|$.*

Theorem 7 shows that by choosing an appropriate α, `ScaledGD(λ)` finds an ϵ-accurate solution, i.e., $\|\boldsymbol{L}_T \boldsymbol{L}_T^\top - \boldsymbol{X}_\star\|_{\mathsf{F}} \le \epsilon \|\boldsymbol{X}_\star\|$, in no more than an order of

$$\log \kappa \cdot \log(\kappa n) + \log(1/\epsilon)$$

iterations. Roughly speaking, this asserts that `ScaledGD(λ)` converges at a constant linear rate independent of the condition number κ after an initial phase of approximately $O(\log \kappa \cdot \log(\kappa n))$ iterations. In contrast, overparameterized `GD` requires $O(\kappa^4 + \kappa^3 \log(\kappa n/\epsilon))$ iterations to converge from a small random initialization to ϵ-accuracy; see [39, 51]. Thus, the convergence of overparameterized `GD` is much slower than `ScaledGD(λ)` even for mildly ill-conditioned matrices. Furthermore, our sample complexity depends only on the true rank r, but not on the overparameterized rank r'—a crucial feature in order to provide meaningful guarantees when the overparameterized rank r' is uninformative of the true rank, i.e., close to the full dimension n. The dependency on κ in the sample complexity, on the other end, has been generally unavoidable in nonconvex low-rank estimation [19].

5 Numerical Experiments

We illustrate the performance of ScaledGD and ScaledGD(λ) via a real data experiment to highlight the consideration of rank selection, when the data matrix is approximately low rank, a scenario which occurs frequently in practice. We consider a dataset[2] that measures chlorine concentrations in a drinking water distribution system, over different junctions and recorded once every 5 minutes during 15 days. The data matrix of interest, $\boldsymbol{X}_\star \in \mathbb{R}^{120\times 180}$, corresponds to the data extracted at 120 junctions over 15 hours. Figure 3 plots the spectrum of $\boldsymbol{X}_\star$, where its singular values decay rapidly, suggesting it can be well approximated by a low-rank matrix.

Since the data matrix $\boldsymbol{X}_\star$ is not exactly low rank, the choice of the rank r determines how good the low-rank approximation is as well as the condition number κ: Choosing a larger r leads to a lower approximation error but also a higher κ. We explore the behavior of ScaledGD and ScaledGD(λ) in comparison with vanilla gradient descent under different choices of r in Fig. 4. Moreover, apart from the spectral initialized ScaledGD, we consider yet another variant of ScaledGD which we found useful in practice: We start with ScaledGD(λ) for a few iterations and switch to ScaledGD after it is detected $\sigma_{\min}^2(\boldsymbol{L}_t) \gtrsim \lambda$. We call this variant ScaledGD with *mixed initialization*. The philosophy of this variant will be introduced after we discuss the results in Fig. 4.

We consider the matrix completion setting, where we randomly observe 80% of the entries in the data matrix. Figure 4a illustrates the performance of different algorithms when $r = 5$. ScaledGD with spectral initialization achieves the fastest convergence, while ScaledGD(λ) and ScaledGD with mixed initialization take a few more iterations at the beginning to warm up. All variants of ScaledGD converge considerably faster than vanilla GD and approach the optimal rank-r approximation error. Figure 4b illustrates the performance of different algorithms when $r = 20$,

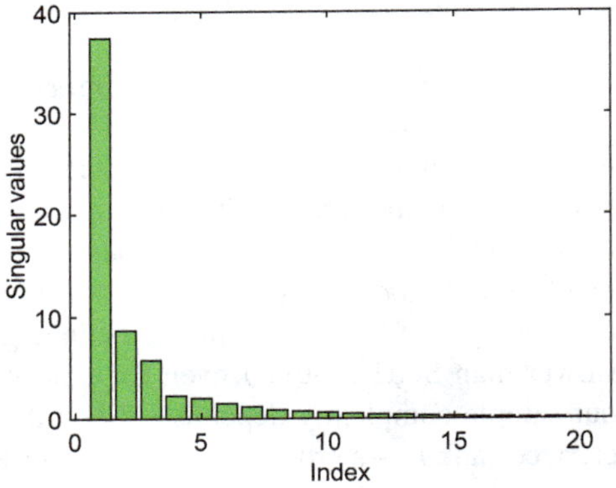

Fig. 3 The spectrum of the chlorine concentration data matrix, where its singular values decay rapidly

[2] The dataset can be accessed from http://www.cs.cmu.edu/afs/cs/project/spirit-1/www/.

Fig. 4 Performance comparison of different algorithms for matrix completion on the chlorine concentration dataset, under $r = 5$ (**a**) and $r = 20$ (**b**)

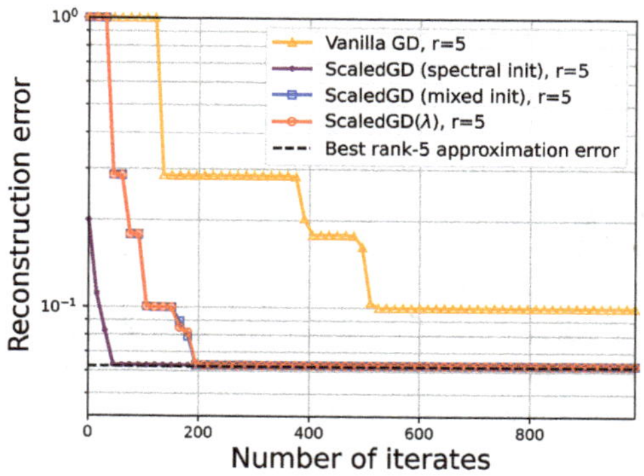

(a) $r = 5$

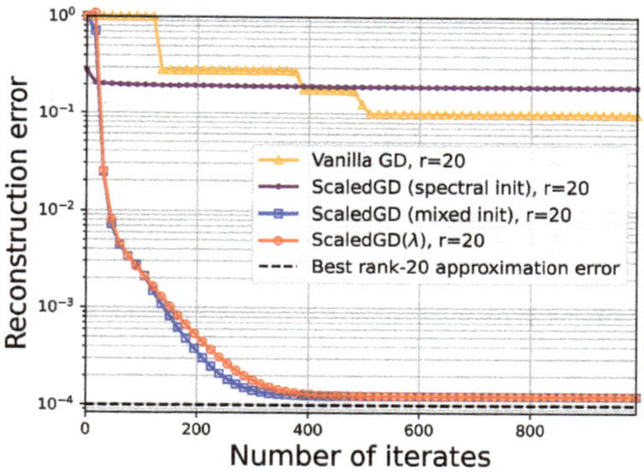

(b) $r = 20$

where the situation becomes different: `ScaledGD` with spectral initialization no longer converges, while `ScaledGD(λ)` and `ScaledGD` with mixed initialization still demonstrate fast convergence to the optimal rank-r approximation error. Vanilla GD, on the other hand, is still significantly slower.

The experimental results help to explain the motivation of using `ScaledGD` with mixed initialization. The reason for the instability of spectrally initialized `ScaledGD` for large r stems from the fact that spectral initialization does not cope well with overparameterization. On the other hand, small random initialization is known to help stabilize with overparameterization [39], but does not integrate with `ScaledGD` since the preconditioner $(\boldsymbol{L}_0^\top \boldsymbol{L}_0)^{-1}$ at the first iteration would be extremely large if the initialization $\boldsymbol{L}_0$ were small. Therefore, initializing with `ScaledGD(λ)`, which is provably robust against overparameterization and can be integrated perfectly into `ScaledGD`, becomes a reasonable choice.

6 Conclusions

This chapter highlights a novel approach to provably accelerate ill-conditioned low-rank estimation via ScaledGD. Its fast convergence, together with low computational and memory costs by operating in the factor space, makes it a highly scalable and desirable method in practice. The performance of ScaledGD is also robust when the data matrix is only approximately low rank and the observations are noisy; we refer interested readers to [66] for further details. In addition, the design methodology is also amenable to nonsmooth optimization problems, where similar preconditioners can be applied to precondition the subgradients [56, 57]. In terms of future directions, it is of great interest to explore the design of effective preconditioners for other statistical estimation and learning tasks such as superresolution [23], as well as further understand the implications of preconditioning in the presence of overparameterization for the asymmetric setting.

Acknowledgments This work is supported in part by the Office of Naval Research under N00014-19-1-2404 and by the National Science Foundation under CCF-1901199, DMS-2134080, and ECCS-2126634 to Y. Chi and by the National Science Foundation under DMS-2311127 to C. Ma.

References

1. Pierre Baldi and Kurt Hornik. Neural networks and principal component analysis: Learning from examples without local minima. *Neural Networks*, 2(1):53–58, 1989.
2. Boaz Barak and Ankur Moitra. Noisy tensor completion via the sum-of-squares hierarchy. In *Conference on Learning Theory*, pages 417–445. PMLR, 2016.
3. Srinadh Bhojanapalli, Behnam Neyshabur, and Nati Srebro. Global optimality of local search for low rank matrix recovery. In *Advances in Neural Information Processing Systems*, pages 3873–3881, 2016.
4. Léon Bottou, Frank E Curtis, and Jorge Nocedal. Optimization methods for large-scale machine learning. *SIAM Review*, 60(2):223–311, 2018.
5. Changxiao Cai, Gen Li, Yuejie Chi, H Vincent Poor, and Yuxin Chen. Subspace estimation from unbalanced and incomplete data matrices: $\ell_{2,\infty}$ statistical guarantees. *The Annals of Statistics*, 49(2):944–967, 2021.
6. Jian-Feng Cai, Jingyang Li, and Dong Xia. Generalized low-rank plus sparse tensor estimation by fast Riemannian optimization. *Journal of the American Statistical Association*, 0(0):1–17, 2022.
7. Emmanuel Candès, Xiaodong Li, and Mahdi Soltanolkotabi. Phase retrieval via Wirtinger flow: Theory and algorithms. *IEEE Transactions on Information Theory*, 61(4):1985–2007, 2015.
8. Emmanuel J. Candès, Xiaodong Li, Yi Ma, and John Wright. Robust principal component analysis? *Journal of the ACM*, 58(3):11:1–11:37, 2011.
9. Emmanuel J Candès and Yaniv Plan. Tight oracle inequalities for low-rank matrix recovery from a minimal number of noisy random measurements. *IEEE Transactions on Information Theory*, 57(4):2342–2359, 2011.
10. Venkat Chandrasekaran, Sujay Sanghavi, Pablo Parrilo, and Alan Willsky. Rank-sparsity incoherence for matrix decomposition. *SIAM Journal on Optimization*, 21(2):572–596, 2011.

11. Vasileios Charisopoulos, Yudong Chen, Damek Davis, Mateo Díaz, Lijun Ding, and Dmitriy Drusvyatskiy. Low-rank matrix recovery with composite optimization: good conditioning and rapid convergence. *Foundations of Computational Mathematics*, pages 1–89, 2021.
12. Ji Chen and Xiaodong Li. Model-free nonconvex matrix completion: Local minima analysis and applications in memory-efficient kernel PCA. *Journal of Machine Learning Research*, 20(142):1–39, 2019.
13. Ji Chen, Dekai Liu, and Xiaodong Li. Nonconvex rectangular matrix completion via gradient descent without $\ell_{2,\infty}$ regularization. *IEEE Transactions on Information Theory*, 66(9):5806–5841, 2020.
14. Yudong Chen. Incoherence-optimal matrix completion. *IEEE Transactions on Information Theory*, 61(5):2909–2923, 2015.
15. Yudong Chen and Yuejie Chi. Harnessing structures in big data via guaranteed low-rank matrix estimation: Recent theory and fast algorithms via convex and nonconvex optimization. *IEEE Signal Processing Magazine*, 35(4):14–31, 2018.
16. Yudong Chen and Martin J Wainwright. Fast low-rank estimation by projected gradient descent: General statistical and algorithmic guarantees. *arXiv preprint arXiv:1509.03025*, 2015.
17. Yuxin Chen, Yuejie Chi, Jianqing Fan, Cong Ma, and Yuling Yan. Noisy matrix completion: Understanding statistical guarantees for convex relaxation via nonconvex optimization. *SIAM Journal on Optimization*, 30(4):3098–3121, 2020.
18. Yuxin Chen, Jianqing Fan, Cong Ma, and Yuling Yan. Bridging convex and nonconvex optimization in robust PCA: Noise, outliers, and missing data. *The Annals of Statistics*, 49(5):2948–2971, 2021.
19. Yuejie Chi, Yue M Lu, and Yuxin Chen. Nonconvex optimization meets low-rank matrix factorization: An overview. *IEEE Transactions on Signal Processing*, 67(20):5239–5269, 2019.
20. Damek Davis, Dmitriy Drusvyatskiy, and Courtney Paquette. The nonsmooth landscape of phase retrieval. *IMA Journal of Numerical Analysis*, 40(4):2652–2695, 2020.
21. Harry Dong, Tian Tong, Cong Ma, and Yuejie Chi. Fast and provable tensor robust principal component analysis via scaled gradient descent. *Information and Inference: A Journal of the IMA*, 12(3):1716–1758, 06 2023.
22. Simon S Du, Wei Hu, and Jason D Lee. Algorithmic regularization in learning deep homogeneous models: Layers are automatically balanced. In *Advances in Neural Information Processing Systems*, pages 384–395, 2018.
23. Maxime Ferreira Da Costa and Yuejie Chi. Local geometry of nonconvex spike deconvolution from low-pass measurements. *IEEE Journal on Selected Areas in Information Theory*, 4:1–15, 2023.
24. Abraham Frandsen and Rong Ge. Optimization landscape of Tucker decomposition. *Mathematical Programming*, pages 1–26, 2020.
25. Rong Ge, Furong Huang, Chi Jin, and Yang Yuan. Escaping from saddle points-online stochastic gradient for tensor decomposition. In *Conference on Learning Theory (COLT)*, pages 797–842, 2015.
26. Rong Ge, Chi Jin, and Yi Zheng. No spurious local minima in nonconvex low rank problems: A unified geometric analysis. In *International Conference on Machine Learning*, pages 1233–1242, 2017.
27. Rong Ge, Jason D Lee, and Tengyu Ma. Matrix completion has no spurious local minimum. In *Advances in Neural Information Processing Systems*, pages 2973–2981, 2016.
28. Rungang Han, Rebecca Willett, and Anru R Zhang. An optimal statistical and computational framework for generalized tensor estimation. *The Annals of Statistics*, 50(1):1–29, 2022.
29. Moritz Hardt and Mary Wootters. Fast matrix completion without the condition number. In *Proceedings of The 27th Conference on Learning Theory*, pages 638–678, 2014.
30. Bo Huang, Cun Mu, Donald Goldfarb, and John Wright. Provable models for robust low-rank tensor completion. *Pacific Journal of Optimization*, 11(2):339–364, 2015.
31. Prateek Jain and Purushottam Kar. Non-convex optimization for machine learning. *Foundations and Trends® in Machine Learning*, 10(3–4):142–336, 2017.

32. Prateek Jain, Raghu Meka, and Inderjit S Dhillon. Guaranteed rank minimization via singular value projection. In *Advances in Neural Information Processing Systems*, pages 937–945, 2010.
33. Prateek Jain, Praneeth Netrapalli, and Sujay Sanghavi. Low-rank matrix completion using alternating minimization. In *Proceedings of the 45th Annual ACM Symposium on Theory of Computing*, pages 665–674, 2013.
34. Chi Jin, Rong Ge, Praneeth Netrapalli, Sham M Kakade, and Michael I Jordan. How to escape saddle points efficiently. In *International Conference on Machine Learning*, pages 1724–1732, 2017.
35. Hiroyuki Kasai and Bamdev Mishra. Low-rank tensor completion: a Riemannian manifold preconditioning approach. In *International Conference on Machine Learning*, pages 1012–1021, 2016.
36. Kenji Kawaguchi. Deep learning without poor local minima. In *Advances in Neural Information Processing Systems*, pages 586–594, 2016.
37. Xiaodong Li, Shuyang Ling, Thomas Strohmer, and Ke Wei. Rapid, robust, and reliable blind deconvolution via nonconvex optimization. *Applied and Computational Harmonic Analysis*, 47(3):893–934, 2019.
38. Yuanxin Li, Cong Ma, Yuxin Chen, and Yuejie Chi. Nonconvex matrix factorization from rank-one measurements. *IEEE Transactions on Information Theory*, 67(3):1928–1950, 2021.
39. Yuanzhi Li, Tengyu Ma, and Hongyang Zhang. Algorithmic regularization in over-parameterized matrix sensing and neural networks with quadratic activations. In *Conference On Learning Theory*, pages 2–47. PMLR, 2018.
40. Yuetian Luo and Anru R Zhang. Low-rank tensor estimation via Riemannian Gauss-Newton: Statistical optimality and second-order convergence. *arXiv preprint arXiv:2104.12031*, 2021.
41. Cong Ma, Yuanxin Li, and Yuejie Chi. Beyond Procrustes: Balancing-free gradient descent for asymmetric low-rank matrix sensing. *IEEE Transactions on Signal Processing*, 69:867–877, 2021.
42. Cong Ma, Kaizheng Wang, Yuejie Chi, and Yuxin Chen. Implicit regularization in nonconvex statistical estimation: Gradient descent converges linearly for phase retrieval, matrix completion, and blind deconvolution. *Foundations of Computational Mathematics*, pages 1–182, 2019.
43. Song Mei, Yu Bai, and Andrea Montanari. The landscape of empirical risk for nonconvex losses. *The Annals of Statistics*, 46(6A):2747–2774, 2018.
44. Yurii Nesterov and Boris T Polyak. Cubic regularization of Newton method and its global performance. *Mathematical Programming*, 108(1):177–205, 2006.
45. Praneeth Netrapalli, UN Niranjan, Sujay Sanghavi, Animashree Anandkumar, and Prateek Jain. Non-convex robust PCA. In *Advances in Neural Information Processing Systems*, pages 1107–1115, 2014.
46. Dohyung Park, Anastasios Kyrillidis, Constantine Carmanis, and Sujay Sanghavi. Non-square matrix sensing without spurious local minima via the Burer-Monteiro approach. In *Artificial Intelligence and Statistics*, pages 65–74, 2017.
47. Garvesh Raskutti, Ming Yuan, and Han Chen. Convex regularization for high-dimensional multiresponse tensor regression. *The Annals of Statistics*, 47(3):1554–1584, 2019.
48. Holger Rauhut, Reinhold Schneider, and Željka Stojanac. Low rank tensor recovery via iterative hard thresholding. *Linear Algebra and its Applications*, 523:220–262, 2017.
49. Benjamin Recht, Maryam Fazel, and Pablo A Parrilo. Guaranteed minimum-rank solutions of linear matrix equations via nuclear norm minimization. *SIAM Review*, 52(3):471–501, 2010.
50. Laixi Shi and Yuejie Chi. Manifold gradient descent solves multi-channel sparse blind deconvolution provably and efficiently. *IEEE Transactions on Information Theory*, 67(7):4784–4811, 2021.
51. Dominik Stöger and Mahdi Soltanolkotabi. Small random initialization is akin to spectral learning: Optimization and generalization guarantees for overparameterized low-rank matrix reconstruction. *Advances in Neural Information Processing Systems*, 34:23831–23843, 2021.
52. Ju Sun, Qing Qu, and John Wright. Complete dictionary recovery using nonconvex optimization. In *Proceedings of the 32nd International Conference on Machine Learning*, pages 2351–2360, 2015.

53. Ju Sun, Qing Qu, and John Wright. A geometric analysis of phase retrieval. *Foundations of Computational Mathematics*, 18(5):1131–1198, 2018.
54. Ruoyu Sun and Zhi-Quan Luo. Guaranteed matrix completion via non-convex factorization. *IEEE Transactions on Information Theory*, 62(11):6535–6579, 2016.
55. Tian Tong, Cong Ma, and Yuejie Chi. Accelerating ill-conditioned low-rank matrix estimation via scaled gradient descent. *Journal of Machine Learning Research*, 22(150):1–63, 2021.
56. Tian Tong, Cong Ma, and Yuejie Chi. Low-rank matrix recovery with scaled subgradient methods: Fast and robust convergence without the condition number. *IEEE Transactions on Signal Processing*, 69:2396–2409, 2021.
57. Tian Tong, Cong Ma, and Yuejie Chi. Accelerating ill-conditioned robust low-rank tensor regression. In *2022 IEEE International Conference on Acoustics, Speech and Signal Processing (ICASSP)*, pages 9072–9076. IEEE, 2022.
58. Tian Tong, Cong Ma, Ashley Prater-Bennette, Erin Tripp, and Yuejie Chi. Scaling and scalability: Provable nonconvex low-rank tensor estimation from incomplete measurements. *Journal of Machine Learning Research*, 23(163):1–77, 2022.
59. Stephen Tu, Ross Boczar, Max Simchowitz, Mahdi Soltanolkotabi, and Benjamin Recht. Low-rank solutions of linear matrix equations via Procrustes flow. In *International Conference Machine Learning*, pages 964–973, 2016.
60. Ledyard R Tucker. Some mathematical notes on three-mode factor analysis. *Psychometrika*, 31(3):279–311, 1966.
61. Haifeng Wang, Jinchi Chen, and Ke Wei. Implicit regularization and entrywise convergence of Riemannian optimization for low tucker-rank tensor completion. *arXiv preprint arXiv:2108.07899*, 2021.
62. Ke Wei, Jian-Feng Cai, Tony F Chan, and Shingyu Leung. Guarantees of Riemannian optimization for low rank matrix recovery. *SIAM Journal on Matrix Analysis and Applications*, 37(3):1198–1222, 2016.
63. Dong Xia and Ming Yuan. On polynomial time methods for exact low-rank tensor completion. *Foundations of Computational Mathematics*, 19(6):1265–1313, 2019.
64. Dong Xia, Ming Yuan, and Cun-Hui Zhang. Statistically optimal and computationally efficient low rank tensor completion from noisy entries. *The Annals of Statistics*, 49(1):76–99, 2021.
65. Dong Xia, Anru R Zhang, and Yuchen Zhou. Inference for low-rank tensors–no need to debias. *arXiv preprint arXiv:2012.14844*, 2020.
66. Xingyu Xu, Yandi Shen, Yuejie Chi, and Cong Ma. The power of preconditioning in overparameterized low-rank matrix sensing. *arXiv preprint arXiv:2302.01186*, 2023.
67. Xinyang Yi, Dohyung Park, Yudong Chen, and Constantine Caramanis. Fast algorithms for robust PCA via gradient descent. In *Advances in Neural Information Processing Systems*, pages 4152–4160, 2016.
68. Ming Yuan and Cun-Hui Zhang. On tensor completion via nuclear norm minimization. *Foundations of Computational Mathematics*, 16(4):1031–1068, 2016.
69. Anru Zhang and Dong Xia. Tensor SVD: Statistical and computational limits. *IEEE Transactions on Information Theory*, 64(11):7311–7338, 2018.
70. Qinqing Zheng and John Lafferty. A convergent gradient descent algorithm for rank minimization and semidefinite programming from random linear measurements. In *Advances in Neural Information Processing Systems*, pages 109–117, 2015.
71. Qinqing Zheng and John Lafferty. Convergence analysis for rectangular matrix completion using Burer-Monteiro factorization and gradient descent. *arXiv preprint arXiv:1605.07051*, 2016.
72. Zhihui Zhu, Qiuwei Li, Gongguo Tang, and Michael B Wakin. Global optimality in low-rank matrix optimization. *IEEE Transactions on Signal Processing*, 66(13):3614–3628, 2018.

CLAIRE: Scalable GPU-Accelerated Algorithms for Diffeomorphic Image Registration in 3D

Andreas Mang

Abstract We present our work on scalable, GPU-accelerated algorithms for diffeomorphic image registration. The associated software package is termed CLAIRE. Image registration is a nonlinear inverse problem. It is about computing a spatial mapping from one image of the same object or scene to another. In diffeomorphic image registration, the set of admissible spatial transformations is restricted to maps that are smooth, are one-to-one, and have a smooth inverse. We formulate diffeomorphic image registration as a variational problem governed by transport equations. We use an inexact, globalized (Gauss–)Newton–Krylov method for numerical optimization. We consider semi-Lagrangian methods for numerical time integration. Our solver features mixed-precision, hardware-accelerated computational kernels for optimal computational throughput. We use the message-passing interface for distributed-memory parallelism and deploy our code on modern high-performance computing architectures. Our solver allows us to solve clinically relevant problems in under four seconds on a single GPU. It can also be applied to large-scale 3D imaging applications with data that is discretized on meshes with billions of voxels. We demonstrate that our numerical framework yields high-fidelity results in only a few seconds, even if we search for an optimal regularization parameter.

1 Introduction

In the present work, we discuss scalable, hardware-accelerated algorithms for diffeomorphic image registration. We review our past contributions and showcase results for a software framework termed CLAIRE [1–3]. Image registration is an ill-posed inverse problem [4]. It is a key methodology in medical image analysis. The inputs are two (or more, noisy) images $m_i \in \mathfrak{I}$, $i = 0, 1$, $\mathfrak{I} \subset \{u : \Omega \to \mathbb{R}\}$, of the same object or scene, compactly supported on some domain $\Omega \subseteq \mathbb{R}^d$, where

A. Mang (✉)
Department of Mathematics, University of Houston, Houston, TX, USA
e-mail: andreas@math.uh.edu

S. Foucart, S. Wojtowytsch (eds.), *Explorations in the Mathematics of Data Science*, Applied and Numerical Harmonic Analysis,
https://doi.org/10.1007/978-3-031-66497-7_8

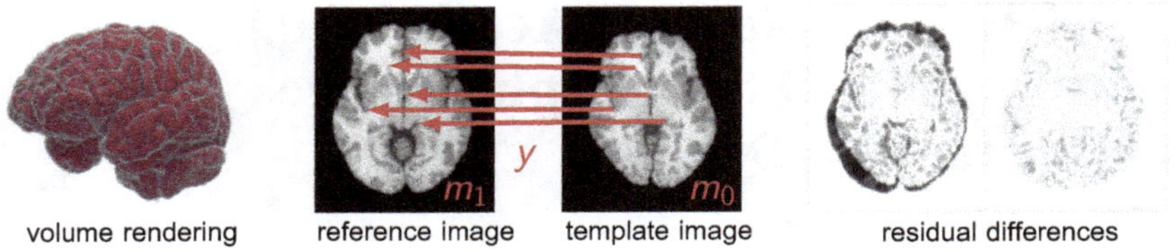

Fig. 1 Image registration problem. On the left, we show a volume rendering of a 3D brain MRI. The figures in the middle show an axial slice of two MRI brain scans of different individuals. In image registration, we seek a map $\boldsymbol{y} \in \mathfrak{Y}_{\text{ad}} \subset \{\boldsymbol{\phi} : \mathbb{R}^d \to \mathbb{R}^d\}$, $d \in \{2, 3\}$, that establishes a *plausible* spatial correspondence between these two images. In this work, we restrict the set of admissible spatial transformations $\mathfrak{Y}_{\text{ad}}$ to $\mathbb{R}^d$-diffeomorphisms. On the right, we show the residual differences between the axial slices of the images shown in the middle before (left) and after (right) diffeomorphic (deformable) registration. Here, white represents small residual differences, and black indicates large residual differences. We note that the registration of two brains from different individuals is a common application for diffeomorphic image registration in computational anatomy. However, strictly speaking this example is in violation with our underlying assumptions; we do not register the "same object"—we register images of brains of different individuals (in an attempt to study anatomical variability)

$d \in \{2, 3\}$. In image registration, we seek a *plausible* spatial transformation $\boldsymbol{y} \in \mathfrak{Y}_{\text{ad}}$, $\mathfrak{Y}_{\text{ad}} \subset \{\boldsymbol{\phi} : \mathbb{R}^d \to \mathbb{R}^d\}$, that maps points in the so-called template or source image m_0 to its corresponding points in the so-called reference or target image m_1 [4–6]. The notion of the plausibility of the map $\boldsymbol{y}$ depends on the particular application. In the present work, we restrict the set $\mathfrak{Y}_{\text{ad}}$ of admissible maps $\boldsymbol{y}$ to $\mathbb{R}^d$-diffeomorphisms [7]. That is, $\mathfrak{Y}_{\text{ad}} \subseteq \text{diff}(\mathbb{R}^d)$, where $\text{diff}(\mathbb{R}^d)$ is the set of $\mathbb{R}^d$-diffeomorphisms, i.e., smooth maps from $\mathbb{R}^d$ to $\mathbb{R}^d$ that are one-to-one and onto, with a smooth inverse. The set $\text{diff}(\mathbb{R}^d)$ is closed under composition and taking the inverse; it forms a group. In this framework, deforming the template image m_0 corresponds to a change of coordinates $m_0 \circ \boldsymbol{y}^{-1}$; the image intensity in the transformed image $m_0 \circ \boldsymbol{y}^{-1}$ at coordinate $\boldsymbol{y}(\boldsymbol{x}) \in \mathbb{R}^d$ is identical to the value at the location $\boldsymbol{x} \in \mathbb{R}^d$ in the original image. Using this notation, the diffeomorphic image registration problem can be formulated as the problem of finding $\boldsymbol{y} \in \text{diff}(\mathbb{R}^d)$ such that $m_0 \circ \boldsymbol{y}^{-1} = m_1$. We illustrate this in Fig. 1. We summarize the main notation and acronyms in Table 1.

1.1 Outline of the Method

The approach considered in the present work is related to a mathematical framework referred to as LDDMM [7–11]. We consider PDE-constrained optimization problems [12–15] governed by transport equations for diffeomorphic image registration. The transport map is parameterized by a smooth space-time field $\boldsymbol{v} \in \mathfrak{V}$, $\mathfrak{V} := L^q([0, 1], \mathfrak{H})$, $q \in \mathbb{N}$, where $\mathfrak{H}$ is a Sobolev space of suitable regularity, i.e.,

Table 1 Notation, symbols, and acronyms

Symbol/acronym	Meaning
$d \in \mathbb{N}$	Dimensionality of the ambient space
$\Omega \subset \mathbb{R}^d$	Spatial domain
$m_0 : \bar{\Omega} \to \mathbb{R}$	Template image
$m_1 : \bar{\Omega} \to \mathbb{R}$	Reference image
$m : [0, 1] \times \bar{\Omega} \to \mathbb{R}$	State variable (transported image intensities)
$\boldsymbol{v} : \bar{\Omega} \to \mathbb{R}^d$	Control variable (stationary velocity field)
$\lambda : [0, 1] \times \bar{\Omega} \to \mathbb{R}$	Dual variable
$\text{dist} : \mathfrak{I} \times \mathfrak{I} \to \mathbb{R}$	Distance functional
$\text{reg} : \mathfrak{V} \to \mathbb{R}$	Regularization functional
$\mathcal{L} : \mathfrak{V} \to \mathfrak{V}^*$	Regularization operator
$\alpha \in \mathbb{R}$	Regularization parameter
$\mathfrak{I}$	Orbit
$\mathfrak{G}$	Group of diffeomorphisms
$\text{diff}(\mathbb{R}^d)$	Set of $\mathbb{R}^d$-diffeomorphisms
FFT	Fast Fourier transform
GMRES	Generalized minimal residual (method)
GPU	Graphics processing unit
HPC	High-performance computing
KKT	Karush–Kuhn–Tucker (conditions)
LDDMM	Large deformation diffeomorphic metric mapping
MRI	Magnetic resonance imaging
MPI	Message passing interface
ODE	Ordinary differential equation
PCG	Preconditioned conjugate gradient (method)
PDE	partial differential equation
RK2	Second-order Runge–Kutta (method)
SL	Semi-Lagrangian (method)

$\mathfrak{H} = W^{p,s}(\Omega, \mathbb{R}^d)$, $p \in \mathbb{N}$, $s \in \mathbb{N}$. Our problem formulation is of the form

$$\underset{m \in \mathfrak{M}_{\text{ad}},\, \boldsymbol{v} \in \mathfrak{V}_{\text{ad}}}{\text{minimize}} \quad \text{dist}(m(1), m_1) + \text{reg}(\boldsymbol{v}) \tag{1a}$$

$$\text{subject to} \quad c(m, \boldsymbol{v}) = 0. \tag{1b}$$

Here, $c : \mathfrak{M} \times \mathfrak{V} \to \mathfrak{Q}$ represents a PDE constraint. It is of the general form $c(m, \boldsymbol{v}) = \mathcal{A}(m, \boldsymbol{v}) - q$. The *parameter-to-observation map* $f : \mathfrak{V} \to \mathfrak{I}$ (i.e., the solution operator for the constraint) is formally given by $f(\boldsymbol{v}) = Q\mathcal{A}^{-1}(m, \boldsymbol{v})q$. Here, Q denotes the observation operator, i.e., a mapping that takes the output of $\mathcal{A}^{-1}$ and maps it to "locations" at which data is available. The functional dist : $\mathfrak{I} \times \mathfrak{I} \to \mathbb{R}$ measures the discrepancy between the deformed template image $m(1) =$

$f(\boldsymbol{v})$ and the reference image m_1. The functional reg : $\mathfrak{V} \rightarrow \mathbb{R}$ is a regularization functional. We specify the precise choices in greater detail below.

We use the method of Lagrange multipliers to solve (1). We consider an *optimize-then-discretize* approach. We use a globalized, inexact reduced-space (Gauss–)Newton–Krylov method for numerical optimization. We solve the PDEs that appear in the optimality conditions based on a SL method. The main computational kernels of our algorithm are interpolation and numerical differentiation. For interpolation, we use a Lagrange polynomial. For numerical differentiation, we consider a mixture of high-order finite difference operators and a pseudo-spectral method. We use MPI for distributed-memory parallelism and deploy our code on dedicated GPU architectures.

1.2 Related Work

We consider a PDE-constrained optimization problem for velocity-based diffeomorphic image registration. We refer to [12–14, 16–18] for insights into theory and algorithmic developments related to PDE-constrained optimization. Additional information about image registration and related work can be found in [4–7, 19, 20]. As we mentioned above, we restrict ourselves to diffeomorphic image registration. An intuitive approach to safeguard against non-diffeomorphic maps $\boldsymbol{y}$ is to add hard and/or soft constraints to the variational problem [21–24]. An alternative strategy is to introduce a pseudo-time variable t and invert for a smooth velocity field $\boldsymbol{v}$ that parameterizes $\boldsymbol{y}$ [7, 8, 10, 11, 25–29]; our approach falls into this category. In [8–11, 26, 29], the flow of the sought-after diffeomorphism $\boldsymbol{y}$ is modelled as the solution of the ODE $\partial_t \boldsymbol{\phi} = \boldsymbol{v} \circ \boldsymbol{\phi}$ for $t \in (0, 1]$ with initial condition $\boldsymbol{\phi} = \mathrm{id}_{\mathbb{R}^d}$ at time $t = 0$, where $\boldsymbol{v}$ is a smooth, time-dependent vector field from $\mathbb{R}^d$ to $\mathbb{R}^d$ and $\mathrm{id}_{\mathbb{R}^d} : \mathbb{R}^d \rightarrow \mathbb{R}^d$, $\mathrm{id}_{\mathbb{R}^d}(\boldsymbol{x}) = \boldsymbol{x}$, is the identity transformation in $\mathbb{R}^d$. This ODE enters the variational problem as a constraint; we arrive at a nonlinear optimal control problem with state variable $\boldsymbol{\phi}$ and control $\boldsymbol{v}$. The sought-after diffeomorphism $\boldsymbol{y}$ that maps one image to another corresponds to the endpoint of the flow $\boldsymbol{\phi}$, i.e., $\boldsymbol{y} = \boldsymbol{\phi}(t = 1)$. This approach is commonly referred to as LDDMM [8]. We describe it in greater detail in the main part of this manuscript. In our formulation, the diffeomorphism $\boldsymbol{y}$ no longer appears; we do *not* model the deformed template image as the application of $\boldsymbol{y}$ to m_0. Instead, we transport the intensities of the template image m_0 given some candidate $\boldsymbol{v}$ based on a hyperbolic transport equation [30–32]. Unlike most existing approaches, our framework features explicit control on volume change introduced by the mapping by controlling the divergence of $\boldsymbol{v}$. This formulation was originally proposed in [33]; a similar approach is described in [32]. Works of other groups that consider divergence-free velocities $\boldsymbol{v}$ in similar contexts have been described in [34–38].

Our formulation has been introduced in [30, 33]. The work most closely related to ours in terms of the problem formulation is [31, 32, 39–44]. Related formulations for optimal mass transport are discussed in [44–48]. In contrast to optimal mass transport, our formulation keeps the transported quantities constant along the characteristics, i.e., mass is *not* preserved. Our formulation is related to traditional optical flow formulations [37, 49, 50]. The main difference is that the transport equation for the image intensities of m_0 enters our formulation as a hard constraint. PDE-constrained formulations for optical flow that are equivalent to our formulation are described in [32, 34, 51, 52].

Among the most popular packages for diffeomorphic registration are `Demons` [27, 53], `ANTs` [54, 55], `Deformetrica` [56, 57], or `DARTEL` [58]. There are only few works on effective numerical methods for velocity-based diffeomorphic image registration and even fewer on scalable algorithms. Works of other groups on numerical algorithms for the solution of diffeomorphic association problems (for images as well as surface representations) are, e.g., described in [8, 31, 43, 59–64]. The majority of existing works consider an *optimize-then-discretize* approach for solving the variational problem [8, 31, 43, 64–67]; *discretize-then-optimize* approaches for related problem formulations can be found in [48, 62, 68–70]. In the work discussed in this exposition, we also consider an *optimize-then-discretize* approach [30, 33, 71]; an implementation for a *discretize-then-optimize* approach for problem formulations similar to the one considered here can be found in [48].

Despite the fact that first-order methods for optimization have poor convergence rates for nonlinear, ill-posed inverse problems, most work on algorithms for formulations similar to ours, with the exception of ours [15, 30, 33, 48, 71, 72] and [27, 44, 45, 61, 66, 73, 74], use first-order gradient descent-type approaches. Work on operator-splitting algorithms for LDDMM (and related problems) can be found in [69, 70, 75, 76]. Other recent works that do not explicitly derive optimality conditions based on variational principles but rely on automatic differentiation can be found in [56, 77–80]. Lastly, we note that the success of machine learning in various scientific disciplines has led to several recent works that attempt to solve the inverse problem of diffeomorphic registration based on machine learning techniques [80–89]. As we will show, our dedicated hardware-accelerated implementation [90, 91] allows us to solve diffeomorphic image registration problems in 3 to 4 seconds on a single GPU without considering machine learning approaches.

We consider a globalized, reduced-space (Gauss–)Newton–Krylov method [1, 30]. For these methods to be effective, it is crucial to design a good preconditioner for solving the reduced-space KKT system [92]. Related work on designing preconditioners for problems similar to ours can be found in [44, 45, 74]. Another key ingredient is fast algorithms to solve the PDEs that appear in the optimality systems. In our case, the most expensive PDE operators are (hyperbolic) transport equations. We refer to [8, 31, 32, 34, 45, 48, 61, 74] for different numerical methods to solve these types of PDEs in the context of PDE-constrained optimization. We use a SL method [8, 34, 71, 72].

What separates CLAIRE [1, 2] from most existing packages for velocity-based diffeomorphic image registration, aside from the numerics, is that it features hardware-accelerated computational kernels and that it has been deployed to dedicated HPC architectures [1, 71, 90, 91, 93]. Examples for parallel algorithms for PDE-constrained optimization problems can be found in [16, 94–100]. Surveys for parallel implementations of image registration algorithms are [101–104]. Many of these works consider low-dimensional parameterizations based on an expansion of the deformation map $\boldsymbol{y}$ in terms of smooth basis functions. Examples of GPU implementations of these approaches are [105–107]. GPU implementations of formulations similar to ours are described in [47, 56, 108–111]. Our memory-distributed implementation uses MPI for parallelism and allows us to solve problems of unprecedented scale [33, 90, 93, 112]. The linear solvers and the optimizer are built on top of `PETSc/TAO` [113–116]. Our CPU implementation [1, 33, 93] allows us to solve problems with 3,221,225,472 unknowns in 2 min on 22 compute nodes (256 MPI tasks) and in less than 5 s if we use 342 compute nodes (4096 MPI tasks). Our GPU implementation [90, 91] allows us to solve clinically relevant problems (50,000,000 unknowns) in less than 5 s on a single GPU. Our multi-GPU implementation for large-scale problems is described in [90] and applied to large-scale (biomedical) imaging data in [112]. We limit the numerical results reported in this study to our GPU implementation [90, 91].

1.3 Contributions

Our contributions are as follows:

- We overview our work on CLAIRE—a memory-distributed algorithmic framework for diffeomorphic image registration based on variational optimization problems governed by transport equations. In particular, we recapitulate our contributions presented in [1, 3, 15, 30, 33, 71, 72, 90, 91, 93].
- We report new results and study the performance of CLAIRE for real-world medical imaging data in 3D [1–3]. We include results for different similarity measures—normalized cross correlation and the default squared L^2-distance. In addition, we present results for an improved implementation in which we store the state variable and its gradient, reducing the runtime from roughly five seconds reported in prior work to slightly more than three seconds for clinically relevant problems (50,000,000 unknowns).

1.4 Limitations

CLAIRE has several limitations. First, CLAIRE only supports stationary velocity fields. Stationary velocities yield similar residuals in practical applications [30]. However, they are less expressive [48]; they only allow us to model a subset of the diffeomorphisms that can be modeled by using nonstationary $\boldsymbol{v}$. Second, CLAIRE only supports the registration of images acquired from the same modality. Implementing distance measures that allow for more complicated intensity relationships between images to be registered requires more work. Third, while our schemes for preconditioning the reduced-space Hessian are effective and, in general, mesh-independent, they are not independent of the choice of the regularization parameter. Fourth, the GPU implementation only supports single precision. We trade numerical accuracy for computational throughput. For example, applying (the inverse of) high-order differential operators (e.g., biharmonic operators) results in significant numerical roundoff errors and, consequently, is currently not supported. This, in conjunction with other algorithmic choices, does not allow us to solve the optimization problem to arbitrary accuracy, in particular for practical parameter choices. Nonetheless, we can see that in practice we obtain an excellent agreement between the registered datasets even if the gradient of our problem is not driven to zero.

1.5 Outline

We present the formulation and numerical methods in Sect. 2. This includes a discussion of the mathematical framework that motivates our approach (see Sect. 2.1), a brief recapitulation of the general problem formulation (see Sect. 2.2), the optimality conditions (see Sect. 2.3), and the Newton step (see Sect. 2.4), followed by a presentation of our numerical approach (see Sect. 2.5). We present some numerical results in Sect. 3 and conclude with Sect. 4.

2 Methods

In the following, we present the problem formulation as well as our numerical approach and implementation aspects. We start with discussing some background material related to the considered problem formulation.

2.1 Mathematical Foundations

Our problem formulation is related to LDDMM [7–10, 117, 118]—a mathematical framework for diffeomorphic image registration and shape matching. It builds upon the seminal work [119–121].

Let $\boldsymbol{k} = (k_1, \ldots, k_d) \in \mathbb{N}^d$, $d \in \{1, 2, 3\}$, denote a multi-index, and let

$$\partial^{\boldsymbol{k}} := \frac{\partial^{|\boldsymbol{k}|}}{\partial_1^{k_1} \cdots \partial_d^{k_d}}$$

denote the differential operator of order $|\boldsymbol{k}| = \sum_{i=1}^{d} k_i$. Here, ∂_i denotes the partial derivative with respect to the coordinate direction x_i with $\boldsymbol{x} := (x_1, \ldots, x_d) \in \Omega$

defined on some domain $\Omega \subseteq \mathbb{R}^d$. Moreover, let $q \in \mathbb{N}$, $1 \leq p \leq \infty$. We denote by

$$C^q(\Omega) := \left\{u : \Omega \to \mathbb{R} : \partial^{\boldsymbol{k}} u \text{ is continuous for } |\boldsymbol{k}| \leq q\right\}$$

the space of q-times continuously differentiable functions on Ω. Moreover, let

$$W^{q,p}(\Omega) := \left\{u \in L^p(\Omega) : \partial^{\boldsymbol{k}} u \in L^p(\Omega) \text{ for } 0 \leq |\boldsymbol{k}| \leq q\right\}$$

denote the Sobolev space with norm

$$\|u\|_{q,p} := \begin{cases} \left(\sum_{0\leq|\boldsymbol{k}|\leq q} \|\partial^{\boldsymbol{k}} u\|_p^p\right)^{1/p} & \text{if } 1 \leq p < \infty, \\ \max_{0\leq|\boldsymbol{k}|\leq q} \|\partial^{\boldsymbol{k}} u\|_\infty & \text{if } p = \infty. \end{cases} \tag{2}$$

Here, $\|\cdot\|_\infty$ denotes the standard supremum norm. Using these definitions, we denote by $C_0^q(\Omega)^d \subset C^q(\Omega)^d$ with $q \in \mathbb{N}$ the completion of the space of vector fields of class C^q which along with their derivatives of order less than or equal to q converge to zero at infinity. The space $C_0^q(\Omega)^d$ is a Banach space for the norm $\|u\|_{q,\infty}$. Similarly, we define the Sobolev space $W_0^{q,p}(\Omega)^d$ as a space that consists of elements with compact support on $\Omega \subseteq \mathbb{R}^d$.

We introduce a pseudo-time variable $t \in [0, 1]$, a suitable Hilbert space $\mathfrak{H}$ of smooth vector fields in $\mathbb{R}^d$, and parameterize diffeomorphisms using smooth vector fields $\boldsymbol{v} \in \mathfrak{V}$, $\mathfrak{V} := L^r([0, 1], \mathfrak{H})$, $1 \leq r \leq \infty$, $t \mapsto \boldsymbol{v}_t := \boldsymbol{v}(t, \cdot)$, $\boldsymbol{v}_t \in \mathfrak{H}$. This allows us to model the flow of $\mathbb{R}^d$-diffeomorphisms $\boldsymbol{\phi}_t := \boldsymbol{\phi}(t, \cdot)$ as the solution of the ODE

$$\begin{aligned} \partial_t \boldsymbol{\phi}_t &= \boldsymbol{v}_t \circ \boldsymbol{\phi}_t && \text{for } t \in (0, 1], \\ \boldsymbol{\phi}_t &= \mathrm{id}_{\mathbb{R}^d} && \text{for } t = 0, \end{aligned} \tag{3}$$

where $\mathrm{id}_{\mathbb{R}^d} : \mathbb{R}^d \to \mathbb{R}^d$, $\mathrm{id}_{\mathbb{R}^d}(\boldsymbol{x}) = \boldsymbol{x}$, is the identity transformation in $\mathbb{R}^d$ and the vector field $\boldsymbol{v}_t$ tends to zero as $\boldsymbol{x} \to \infty$; that is, we assume $\boldsymbol{v}_t \in C_0^q(\Omega)^d \supseteq \mathfrak{H}$ for any $t \in [0, 1]$. This assumption, along with suitable regularity requirements in time, guarantees that (3) admits a unique solution. Moreover, it is ensured that solutions of (3) are $\mathbb{R}^d$-diffeomorphisms [7].

We assume L^1-integrability in time, i.e., $r = 1$ [7]. The differentiability class q and the integrability order p of the Sobolev norm (2) are chosen to stipulate adequate regularity requirements in space. A common choice for p is $p = 2$. The choice of q depends on the dimension d of the ambient space Ω. In general, we have $d \in \{1, 2, 3\}$. Based on the Sobolev embedding theorem [122] we observe that for $p = 2$ and $q > (d/2) + 1$ the embedding $W_0^{q,2}(\Omega) \hookrightarrow C^1(\bar{\Omega})$ is compact. Since this embedding holds for all components of $\boldsymbol{v}_t$, we have that $\boldsymbol{v}_t \in \mathfrak{H} = W^{q,2}(\Omega)^d$ with $q > 5/2$ for $d = 3$ is an admissible space that yields a diffeomorphic flow $\boldsymbol{\phi}_t$ of smoothness class $1 \leq s < q - (3/2)$, $s \in \mathbb{N}$. We refer to [7, 11, 118] for a more rigorous discussion.

The set of all endpoints $\boldsymbol{y} := \boldsymbol{\phi}_1$ at time $t = 1$ of admissible flows $\boldsymbol{\phi}_t$ is a subgroup

$$\mathfrak{G} := \left\{ \boldsymbol{\phi}_1 : \int_0^1 \|\boldsymbol{v}_t\|_{\mathfrak{H}} \, \mathrm{d}t < \infty \right\}$$

of C^s-diffeomorphisms in $\mathbb{R}^d$. This subgroup can be equipped with a right-invariant metric defined as the minimal path length of all geodesics joining two elements in $\mathfrak{G}$ [9, 25, 123, 124]. The geodesic distance between $\mathrm{id}_{\mathbb{R}^d}$ and a mapping $\boldsymbol{y} \in \mathfrak{G}$ corresponds to the square root of the kinetic energy:

$$\mathrm{kin}(\boldsymbol{v}) := \|\boldsymbol{v}\|^2_{L^2([0,1],\mathfrak{H})} = \int_0^1 \|\boldsymbol{v}_t\|^2_{\mathfrak{H}} \, \mathrm{d}t \tag{4}$$

subject to the constraint that $\boldsymbol{y}$ is equal to the solution $\boldsymbol{\phi}$ of (3) at time $t = 1$ for the energy-minimizing velocity $\boldsymbol{v}$. We denote this geodesic distance by $\mathrm{dist}_{\mathfrak{G}}(\mathrm{id}_{\mathbb{R}^d}, \boldsymbol{y})$,

$$\begin{aligned} &\mathrm{dist}_{\mathfrak{G}}(\mathrm{id}_{\mathbb{R}^d}, \boldsymbol{y})^2 \\ &:= \inf_{\boldsymbol{v} \in \mathfrak{V}} \left\{ \int_0^1 \|\boldsymbol{v}_t\|^2_{\mathfrak{H}} \, \mathrm{d}t : \boldsymbol{v} \in \mathfrak{V},\ \boldsymbol{y} = \boldsymbol{\phi}_1,\ \partial_t \boldsymbol{\phi}_t = \boldsymbol{v}_t \circ \boldsymbol{\phi}_t,\ \boldsymbol{\phi}_0 = \mathrm{id}_{\mathbb{R}^d} \right\}. \end{aligned}$$

The geodesic distance between two maps $\boldsymbol{y}$ and $\boldsymbol{\psi}$ is given by $\mathrm{dist}_{\mathfrak{G}}(\mathrm{id}_{\mathbb{R}^d}, \boldsymbol{y} \circ \boldsymbol{\psi}^{-1})$.

Similarly, we can measure the geodesic distance between two images $m_0, m_1 \in \mathfrak{I}$ in terms of the kinetic energy associated with the energy-minimizing $\boldsymbol{v}$ that gives rise to the diffeomorphic flow $\boldsymbol{\phi}$ that maps m_0 to m_1. To do so, we assume that the image m_1 is in the orbit $\mathfrak{I}$ of the template image m_0 for the group $\mathfrak{G}$ of diffeomorphism, where

$$\mathfrak{I} := \{m : \Omega \to \mathbb{R} : m = m_0 \circ \boldsymbol{y}^{-1},\ \boldsymbol{y} \in \mathfrak{G}\}.$$

Using the geodesic distance $\mathrm{dist}_{\mathfrak{G}}$ introduced above we have

$$\mathrm{dist}_{\mathfrak{I}}(m_0, m_1) := \inf_{\boldsymbol{y} \in \mathfrak{G}} \left\{ \mathrm{dist}_{\mathfrak{G}}(\mathrm{id}_{\mathbb{R}^d}, \boldsymbol{y}) : m_1 = m_0 \circ \boldsymbol{y}^{-1} \right\},$$

where $\boldsymbol{y}$ corresponds to the endpoint of the flow $\boldsymbol{\phi}$. This notion of measuring distances between deformable objects has led to the emergence of a field of study in medical image analysis referred to as *computational anatomy* [123, 125–128].

Putting everything together, we can formulate the diffeomorphic matching of the template image m_0 to the reference image m_1 as a variational optimization problem. We stated initially that we seek a diffeomorphic map $\boldsymbol{y}$ such that $m_0 \circ \boldsymbol{y}^{-1} = m_1$. This is an ill-posed problem; we try to estimate a vector field given scalar data.

Consequently, a solution $\boldsymbol{y}$ may not exist, and if it exists, it may not be unique or depend continuously on the data. To alleviate the ill-posedness we introduce a regularization model that rules out unwanted solutions. For example, we can restrict ourselves to maps $\boldsymbol{y}$ that are close to the identity $\mathrm{id}_{\mathbb{R}^d}$, i.e., we penalize the distance between $\mathrm{id}_{\mathbb{R}^d}$ and $\boldsymbol{y}$. To alleviate existence issues, we relax the exact matching requirement to $m_0 \circ \boldsymbol{y}^{-1} \approx m_1$. We do so by introducing a distance that measures the proximity between the deformed template image $m_0 \circ \boldsymbol{y}^{-1}$ and the reference image m_1. In conclusion, we seek $\boldsymbol{y} \in \mathfrak{G} \subset \mathrm{diff}(\mathbb{R}^d)$ as a solution to

$$\underset{\boldsymbol{y}\in\mathfrak{G}}{\text{minimize}} \quad \mathrm{dist}(m_0 \circ \boldsymbol{y}^{-1}, m_1) + \frac{\alpha}{2}\,\mathrm{dist}_{\mathfrak{G}}(\boldsymbol{y}, \mathrm{id}_{\mathbb{R}^d}).$$

We can reformulate the variational problem above as an optimal control problem governed by (3) [8, 11]. We have

$$\underset{\boldsymbol{\phi}\in\mathfrak{F}_{\mathrm{ad}},\ \boldsymbol{v}\in\mathfrak{V}_{\mathrm{ad}}}{\text{minimize}} \quad \mathrm{dist}(m_0 \circ \boldsymbol{\phi}_1^{-1}, m_1) + \frac{\alpha}{2}\,\mathrm{kin}(\boldsymbol{v}) \tag{5a}$$

$$\begin{aligned} \text{subject to} \quad \partial_t \boldsymbol{\phi}_t &= \boldsymbol{v}_t \circ \boldsymbol{\phi}_t && \text{in } (0, 1], \\ \boldsymbol{\phi}_t &= \mathrm{id}_{\mathbb{R}^d} && \text{for } t = 0, \end{aligned} \tag{5b}$$

where the first term in the objective functional measures the discrepancy between the deformed template image $m_0 \circ \boldsymbol{\phi}_1^{-1}$ and the reference image m_1, the second term denotes the kinetic energy in (4), and the parameter $\alpha > 0$ balances their contribution. The norm $\|\boldsymbol{v}_t\|_{\mathfrak{H}}^2$ in the definition of the kinetic energy in (4) is typically modelled as

$$\|\boldsymbol{v}\|_{\mathfrak{H}}^2 = \langle \boldsymbol{v}, \boldsymbol{v}\rangle_{\mathfrak{H}} = \langle \mathcal{B}\boldsymbol{v}, \mathcal{B}\boldsymbol{v}\rangle_{\mathbb{R}^d}, = \langle \mathcal{L}\boldsymbol{v}, \boldsymbol{v}\rangle_{\mathbb{R}^d},$$

where $\mathcal{L} : \mathfrak{V} \to \mathfrak{V}^*$, $\mathcal{L} = \mathcal{B}^*\mathcal{B}$, is a differential operator of adequate order. A common choice for $\mathcal{B}$ is a symmetric, positive definite Helmholtz operator of the form $\mathcal{B} := (\beta\,\mathrm{id} - \Delta_d)^\gamma$, $\beta, \gamma > 0$ [8], where $\Delta_d \boldsymbol{u}(\boldsymbol{x}) := (\Delta u_1(\boldsymbol{x}), \ldots, \Delta u_d(\boldsymbol{x}))$, $\Delta := \sum_{i=1}^d \partial_{ii}$ for any $\boldsymbol{u} : \bar{\Omega} \to \mathbb{R}^d$.

Other data structures than images that can be registered within this framework are landmarks [129, 130], curves [131, 132], surfaces [64, 68–70, 79, 131, 133, 134], tensor fields [135], or functional data on manifolds. We refer to [8, 62, 68–70] for numerical methods to solve the control problem in (5).

2.2 Variational Problem Formulation

In this section, we review the problem formulation considered in CLAIRE. We assume that $m_i : \Omega \to \mathbb{R}$, $i = 0, 1$, are smooth C^1-functions compactly supported on Ω. As stated in Sect. 1, we formulate diffeomorphic image registration as a PDE-constrained optimization problem of the general form (1). This is different from the ODE-constrained optimization problem (5). Motivated by the formulation discussed above, we introduce a pseudo-time variable $t \in [0, 1]$ and invert for a smooth, time-dependent velocity field $\boldsymbol{v} \in \mathfrak{V}$ [30]. However, to reduce the computational complexity, our hardware-accelerated implementation no longer inverts for a time-dependent velocity $\boldsymbol{v}$ but for a *stationary velocity field* $\boldsymbol{v} : \Omega \to \mathbb{R}^d$. This not only reduces the complexity of the optimization problem but also simplifies the implementation. We note that stationary velocities no longer define a Riemannian metric as described in Sect. 2.1. However, we still generate diffeomorphic transformations. Moreover, we did not observe a deterioration in registration accuracy when comparing results to a nonstationary implementation [30]. Related work by other groups that use stationary $\boldsymbol{v}$ can be found in [27, 58, 136–139].

In its simplest form, the PDE constraint c in (1) for stationary $\boldsymbol{v}$ is given by the *hyperbolic transport equation*:

$$\partial_t m(t, \boldsymbol{x}) + \nabla m(t, \boldsymbol{x}) \cdot \boldsymbol{v}(\boldsymbol{x}) = 0 \qquad \text{in } (0, 1] \times \Omega, \tag{6a}$$

$$m(t, \boldsymbol{x}) = m_0(\boldsymbol{x}) \qquad \text{in } \{0\} \times \Omega. \tag{6b}$$

The solution of this PDE is the transported intensities m of the template image m_0. The endpoint $m(1) := m(1, \cdot)$ at time $t = 1$ corresponds to the deformed template image for some trial velocity $\boldsymbol{v}$.

The second building block of our variational problem formulation is the distance functional $\text{dist} : \mathfrak{I} \times \mathfrak{I} \to \mathbb{R}$ in (1a) that quantifies the discrepancy between the transported template image $m(1) := m(1, \cdot)$ at time $t = 1$ and the reference image m_1. A common choice for this terminal (endpoint) cost is given by the squared L^2-distance:

$$\text{dist}(m(1), m_1) = \frac{1}{2} \int_\Omega (m(1, \boldsymbol{x}) - m_1(\boldsymbol{x}))^2 \, \mathrm{d}\boldsymbol{x}.$$

While this is a common choice in many diffeomorphic image registration packages [1, 8], this distance measure can only be used for registering images acquired using the same imaging modality. We present an alternative in the appendix.

The last building block is the regularization functional $\text{reg} : \mathfrak{V} \to \mathbb{R}$. Motivated by the problem formulation presented in Sect. 2.1, we use

$$\text{reg}(\boldsymbol{v}) = \frac{\alpha}{2} \|\boldsymbol{v}\|_{\mathfrak{H}}^2 = \frac{\alpha}{2} \langle \mathcal{L}\boldsymbol{v}, \boldsymbol{v} \rangle_{\mathbb{R}^d},$$

where $\mathcal{L} : \mathfrak{V} \to \mathfrak{V}^*$ is a differential operator of adequate order. CLAIRE, in general, features H^1-, H^2-, and H^3-norms and seminorms for the regularization of $\boldsymbol{v}$ [1, 30, 33]. The default regularization operator is an H^1-seminorm, i.e., $\mathcal{L} = -\Delta_d$, with an additional H^1-norm that penalizes the divergence of the velocity [1, 90, 91, 112]. We provide additional details in the appendix.

Putting everything together, we arrive at the PDE-constrained optimization problem:

$$\underset{m\in\mathfrak{M}_{\mathrm{ad}},\, \boldsymbol{v}\in\mathfrak{V}_{\mathrm{ad}}}{\text{minimize}} \quad \frac{1}{2}\int_{\Omega}(m(1,\boldsymbol{x})-m_1(\boldsymbol{x}))^2\,\mathrm{d}\boldsymbol{x}+\frac{\alpha}{2}\|\boldsymbol{v}\|_{\mathfrak{H}}^2 \tag{7a}$$

$$\begin{aligned}\text{subject to} \quad \partial_t m(t,\boldsymbol{x})+\nabla m(t,\boldsymbol{x})\cdot\boldsymbol{v}(\boldsymbol{x})&=0 && \text{in } (0,1]\times\Omega,\\ m(t,\boldsymbol{x})&=m_0(\boldsymbol{x}) && \text{in } \{0\}\times\Omega.\end{aligned} \tag{7b}$$

Similar problem formulations have been considered in [31, 32, 39, 40]. For simplicity, we discuss the numerical methods for the problem formulation in (7). However, we note that we considered different variants of this formulation in our past work [1, 30, 33, 90]. We discuss these in greater detail in the appendix.

2.3 Optimality Conditions

In the present work, we consider an *optimize-then-discretize* approach. The advantages of this approach are that the formal optimality conditions are straightforward to derive. They also retain interpretability; for example, we will see that the adjoint equation of the transport equation in (6) represents a continuity equation for the image mismatch (see (10)). Moreover, one can freely decide on the numerical methods to solve the PDEs associated with the optimality conditions. A disadvantage of this approach is that the discrete gradient is (in general) not consistent with the discretized objective functional (contingent on the numerical scheme used for discretization). Consequently, it is not possible to solve the variational optimization problem with arbitrary accuracy (i.e., to machine precision). A *discretize-then-optimize* approach guarantees that the discrete gradient is consistent with the discretized objective functional. However, depending on the discretization this approach also has drawbacks. We refer to [17] for a general discussion and to [48, 62] for examples of *discretize-then-optimize* implementations for a problem of the form (7).

We consider the method of Lagrange multipliers to solve (1). We introduce the dual variable $\lambda : [0,1]\times\bar{\Omega}\rightarrow\mathbb{R}$, $\lambda\in\mathfrak{M}^*$, for the transport Eq. (6). The Lagrangian functional is given by

$$\begin{aligned}\ell(\Xi) &= \frac{1}{2}\int_{\Omega}(m(1,\boldsymbol{x})-m_1(\boldsymbol{x}))^2\,\mathrm{d}\boldsymbol{x}+\frac{\alpha}{2}\langle\mathcal{L}\boldsymbol{v},\boldsymbol{v}\rangle_{\mathbb{R}^d}\\ &\quad+\int_0^1\langle\partial_t m+\nabla m\cdot\boldsymbol{v},\lambda\rangle_{L^2(\Omega)}\mathrm{d}t+\langle m(0)-m_0,\lambda\rangle_{L^2(\Omega)}\end{aligned} \tag{8}$$

where $\boldsymbol{\Xi} := (\boldsymbol{v}, m, \lambda)\in\mathfrak{V}\times\mathfrak{M}\times\mathfrak{M}^*$.

Computing first variations with respect to the control variable $\boldsymbol{v}$ yields the reduced gradient

$$g[\boldsymbol{v}](\boldsymbol{x}) := \alpha\mathcal{L}[\boldsymbol{v}](\boldsymbol{x}) + \int_0^1 \lambda(t,\boldsymbol{x})\nabla m(t,\boldsymbol{x})\,\mathrm{d}t. \tag{9}$$

To be able to evaluate the reduced gradient we require the state variable $m \in \mathfrak{M}$ and the dual variable $\lambda \in \mathfrak{M}^*$. We can find the state variable by solving (6) forward in time. Formally, this equation is obtained by computing the first variations of ℓ in (8) with respect to λ. The dual variable λ can be found by solving the adjoint equations backward in time. Formally, the adjoint equations are found by computing variations of ℓ in (8) with respect to m. We obtain

$$-\partial_t\lambda(t,\boldsymbol{x}) - \nabla\cdot\lambda(t,\boldsymbol{x})\boldsymbol{v}(\boldsymbol{x}) = 0 \quad \text{in } [0,1)\times\Omega, \tag{10a}$$

$$\lambda(t,\boldsymbol{x}) = -(m(1,\boldsymbol{x}) - m_1(\boldsymbol{x})) \quad \text{in } \{1\}\times\Omega, \tag{10b}$$

subject to periodic boundary conditions on $\partial\Omega$. Notice that this equation represents a continuity equation; we transport the mismatch between the deformed template image $m(1,\cdot)$ and the reference image m_1 backward in time. If we change the distance measure dist : $\mathfrak{I}\times\mathfrak{I}\to\mathbb{R}$ in (7), the final conditions in (10b) will change.

2.4 Newton Step

We consider a (Gauss–)Newton–Krylov method for numerical optimization [30]. We provide more details in Sect. 2.5. The PDE operators associated with the Hessian can be found by formally computing second-order variations of the Lagrangian ℓ in (8). The expression for the Hessian matvec—i.e., the application of the Hessian to a vector $\tilde{\boldsymbol{v}}$—is given by

$$\begin{aligned}\mathcal{H}[\tilde{\boldsymbol{v}}](\boldsymbol{x}) &= \mathcal{H}_{\text{reg}}[\tilde{\boldsymbol{v}}](\boldsymbol{x}) + \mathcal{H}_{\text{dat}}[\tilde{\boldsymbol{v}}](\boldsymbol{x})\\ &= \alpha\mathcal{L}[\tilde{\boldsymbol{v}}](\boldsymbol{x}) + \int_0^1 \left\{\tilde{\lambda}(t,\boldsymbol{x})\nabla m(t,\boldsymbol{x}) + \lambda(t,\boldsymbol{x})\nabla\tilde{m}(t,\boldsymbol{x})\right\}\mathrm{d}t.\end{aligned} \tag{11}$$

The variable $\tilde{\boldsymbol{v}} : \bar{\Omega}\to\mathbb{R}^d$, $\boldsymbol{v}\in\mathfrak{V}$, represents the incremental control variable, i.e., the search direction for $\boldsymbol{v}$. The operators $\mathcal{H}_{\text{reg}}$ and $\mathcal{H}_{\text{dat}}$ denote the regularization part and the data part of the reduced-space Hessian, respectively. For the latter, the dependence on $\tilde{\boldsymbol{v}}$ is hidden in the incremental PDE operators. Given a candidate $\boldsymbol{v}$ and a candidate $\tilde{\boldsymbol{v}}$ we require the state variable $m : [0,1]\times\bar{\Omega}\to\mathbb{R}$, the dual variable $\lambda : [0,1]\times\bar{\Omega}\to\mathbb{R}$, the incremental state variable $\tilde{m} : [0,1]\times\bar{\Omega}\to\mathbb{R}$, and the incremental adjoint variable $\tilde{\lambda} : [0,1]\times\bar{\Omega}\to\mathbb{R}$. We can find the state and dual variables during the evaluation of the reduced gradient in (9). The incremental state variable can be found by solving

$$\partial_t \tilde{m}(t, \boldsymbol{x}) + \nabla \tilde{m}(t, \boldsymbol{x}) \cdot \boldsymbol{v}(t, \boldsymbol{x}) + \nabla m(t, \boldsymbol{x}) \cdot \tilde{\boldsymbol{v}}(t, \boldsymbol{x}) = 0 \quad \text{in } (0, 1] \times \Omega, \tag{12a}$$

$$\tilde{m}(t, \boldsymbol{x}) = 0 \quad \text{in } \{0\} \times \Omega, \tag{12b}$$

subject to periodic boundary conditions on $\partial\Omega$ forward in time. We can find the incremental dual variable $\tilde{\lambda}$ by solving

$$-\partial_t \tilde{\lambda}(t, \boldsymbol{x}) + \nabla \cdot (\tilde{\lambda}(t, \boldsymbol{x})\boldsymbol{v}(\boldsymbol{x}) + \lambda(t, \boldsymbol{x})\tilde{\boldsymbol{v}}(\boldsymbol{x})) = 0 \quad \text{in } [0, 1) \times \Omega, \tag{13a}$$

$$\tilde{\lambda}(t, \boldsymbol{x}) = -\tilde{m}(t, \boldsymbol{x}) \quad \text{in } \{1\} \times \Omega, \tag{13b}$$

subject to periodic boundary conditions on $\partial\Omega$ backward in time. Consequently, each time we apply $\mathcal{H}$ to a vector, we have to solve two PDEs.

2.5 Numerics

The numerical implementation discussed below is based on the computational kernels described in [1, 1, 30, 71, 72, 90, 91, 93]. The hardware-accelerated CPU implementation is described in [1, 71, 93]. The GPU implementation is described in [90, 91].

We note that the GPU implementation is only available in single precision. This poses several numerical challenges. In particular, we observed that our scheme does not allow us to solve the optimization problem to arbitrary accuracy. This is due to the accumulation of numerical errors, dominated by the time integration and numerical differentiation. Moreover, the numerical gradient is inconsistent with the objective function. This is caused by our particular choice of a numerical time integrator in conjunction with an optimize-then-discretize approach.

2.5.1 Discretization

We consider a nodal discretization in space. That is, we subdivide the spatial interval $\Omega = [-\pi, \pi)^d \subset \mathbb{R}^d$ into $n_i \in \mathbb{N}$ cells of width $h_i > 0$, $i = 1, \ldots, d$, along each spatial direction x_i, $i = 1, \ldots, d$. The width of the cells $\boldsymbol{h} = (h_1, \ldots, h_d) \in \mathbb{R}^d$ along each spatial direction is given by $h_i = 2\pi/n_i$, $i = 1, \ldots, d$. Let $\boldsymbol{x}_{\boldsymbol{l}} \in \mathbb{R}^d$ denote a mesh point at index $\boldsymbol{l} = (l_1, \ldots, l_d) \in \mathbb{N}^d$, $1 \le l_i \le n_i$. The coordinates $\boldsymbol{x}_{\boldsymbol{l}}$ are computed according to

$$\boldsymbol{x}_{\boldsymbol{l}} = ((\boldsymbol{n}/2) - \boldsymbol{l}) \odot \boldsymbol{h}.$$

Here, $\odot$ denotes an elementwise multiplication between two vectors (Hadamard product) and $\boldsymbol{n} = (n_1, \ldots, n_d) \in \mathbb{N}^d$ represents the number of mesh points along each spatial direction. We denote the resulting mesh by $\boldsymbol{\Omega} = (\mathbf{x}_l) \in \mathbb{R}^{d,n_1,\ldots,n_d}$. Similarly, we subdivide the unit time interval [0, 1] into a uniform mesh with step size $h_t = 1/n_t$. We discretize integrals using a trapezoidal rule.

2.5.2 Time Integration

We use a SL method for numerical time integration [140]. The prototype implementation of this time integrator is described in [72]. Different variants of hardware-accelerated implementations are described in [71, 90, 91]. Other works that consider a SL scheme in a similar context are [8, 78].

The SL scheme is a hybrid between Eulerian and Lagrangian methods. It is unconditionally stable, i.e., we can select the time step size $h_t > 0$ solely based on accuracy considerations. To apply the SL scheme to the PDEs that appear in our optimality system, we need to bring them into the general form:

$$\mathrm{d}_t u = f(t, u, \boldsymbol{v}, \ldots) \tag{14}$$

where $u : [0, 1] \times \Omega \to \mathbb{R}$ denotes an arbitrary scalar function, $\mathrm{d}_t := \partial_t + \boldsymbol{v} \cdot \nabla$ denotes the *material derivative*, and the right-hand-side f represents all remaining terms. To obtain this representation for the equations considered here, we use the vector calculus identity $\nabla \cdot u\boldsymbol{v} = u\nabla \cdot \boldsymbol{v} + \nabla u \cdot \boldsymbol{v}$.

In the first step, we have to compute the characteristic $\boldsymbol{y} : [t^j, t^{j+1}] \to \mathbb{R}^d$ along which particles flow between the timepoint t^j and t^{j+1}, $j = 1, \ldots, n_t$. The question we seek to answer is where particles at time t^{j+1} originate from a given data at time t^j. That is, we compute the departure point $\boldsymbol{y}$ at time $t = t^j$. To compute this characteristic, we solve the ODE $\mathrm{d}_t \boldsymbol{y} = \boldsymbol{v} \circ \boldsymbol{y}$ for $t \in [t^j, t^{j+1})$ with $\boldsymbol{y} = \boldsymbol{x}$ for $t = t^{j+1}$ backward in time. We illustrate the computation of the departure point in Fig. 2.

In our implementation, we compute the characteristics $\boldsymbol{y}$ using an RK2 method. Notice that the velocity $\boldsymbol{v}$ is constant in time; this simplifies the computation of the transported quantities considerably. Let $\mathbf{v}_l := \mathbf{v}(\mathbf{x}_l)$ denote the discretized velocity at a given mesh point $\mathbf{x}_l \in \boldsymbol{\Omega}$. We obtain the lth query point $\mathbf{y}_l \in \mathbb{R}^d$ associated with $\mathbf{x}_l \in \boldsymbol{\Omega}$ according to

$$\begin{aligned}
\tilde{\mathbf{y}}_l &\leftarrow \mathbf{x}_l - h_t \mathbf{v}(\mathbf{x}_l) \\
\mathbf{y}_l &\leftarrow \mathbf{x}_l - \frac{h_t}{2} \left(\mathbf{v}(\mathbf{x}_l) + \mathbf{v}(\tilde{\mathbf{y}}_l) \right).
\end{aligned}$$

The intermediate query points $\tilde{\mathbf{y}}_l$ and the final query point $\mathbf{y}_l$ (i.e., the departure point) are—in general—off-grid locations. Therefore, evaluating quantities of interest at these locations requires interpolation (see Sect. 2.5.4 for details). If (14)

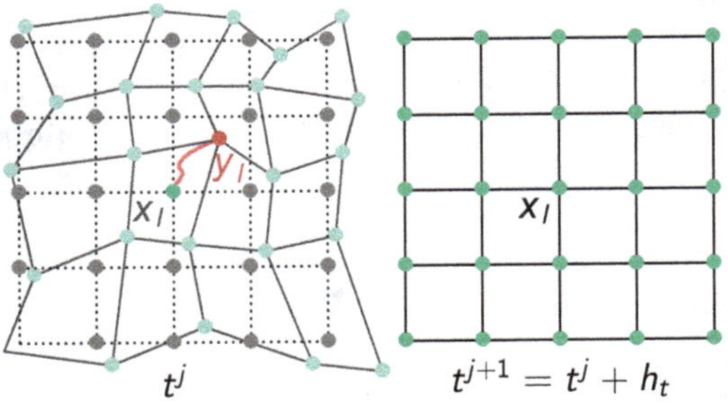

Fig. 2 Illustration of the computation of the characteristic $\boldsymbol{y}$ in the SL scheme. In the SL scheme, we have to compute the departure points at time t^j. To do so, we start with a regular grid at time t^{j+1} and solve for the departure point $\boldsymbol{y}_l$ at a given point $\boldsymbol{x}_l$ at mesh index $\boldsymbol{l}$ backward in time (green line in the graphic on the left). The deformed grid configuration is overlaid onto the initial regular grid at time t^j. (Figure modified from [48, 90])

is homogeneous, i.e., $f = 0$, we only interpolate the transported quantity to obtain its value at the departure point $\mathbf{y}_l$ at time t^j and assign the resulting value to the regular mesh point $\mathbf{x}_l \in \boldsymbol{\Omega}$ at time t^{j+1}. That is,

$$u(t^{j+1}, \mathbf{x}_l) \leftarrow u(t^j, \mathbf{y}_l).$$

If (14) is not homogeneous, i.e., $f \neq 0$, we have to solve the ODE (14) along the characteristic $\boldsymbol{y}$ forward in time. We do so using an RK2 scheme. That is,

$$\begin{aligned} f_0 &\leftarrow f(t^j, u(t^j, \mathbf{y}_l), \mathbf{v}(\mathbf{y}_l), \ldots) \\ \tilde{u}(t^{j+1}, \mathbf{x}_l) &\leftarrow u(t^j, \mathbf{y}_l) + h_t f_0 \\ f_1 &\leftarrow f(t^{j+1}, \tilde{u}(t^{j+1}, \mathbf{x}_l), \mathbf{v}(\mathbf{x}_l), \ldots) \\ u(t^{j+1}, \mathbf{x}_l) &\leftarrow u(t^j, \mathbf{y}_l) - h_t(f_0 + f_1)/2. \end{aligned}$$

Again, quantities evaluated at the query point $\mathbf{y}_l$ at time t^j need to be interpolated. We note that these functions live on a curvilinear mesh (see Fig. 2). Since we use spectral methods with a Fourier basis, we cannot evaluate the differential operators that appear in f on such a mesh. As a remedy, we do not compute the derivative on this curvilinear mesh but interpolate (i.e., transport) the derivatives evaluated on a regular mesh instead.

2.5.3 Differentiation

In our past work, we considered pseudo-spectral methods with a Fourier basis for numerical differentiation [1, 30, 71, 72].

For our GPU implementation [90, 91] we have designed a mixed-precision approach to improve scalability and computational throughput. We consider 8th-order finite differences for first-order derivatives (i.e., the gradient and divergence operators). Higher-order derivative operators (e.g., the Laplacian operator Δ) and their inverse are implemented using a pseudo-spectral discretization with a Fourier basis. That is, we model an arbitrary function $u : \bar{\Omega} \to \mathbb{R}$ discretized on a regular mesh $\mathbf{\Omega}$ at grid points $\mathbf{x}_{\boldsymbol{l}} \in \mathbf{\Omega}$, $\boldsymbol{l} = (l_1, \ldots, l_d) \in \mathbb{N}^d$, $l_i = 1, \ldots, n_i$, as $u_{\boldsymbol{l}} := u(\mathbf{x}_{\boldsymbol{l}})$,

$$u_{\boldsymbol{l}} = \sum_{k_1=-(n_1/2)+1}^{n_1/2} \cdots \sum_{k_d=-(n_d/2)+1}^{n_d/2} \hat{u}_{\boldsymbol{k}} \exp(-\mathrm{i}\langle \boldsymbol{k}, \mathbf{x}_{\boldsymbol{l}} \rangle_{\mathbb{R}^d})$$

with $\boldsymbol{k} \in \mathbb{Z}^d$ and spectral coefficients $\hat{u}_{\boldsymbol{k}}$. This spectral representation is the reason why we assume periodic boundary conditions in our continuous model. We note that images may not necessarily be periodic functions. We can address this by zero-padding the datasets and applying a mollifier close to the boundary $\partial\Omega$. The mapping between the coefficients $\{u_{\boldsymbol{l}}\}$ and $\{\hat{u}_{\boldsymbol{k}}\}$ is done using forward and inverse FFTs. In our CPU implementation, we considered a pencil decomposition [1, 71, 93] (see Fig. 3; right). Here, 1D FFTs along each spatial direction are computed based on the FFTW library. FFTs along other directions are then obtained by transposing the data, resulting in large communication costs. For the single-GPU implementation described in [91] we switched to cuFFT for 3D FFTs. The multi-GPU implementation described in [90] uses a combination of cuFFT and a new 2D slab decomposition (see Fig. 3; middle). This enables us to utilize the highly optimized 2D cuFFT on each GPU. We decompose the spatial domain in the outermost dimension (i.e., x_1) and in the spectral domain in the x_2 direction. Consequently, the innermost x_3 direction remains continuous in memory. This reduces misaligned memory access for the communication of the transpose operations. The real-to-complex transformation is divided into three steps: First, we execute cuFFT's batched 2D FFTs in the plane spanned by the x_2- and x_3-axis. Then, we transpose the complex data to a decomposition in x_2 direction. Then, we apply cuFFT's batched 1D FFTs to the x_1 direction, which is noncontinuous in memory.

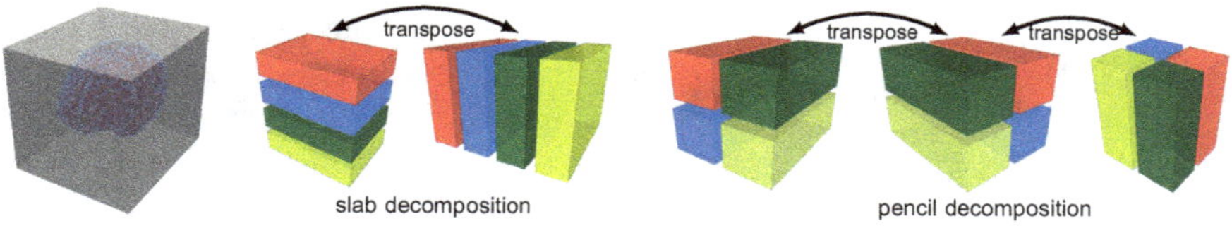

Fig. 3 Domain decomposition for memory-distributed implementation. In each case, we assume that we use four MPI tasks to distribute our data (e.g., four GPUs or four nodes). Left: 3D volume rendering of medical imaging data set (brain image). Middle: Slab decomposition (1D domain decomposition) considered in our GPU implementation [90]. We decompose the spatial domain in the outermost dimension. We transpose the data only once. On the right we illustrate the pencil decomposition (2D domain decomposition) of the data considered in our CPU implementation [1, 71, 93]. The computation in this data layout involves three transposes

For the inverse complex-to-real transformation, these three steps are executed in reverse order, using the respective inverse transformations. For the execution on multiple GPUs, we use CUDA-aware MPI to eliminate expensive on-node host-device transfers [90].

The 8th-order finite difference approximation of the first-order derivative along the ith coordinate direction x_i at a mesh point $\mathbf{x}_l \in \boldsymbol{\Omega}$ is given by

$$\partial_i u(\mathbf{x}_l) \approx \frac{u_i^-(\mathbf{x}_l) + u_i^+(\mathbf{x}_l)}{840 h_i}$$

with $u_i^-(\mathbf{x}_l) := 3u_{l-4e_i} - 32u_{l-3e_i} + 168u_{l-2e_i} - 672u_{l-e_i}$, $u_i^+(\mathbf{x}_l) := 672u_{l+e_i} - 168u_{l+2e_i} + 32u_{l+3e_i} - 3u_{l+4e_i}$, and unit vectors $\boldsymbol{e}_i \in \{0, 1\}^d$, where the jth entry is one for $j = i$ and zero otherwise.

The key limiting factor to obtain optimal strong and weak scalability for our method is the communication costs associated with the FFT. We refer to [90, 91] for additional details.

2.5.4 Interpolation

In our past work [90, 91], we implemented different interpolation models. To obtain the interpolated value of a function u at an arbitrary query point $\mathbf{x} = (x_1, \ldots, x_d) \in \mathbb{R}^d$, we evaluate

$$u(\mathbf{x}) = \sum_{k_1=1}^{p+1} \cdots \sum_{k_d=1}^{p+1} c_{k_1,\ldots,k_d} \prod_{i=1}^{d} \phi_{k_i}(x_i),$$

where $p \in \mathbb{N}$ denotes the polynomial order; $\phi_j : \mathbb{R} \to \mathbb{R}$, $j = 1, \ldots, p+1$, are the polynomial basis functions; and $c_{k_1,\ldots,k_d} \in \mathbb{R}$ are the coefficients. For Lagrange interpolation, the coefficients are identical to the grid values of the discretized function u and ϕ_j are the Lagrange polynomials.

The numerical accuracy and computational performance of different variants of the interpolation kernel are discussed in [91]. The multi-GPU implementation is described in [90]. We use NVIDIA's libraries for texture-based trilinear interpolation [141]. In [91], we also transferred our CPU kernels for cubic Lagrange interpolation [1, 33, 93] to the GPU. We implemented two variants, one that uses texture units for lookup tables and one that implements texture-based interpolation [91]. The latter implementation is similar to [142]; it yields higher computational throughput at lower accuracy. Lastly, we have developed a texture-based B-spline interpolation, the computational kernels of which are inspired by Ruijters et al. [142], Ruijters and Thévenaz [143], and Champagnat and Le Sant [144]. For the execution on multiple GPUs, we use CUDA-aware MPI to eliminate expensive on-node host-device transfers [90]. We utilize the `thrust` library [145] to determine which query points need to be processed by which GPU, thereby completely eliminating host-side computation. We use sparse point-to-point communication to send points to other processors. We adaptively allocate memory for the respective MPI buffers. We do this by computing an estimate of the maximal displacement of grid points along the computed trajectories based on the CFL number of the velocity field. We refer to [90, 91] for additional details.

2.5.5 Optimization

We use an iterative method globalized by an Armijo line search [146, 147]. The outer iterations of our algorithm are summarized in Algorithm 1 in the appendix. At (outer) iteration $k \in \mathbb{N}$, we update the iterate $\mathbf{v}^{(k)}$ according to

$$\mathbf{B}^{(k)}\tilde{\mathbf{v}}^{(k)} = -\mathbf{g}^{(k)}, \quad \mathbf{v}^{(k+1)} = \mathbf{v}^{(k)} + \gamma^{(k)}\tilde{\mathbf{v}}^{(k)},$$

where $\tilde{\mathbf{v}}^{(k)} \in \mathbb{R}^{dn}$ denotes the search direction, $\mathbf{B}^{(k)} \in \mathbb{R}^{dn,dn}$ is a positive-definite matrix, and $\gamma^{(k)} > 0$ is the step size.

For $\mathbf{B}^{(k)} = \mathrm{diag}(1, \ldots, 1) \in \mathbb{R}^{dn,dn}$, the scheme above corresponds to a gradient descent algorithm. In [30], we consider a preconditioned gradient descent algorithm. This scheme is more stable and yields an improved convergence behavior. The preconditioner is the inverse of the regularization operator $\mathcal{L}$. That is, $\mathbf{B}^{(k)} = \alpha\mathbf{L}$, where $\mathbf{L} \succeq \mathbf{0}$ denotes the discretization of $\mathcal{L}$. This scheme can be viewed as a Picard iteration. We note that our spectral discretization allows us to apply the inverse of this operator with vanishing costs; the complexity of inverting $\mathbf{L}$ is $\mathcal{O}(n \log n)$ regardless of the Sobolev norm we consider. If the operator $\mathcal{L}$ has a nontrivial kernel, we set the spectral coefficients that are zero to one before inverting $\alpha\mathbf{L}$. Consequently, the search direction $\tilde{\mathbf{v}}^{(k)}$ is given by $\tilde{\mathbf{v}}^{(k)} = -(\alpha\mathbf{L})^{-1}\mathbf{g}^{(k)}$.

In addition, we have designed a (Gauss)–Newton–Krylov algorithm for numerical optimization [1, 30]. Here, $\mathbf{B}^{(k)}$ corresponds to the Hessian matrix $\mathbf{H}^{(k)}$ at (outer) iteration k. Consequently, we have to invert $\mathbf{H}^{(k)}$ at each iteration. We note that forming and storing $\mathbf{H}^{(k)}$ results in prohibitive computational costs and memory requirements. As such, we cannot use direct methods [148, 149]. Instead, we use iterative methods to invert $\mathbf{H}^{(k)}$. In particular, we use matrix-free Krylov subspace methods—more precisely, a PCG algorithm [150]—to compute the action of the inverse of $\mathbf{H}^{(k)}$ on the vector $-\mathbf{g}^{(k)}$. As such, we only require an expression for the Hessian matvec. This is precisely what is given by (11). Thus, we need to evaluate (11) at every inner iteration of our Krylov subspace method. This involves solving the PDEs (12) and (13) at every inner iteration of the PCG algorithm. These matvecs constitute the most expensive part of our algorithm. We summarize this algorithm in Algorithm 2 in the appendix.

We note that we can use other iterative methods to compute the action of the inverse of the Hessian. In fact, we have tested different methods. In our experiments, we did not observe any issues with the PCG algorithm nor did we see any benefits from using different iterative methods. Since the Hessian is (also for all practical purposes, in computation) a symmetric positive definite operator, we prefer to use the PCG method over, e.g., GMRES. Having said this, we note that CLAIRE supports different Krylov subspace methods via PETSc [115, 116]. We discuss this in greater detail in the next subsection.

Since the considered optimization problem is, in general, nonconvex, one additional challenge that arises is that the Hessian is not guaranteed to be positive definite, especially far away from a (local) minimizer. One approach to address this issue is to terminate the PCG algorithm as soon as one detects negative curvature. In this case, we use the former iterate of the PCG algorithm as a search direction. We consider a Gauss–Newton approximation to $\mathbf{H}^{(k)}$ [1, 30, 48] as an alternative to this approach. This approximation is guaranteed to be positive semi-definite. On the downside, we can expect the convergence to drop from quadratic to superlinear. This Gauss–Newton approximation is obtained by dropping all terms that involve the dual variable λ in (11) and (13), respectively. Notice that the final condition for the dual variable λ in (10) corresponds to the mismatch between the transported intensities of the template image m_0 and the reference image m_1. Thus, as we approach a (local) minimizer of our problem, we can expect that λ tends to zero; our Gauss–Newton approximation becomes exact and we recover quadratic convergence.

To further amortize computational costs, we do not invert $\mathbf{H}^{(k)}$ exactly. We consider an inexact scheme [146, 151, 152]. This is accomplished by selecting the stopping condition for the PCG method to be proportional to the norm of the reduced gradient; as we approach a (local) minimizer, the tolerance decreases and we solve for the search direction more accurately. That is, we terminate the algorithm if

$$\|\mathbf{r}^{(k)}\|_\infty \leq \eta^{(k)}\|\mathbf{g}^{(k)}\|_\infty, \quad \mathbf{r}^{(k)} := \mathbf{H}^{(k)}\tilde{\mathbf{v}}^{(k)} + \mathbf{g}^{(k)},$$

with forcing sequence $\eta^{(k)} = \min(1/2, \sqrt{\|\mathbf{g}^{(k)}\|_\infty})$ or $\eta^{(k)} = \min(1/2, \|\mathbf{g}^{(k)}\|_\infty)$ for superlinear or quadratic convergence, respectively. See Algorithm 2, line 13, in the appendix.

In [30] we demonstrate that the preconditioned gradient descent scheme is less effective than our (Gauss–)Newton–Krylov scheme. As such, we only consider our (Gauss–)Newton–Krylov algorithm here.

We terminate the optimization if we reduce the gradient by $\epsilon_{\text{opt}} > 0$, i.e.,

$$\|\mathbf{g}^{(k)}\|_\infty \leq \epsilon_{\text{opt}}\|\mathbf{g}^{(0)}\|_\infty$$

or if $\|\mathbf{g}^{(k)}\|_\infty \leq 1e\text{–}6$. We have implemented alternative stopping criteria [30] but do not consider them here.

2.5.6 Preconditioning

The main cost of the (Gauss–)Newton–Krylov algorithm is the solution of the linear system:

$$\mathbf{H}^{(k)}\tilde{\mathbf{v}}^{(k)} = -\mathbf{g}^{(k)}, \quad k = 1, 2, 3, \ldots, \tag{15}$$

at each outer iteration k, with $\mathbf{H}^{(k)} = \mathbf{H}_{\text{reg}}^{(k)} + \mathbf{H}_{\text{dat}}^{(k)}$, where $\mathbf{H}_{\text{reg}}^{(k)} = \mathbf{H}_{\text{reg}} \in \mathbb{R}^{dn,dn}$ is a discrete representation of the regularization operator $\mathcal{H}_{\text{reg}} = \alpha\mathcal{L}$ and $\mathbf{H}_{\text{dat}}^{(k)} \in \mathbb{R}^{dn,dn}$ is the discrete version of $\mathcal{H}_{\text{dat}}$ in (11). For the model outlined in Sect. 2.2 the Hessian behaves like a compact operator; large eigenvalues are associated with smooth eigenvectors and the eigenvectors become more oscillatory as the eigenvalues decrease [30].

To amortize the computational costs of our algorithm and make it competitive with gradient descent schemes that consider first-order derivative information only, we have to design effective methods for preconditioning the linear system given above. That is, we seek a matrix $\mathbf{M}^{(k)} \succ \mathbf{0}$ such that, ideally, $(\mathbf{M}^{(k)})^{-1}\mathbf{H}^{(k)} \approx \mathbf{I}_{dn}$, $\mathbf{I}_{dn} := \text{diag}(1, \dots, 1) \in \mathbb{R}^{dn,dn}$. This makes approximations to $\mathbf{H}^{(k)}$ (which are "easy" to invert) an obvious choice.

Regularization Preconditioner

A common choice in PDE-constrained optimization is to consider the regularization operator $\mathbf{H}_{\text{reg}}$ as a preconditioner $\mathbf{M}$ [153–155]. The preconditioned Hessian is a perturbation of the identity, i.e.,

$$\big(\mathbf{H}_{\text{reg}}\big)^{-1}\mathbf{H}^{(k)} = \mathbf{I}_{dn} + \big(\mathbf{H}_{\text{reg}}\big)^{-1}\mathbf{H}_{\text{dat}}^{(k)}$$

with $\mathbf{I}_{dn} := \text{diag}(1, \dots, 1) \in \mathbb{R}^{dn,dn}$. Since $\mathbf{H}_{\text{reg}} \succeq \mathbf{0}$ is a (high-order) differential operator (typically, a Helmholtz-type operator), its inverse acts as a smoother. We note that applying the inverse of $\mathbf{H}_{\text{reg}}$ has a complexity of $O(n \log n)$ in our implementation, i.e., we have to compute two FFTs and a diagonal scaling in the spectral domain using the appropriate Fourier coefficients. As such, this strategy for preconditioning the reduced-space Hessian has vanishing costs. This preconditioner has been considered in [15, 30, 33, 71]. The performance of this preconditioner is mesh independent (assuming we can entirely resolve the problem on the coarsest mesh). However, it deteriorates significantly as we decrease the regularization parameter α.

Two-Level Preconditioner

Inspired by multigrid approaches, we designed a two-level preconditioner for the reduced-space Hessian [1, 48]. We use a coarse grid approximation of the inverse of the reduced-space Hessian as a preconditioner. The basic idea is to iterate only on the low-frequency part and ignore the high-frequency components. That is, we use the inverse of the reduced-space Hessian $\mathbf{H}^{(k)}$, inverted on a coarser grid, as a preconditioner. This idea is motivated by the work in [156–161]. For simplicity of notation, we drop the dependence of the Hessian on the outer iteration index k.

We decompose the Hessian into two operators $\mathbf{H}_L$ and $\mathbf{H}_H$—one acting on low and the other acting on high frequencies, respectively. We denote the operators that project on the low- and high-frequency subspaces by $\mathbf{P}_L : \mathbb{R}^{dn} \to \mathbb{R}^{dn}$ and $\mathbf{P}_H : \mathbb{R}^{dn} \to \mathbb{R}^{dn}$. Let $\mathbf{e}_j \in \mathbb{R}^n$, $(\mathbf{e}_j)_i = 1$ if $j = i$ and $(\mathbf{e}_j)_i = 0$ for $i \neq j$, $i, j = 1, \ldots, n$, denote an eigenvector of $\mathbf{H}$ with $(\mathbf{P}_L\mathbf{H}\mathbf{P}_H)\mathbf{e}_j = (\mathbf{P}_H\mathbf{H}\mathbf{P}_L)\mathbf{e}_j = \mathbf{0}$. Then, with $\mathbf{P}_H + \mathbf{P}_L = \mathbf{I}_{dn}$, $\mathbf{I}_{dn} := \operatorname{diag}(1, \ldots, 1) \in \mathbb{R}^{dn,dn}$, we have

$$\mathbf{H}\mathbf{e}_j = (\mathbf{P}_H + \mathbf{P}_L)\mathbf{H}(\mathbf{P}_H + \mathbf{P}_L)\mathbf{e}_j = \mathbf{P}_H\mathbf{H}\mathbf{P}_H\mathbf{e}_j + \mathbf{P}_L\mathbf{H}\mathbf{P}_L\mathbf{e}_j.$$

In general, this equality will not hold. However, we are not interested in using this model as a surrogate for the Hessian $\mathbf{H}$; we are merely interested in designing an effective preconditioner $\mathbf{M}$ so that $\operatorname{cond}(\mathbf{M}^{-1}\mathbf{H}) \ll \operatorname{cond}(\mathbf{H})$.

Suppose we can decompose $\tilde{\mathbf{v}} \in \mathbb{R}^{dn}$ into a smooth component $\tilde{\mathbf{v}}_L \in \mathbb{R}^{dn}$ and a high-frequency component $\tilde{\mathbf{v}}_H \in \mathbb{R}^{dn}$, where each of these vectors can be found by solving

$$\mathbf{H}_L\tilde{\mathbf{v}}_L = (\mathbf{P}_L\mathbf{H}\mathbf{P}_L)\tilde{\mathbf{v}}_L = -\mathbf{P}_L\mathbf{g} \qquad \text{and} \qquad \mathbf{H}_H\tilde{\mathbf{v}}_H = (\mathbf{P}_H\mathbf{H}\mathbf{P}_H)\tilde{\mathbf{v}}_H = -\mathbf{P}_H\mathbf{g},$$

respectively. We use this construction to design an effective preconditioner for the smooth spectrum of our problem. Let $\mathbf{r} \in \mathbb{R}^{dn}$ denote the vector we apply the inverse of our preconditioner $\mathbf{M} \in \mathbb{R}^{dn,dn}$ to. Since our implementation is matrix-free, we iteratively solve $\mathbf{M}\mathbf{s} = \mathbf{r}$ to obtain the action of the inverse of $\mathbf{M}$ on $\mathbf{r}$. In the spirit of the conceptual idea introduced above, we assume that we can decompose $\mathbf{s}$ into a smooth component $\mathbf{s}_L$ and a high-frequency component $\mathbf{s}_H$. Let $\mathbf{Q}_R \in \mathbb{R}^{dn/2,dn}$ denote a restriction operator and $\mathbf{Q}_P \in \mathbb{R}^{dn,dn/2}$ denote prolongation operator. Moreover, let $\mathbf{F}_L \in \mathbb{R}^{dn,dn}$ and $\mathbf{F}_H \in \mathbb{R}^{dn,dn}$ denote a low- and high-pass filter, respectively. We project the vector $\mathbf{r} \in \mathbb{R}^{dn}$ to a vector $\mathbf{r}_L \in \mathbb{R}^{dn/2}$ by filtering the high-frequency components and restricting the resulting vector to a coarser mesh, i.e., $\mathbf{r}_L = \mathbf{Q}_R\mathbf{F}_L\mathbf{r}$. Subsequently, we obtain the smooth component $\mathbf{s}_L$ by solving

$$\tilde{\mathbf{M}}_L\tilde{\mathbf{s}}_L = \mathbf{Q}_R\mathbf{F}_L\mathbf{r}$$

where $\tilde{\mathbf{M}}_L \in \mathbb{R}^{dn/2,dn/2}$ is a coarse grid approximation of the low-frequency part of the reduced-space Hessian $\mathbf{H}$ and $\tilde{\mathbf{s}}_L \in \mathbb{R}^{dn/2}$. This allows us to precondition the smooth part of $\mathbf{r}$. We note that we do not precondition the high-frequency components $\mathbf{r}_H := \mathbf{F}_H\mathbf{r}$ of $\mathbf{r}$, where $\mathbf{F}_H \in \mathbb{R}^{dn,dn}$ is a high-pass filter with $\mathbf{F}_H + \mathbf{F}_L = \mathbf{I}_{dn}$. Consequently, $\mathbf{s}_H = \mathbf{F}_H\mathbf{r}$. In summary, the solution of $\mathbf{M}\mathbf{s} = \mathbf{r}$ is given by

$$\mathbf{s} = \mathbf{s}_L + \mathbf{s}_H \approx \mathbf{Q}_P\mathbf{F}_L\tilde{\mathbf{s}}_L + \mathbf{F}_H\mathbf{r} \approx \mathbf{Q}_P\mathbf{F}_L(\tilde{\mathbf{M}}_L)^{-1}\mathbf{Q}_R\mathbf{F}_L\mathbf{r} + \mathbf{F}_H\mathbf{r}.$$

To counter the fact that we leave the high-frequency components untouched, we do not directly apply this preconditioner to the reduced-space KKT system in (15) but the regularization preconditioned system:

$$(\mathbf{I} + \mathbf{H}_{\text{reg}}^{-1/2}\mathbf{H}_{\text{data}}\mathbf{H}_{\text{reg}}^{-1/2})\mathbf{w} = -\mathbf{H}_{\text{reg}}^{-1/2}\mathbf{g}$$

where $\mathbf{w} := \mathbf{H}_{\text{reg}}^{1/2}\tilde{\mathbf{v}}$. Notice that the square root of the inverse of $\mathbf{H}_{\text{reg}}$ acts as a smoother. This scheme can be viewed as an approximation of a two-level multigrid V-cycle with an explicit (algebraic) smoother $\mathbf{H}_{\text{reg}}^{-1/2}$.

Before we explore extensions of this idea, we present some implementation aspects. We use spectral restriction and prolongation operators $\mathbf{Q}_R$ and $\mathbf{Q}_P$ [1, 48]. The operators $\mathbf{F}_L$ and $\mathbf{F}_H$ are implemented as cutoff filters in the frequency domain [1, 48]. For the implementation of the coarse grid operator $\tilde{\mathbf{M}}_L$ we have two choices. First, we can use a Galerkin discretization, which is formally given by $\tilde{\mathbf{M}}_L = \mathbf{Q}_R\mathbf{H}\mathbf{Q}_P$ [162]. The drawback of this approach is that we do not significantly reduce the computational costs compared to inverting the fine-grid Hessian, since each matvec necessitates the solution of the incremental state and adjoint equation at full resolution. Conversely, we can directly discretize the Hessian on a coarse grid to obtain $\tilde{\mathbf{M}}_L$. This makes the implementation slightly more involved but reduces the computational costs drastically. We opt for the latter approach [1, 48].

To invert the matrix $\tilde{\mathbf{M}}_L$ we have several options. Again, traditional direct methods are out of the question. However, we can use a nested Krylov subspace method. If we use a Krylov subspace method as an outer method (i.e., for computing the search direction), we have to select a tolerance for the inner Krylov subspace method that is a fraction of the tolerance used to compute the search direction. Alternatively, we can replace the solver for the Newton step with a flexible Krylov subspace method [163, 164] and use a fixed number of iterations for the nested (inner) Krylov subspace method. Alternatively, we can use a semi-iterative Chebyshev method [165] with a fixed number of iterations on the inside. This yields a fixed linear operator for a particular choice of eigenvalue bounds [166]. These bounds can be estimated using a Lanczos method. We have tested and compared these approaches in [1, 48]. This also includes the use of different Krylov subspace methods for not only applying the preconditioner but also solving for the Newton step such as the standard and flexible GMRES method, the standard and flexible PCG method, or the Chebyshev method (some of which have been mentioned above). In [1], we observed that the nested PCG method converges most quickly in the 3D setting.

Zero-Velocity Approximation

The preconditioner introduced in the former section requires a repeated evaluation of the incremental state and adjoint equations. The savings come from discretizing the reduced-space Hessian on a mesh of half the resolution. In [90] we developed a preconditioner that does not require solving any PDEs; the Hessian operator is fixed across all iterations. This is accomplished by fixing $\boldsymbol{v}$ to $\boldsymbol{v} = \mathbf{0}$ (our initial guess for the optimization problem). Under the assumption, the state equation simplifies to $\partial_t m = 0$, i.e., $m(t, \boldsymbol{x}) = m_0(\boldsymbol{x})$ for all $\boldsymbol{x} \in \Omega$ and $t \in [0, 1]$. Likewise, we have $\partial_t \lambda = 0$, i.e., $\lambda(t, \boldsymbol{x}) = -(m_0(\boldsymbol{x}) - m_1(\boldsymbol{x}))$, for all $\boldsymbol{x} \in \Omega$ and $t \in [0, 1]$. Inserting these expressions into the incremental state equation we have $\partial_t \tilde{m} = -\nabla m_0 \cdot \tilde{\boldsymbol{v}}$, which implies that $\tilde{m}(t = 1) = -\nabla m_0 \cdot \tilde{\boldsymbol{v}}$. The incremental adjoint equation for the Gauss–Newton approximation for $\boldsymbol{v} = \boldsymbol{0}$ is given by $\partial_t \tilde{\lambda} = 0$, i.e., $\tilde{\lambda}(t, \boldsymbol{x}) = \nabla m_0(\boldsymbol{x}) \cdot \tilde{\boldsymbol{v}}(\boldsymbol{x})$ for all $\boldsymbol{x} \in \Omega$ and $t \in [0, 1]$. Consequently, the Gauss–Newton approximation of the Hessian matvec for $\boldsymbol{v} = \boldsymbol{0}$ is given by

$$\mathcal{H}_0[\tilde{\boldsymbol{v}}](\boldsymbol{x}) = \alpha \mathcal{L} \tilde{\boldsymbol{v}}(\boldsymbol{x}) + (\nabla m_0(\boldsymbol{x}) \otimes \nabla m_0(\boldsymbol{x})) \tilde{\boldsymbol{v}}(\boldsymbol{x}).$$

This approximation deteriorates as we move away from our initial guess $\mathbf{v}^{(0)} = \mathbf{0}$. As a remedy, we replace m_0 in the expression above with our current estimate m at $t = 1$ at each outer iteration k for a trial velocity $\mathbf{v}^{(k)}$. Like in previous sections, we do not form or store $\mathbf{H}_0 \in \mathbb{R}^{dn,dn}$ (the discrete version of $\mathcal{H}_0$); we invert the matrix iteratively using a nested PCG method. To further reduce the computational costs, we combine the $\mathbf{H}_0$ approximation with the two-level scheme discussed above. That is, we replace the coarse grid preconditioner $\tilde{\mathbf{M}}_L$ with a coarse grid approximation of $\mathcal{H}_0$.

2.6 Parameter Selection

Based on empirical observations, we fix most of our numerical parameters. For the number of time steps n_t in the numerical time integration, we found that $n_t = 4$ provides sufficient accuracy to obtain a good matching between images at resolutions at the order of 256^3 (standard size for brain images acquired in clinical practice). We set the tolerance for the relative reduction of the gradient (stopping condition for optimization) to $\epsilon_{\text{opt}} = 5e\text{–}2$. We use a superlinear forcing sequence to compute the tolerance for the outer PCG algorithm. We use a two-level implementation of the zero-velocity approximation of the reduced-space Hessian as a preconditioner. The tolerance for the inner PCG to invert $\mathbf{H}_0$ is 10 times smaller than the outer tolerance of the PCG. The formulation we consider for diffeomorphic image registration is an extension of what we discussed so far; it considers near-incompressible velocities. We describe this formulation in greater detail in the appendix. The regularization parameter for the H^1 penalty for the divergence of the velocity field is fixed and set to $\beta = 1e\text{–}4$. We compute an optimal regularization parameter α as described below.

Several methods exist to estimate an optimal regularization parameter for inverse problems (see, e.g., [167] for examples). All of these methods have in common that the estimation of an optimal regularization parameter is expensive. Methods that assume that the differences between model output and observed data are random (such as, e.g., generalized cross validation) are not necessarily reliable in the context of image registration, since imaging noise is prone to be highly structured [168]. In our work, we consider a binary search for identifying an optimal value α for the regularization model for the velocity field $\boldsymbol{v}$ [1, 30, 112]. This approach is in spirit similar to an L-curve strategy. Related parameter continuation strategies have been considered in [68, 168, 169]. As a measure for optimality, we select bounds on the determinant of the deformation gradient $\det \nabla \boldsymbol{y}$. Notice that we do not compute $\boldsymbol{y}$ to obtain this quantity but solve a transport problem (see appendix). Assuming that we start from an identity map $\mathrm{id}_{\mathbb{R}^d}$, the initial value for $\det \nabla \boldsymbol{y}$ is one (this is equivalent to $\boldsymbol{v} = \boldsymbol{0}$ in our formulation). Consequently, we assume that the map is diffeomorphic if $\det \nabla \boldsymbol{y} \geq 0$. This motivates the use of a lower bound $\epsilon_D > 0$. Since the determinant of the deformation gradient of $\boldsymbol{y}$ is inversely proportional to $\det \nabla \boldsymbol{y}^{-1}$, we use $1/\epsilon_D$ as an upper bound. Consequently, we require

$$\epsilon_D < \det \nabla \boldsymbol{y} < 1/\epsilon_D \tag{16}$$

for any admissible $\boldsymbol{y}$. Our approach is as follows: We start with a regularization parameter of $\alpha = 1$ and reduce α by one order of magnitude until the condition in (16) is violated. Subsequently, we perform a binary search in the interval between the last value α for which (16) held and the value for which (16) was violated. For each new trial parameter $\alpha^{(l)}$ at level $l \in \mathbb{N}$, we use the control variable $\boldsymbol{v}_\alpha^{(l-1)} := \boldsymbol{v}(\alpha^{(l-1)})$ obtained for $\alpha^{(l-1)}$ at the prior level $l - 1$ as an initial guess to speed up convergence. More details can be found in [1, 30].

Obviously, this search is expensive since we have to solve the inverse problem for each trial $\alpha^{(l)}$, $l = 0, 1, 2, 3, \ldots$. Once we have identified an adequate regularization parameter $\alpha^\star$ for a particular application, we perform a parameter continuation to speed up convergence. That is, we solve the inverse problem consecutively for different values for α, starting with $\alpha^{(0)} = 1$ and subsequently reducing $\alpha^{(l)}$ by one order of magnitude until we reach the order of $\alpha^\star$. Then, we solve the problem one last time for $\alpha^\star$. For high regularization parameters $\alpha^{(l)}$ we essentially solve a convex problem; we expect quick convergence to a (local) minimizer. We use the estimate for the control variable $\boldsymbol{v}_\alpha^{(l-1)}$ as an initial guess for the next solve at level l. This does not significantly affect the runtime compared to directly solving our problem for $\alpha^\star$. Moreover, it "convexifies" the problem; we anticipate to more quickly converge to a "better" (local) minimizer and/or avoid getting trapped in "less optimal" local minima. We have compared this strategy against multi-scale (scale continuation) and multi-resolution (grid continuation) approaches in [1]. We observed the parameter continuation to be more stable and overall more effective. Combining parameter continuation with scale and/or grid continuation remains subject to future work.

Lastly, we note that machine learning has also recently been considered for regularization operator and parameter tuning [170, 171].

3 Results

We consider a slightly more involved formulation than the one presented in Sect. 2.2. We provide additional details in the appendix. We refer to [1, 33, 90, 93] for weak and strong scaling results of our CPU and GPU implementation of CLAIRE, respectively. In the present work, we limit the performance analysis to a single GPU.

3.1 Data

We report results for the `NIREP` dataset [172]. We refer to [172] for additional information about the datasets, the imaging protocol, and the preprocessing. This repository contains 16 rigidly aligned T1-weighted MRI brain datasets (`na01`–`na16`) of size 256 × 300 × 256 voxels of different individuals. Consequently, we invert for $3(256^2)(300) = 58{,}982{,}400$ unknowns. Each dataset is equipped with 33 labels for anatomical gray matter regions. These labels allow us to assess the performance of the registration; we assess registration accuracy by how well these labels are mapped to one another. To do so, we compute the so-called Dice between the label maps. For a Dice of one, the labels are in perfect agreement. For a Dice of zero, they do not overlap. Notice that the registration software does not consider these labels; registration is solely based on matching corresponding image intensities. That is, we do not explicitly minimize the alignment of the labels but only the mismatch between the data. We show the considered data in Fig. 4. In particular, we show axial slices of all 16 datasets with the associated labels in the overlay.

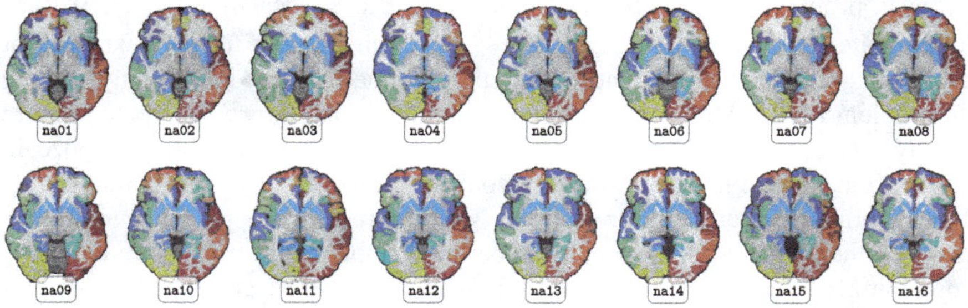

Fig. 4 `NIREP` data repository [172]. We show an axial slice of each dataset (slice number 128). The repository contains 16 rigidly aligned T1-weighted MRI brain datasets (`na01`–`na16`) of size 256 × 300 × 256 voxels of different individuals. Each dataset is equipped with 32 labels of anatomical gray matter regions. We overlay these regions in different colors on the MRI data. We refer to [172] for additional information about the datasets, the imaging protocol, and the preprocessing

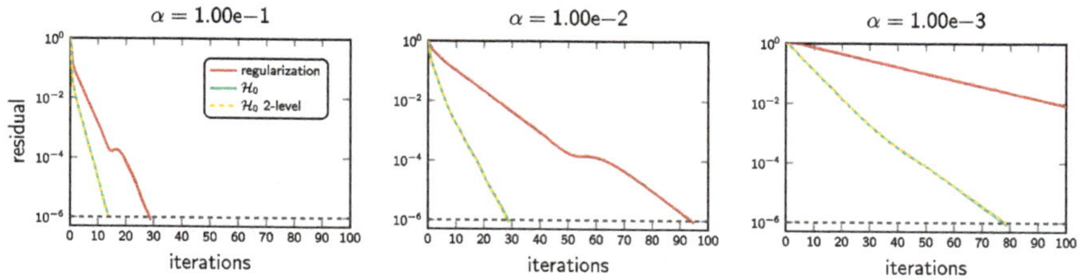

Fig. 5 Convergence of the PCG method for solving for the Newton step. We solve for the search direction at the solution of the registration problem (dataset `na02` registered to `na01`). We consider the squared L^2-distance. We solve this problem at the original resolution of the data. We consider three different preconditioners: the regularization preconditioner, the zero-velocity approximation of the reduced-space Hessian, and the 2-level implementation of the zero-velocity approximation of the reduced-space Hessian. We solve the problem for three different regularization parameter values (from left to right): $\alpha = 1e{-}1$, $\alpha = 1e{-}2$, and $\alpha = 1e{-}3$. The tolerance for the PCG method is $1e{-}6$. We plot the relative residual

3.2 *Preconditioning*

We show representative results for the convergence of different preconditioners in Fig. 5. We consider the regularization preconditioner as well as two variants of the zero-velocity preconditioner—inverting the zero-velocity approximation of the reduced-space Hessian on the fine mesh and a two-level implementation of this preconditioner. To test the performance, we invert the reduced-space Hessian at the true solution of our problem. That is, we solve the registration problem between two images (dataset `na02` registered to `na01`) in our case. We then use the obtained velocity as iterate at which we compute the search direction. We set the tolerance for the PCG method to $1e{-}6$. We consider a squared L^2-distance as a similarity measure. We report results for the full resolution, only; $\boldsymbol{n}_x = (256, 300, 256)$. We report convergence results for three different choices of α; $\alpha = 1e{-}1$, $\alpha = 1e{-}2$, and $\alpha = 1e{-}3$, respectively.

The most important observations are as follows: (i) The convergence of all methods is sensitive with respect to the choice of α. (ii) The zero-velocity approximation yields an improved rate of convergence. (iii) For the zero-velocity approximation, the convergence does not deteriorate as we switch from full resolution to a coarse resolution (2-level implementation).

3.3 *Regularization Parameter Search*

We set the regularization parameter for the divergence of the velocity to $\beta = 1e{-}5$ and search for an optimal regularization parameter α using the scheme described in Sect. 2.6. We register each image with all other images. We also perform the reverse

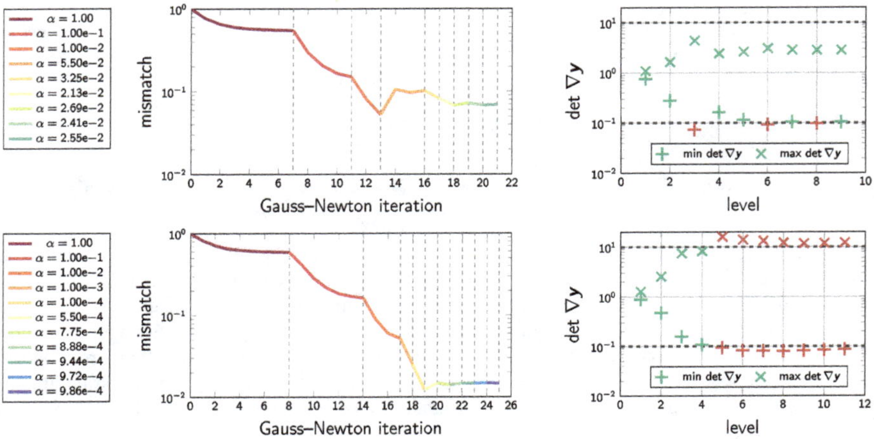

Fig. 6 Illustration of the parameter search for the registration of the dataset with id `na14` to the dataset with id `na01` (top panel) and in the reverse direction (bottom panel). The registration results are shown in Fig. 7 (bottom panel). We show (for each case) the trend of the mismatch for each choice of regularization parameter α (left) and the largest and smallest value of the determinant of the deformation gradient. In the latter plot we also show the lower and upper bound of 0.1 and 10, respectively, for the determinant of the deformation gradient as a dashed line. Whenever these bounds are violated, the marker switches from "green" to "red." For the run shown in the top panel, the optimal regularization parameter is $2.547e{-}2$. For the run at the bottom, the optimal regularization parameter is $1e{-}3$

registration. This results in a total of $16(15) = 240$ registrations. We consider a squared L^2-distance for the similarity measure.

We illustrate the search for an optimal regularization parameter for two registration problems (`na01` to `na14` and `na14` to `na01`) in Fig. 6. We show representative registration results for two images from the considered `NIREP` dataset in Fig. 7. We report statistics for the estimated regularization parameter α in Fig. 8 (left plot). We also compute the minimal, mean, and maximum value of the determinant of the deformation gradient for all registrations. We report the statistics across all 240 registrations for these in Fig. 8 (plots to the right). For the minimum value of the determinant of the deformation gradient, we obtained $1.668e{-}1$ with a standard deviation of $6.119e{-}2$, an overall lowest minimum value of $1.001e{-}1$, and an overall largest minimum value of $4.192e{-}1$. For the mean value of the determinant of the deformation gradient, we obtained $1.028e0$ with a standard deviation of $1.183e{-}2$, an overall lowest minimum value of $1.008e0$, and an overall largest minimum value of $1.084e0$. For the maximum value of the determinant of the deformation gradient, we obtained $7.150e0$ with a standard deviation of $2.324e0$, an overall lowest minimum value of $2.269e0$, and an overall largest minimum value of $9.993e0$. We report the workload for this search in Table 2.

The most important observations are as follows: (i) We can efficiently determine an adequate regularization parameter with an average runtime of 15.135 seconds

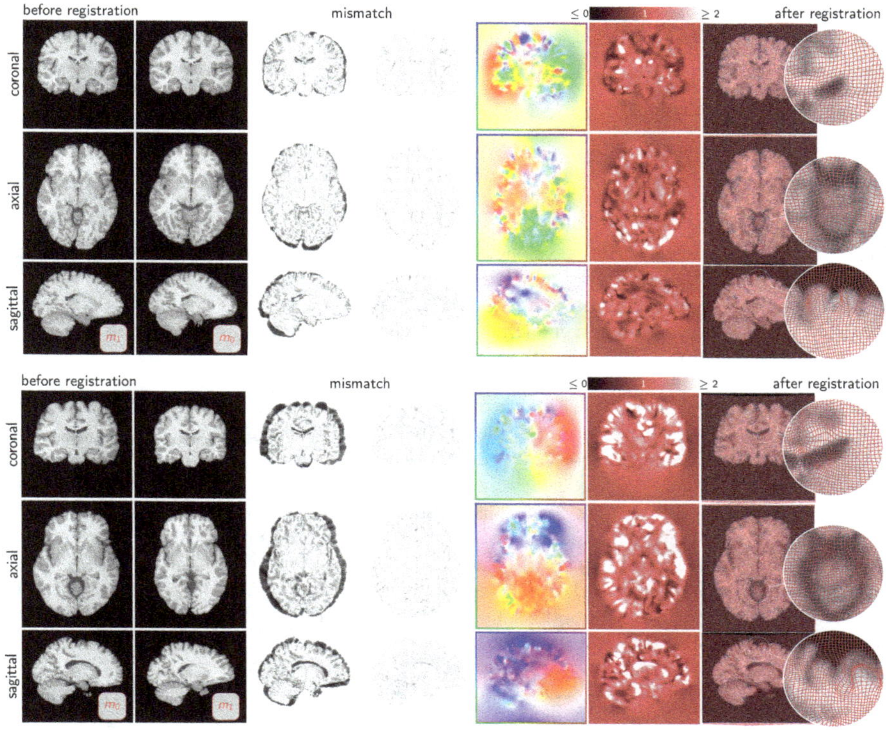

Fig. 7 Representative registration results for CLAIRE. We consider an H^1-seminorm as a regularization model for the velocity field and an H^1-norm to regularize the divergence of the velocity field. We model near incompressible flows. The regularization parameter for the divergence is set to $\beta = 1e{-}5$. The regularization for the velocity is estimated. The bound for the determinant of the deformation gradient is set to $1e{-}1$. We register the dataset with id `na06` to the dataset with id `na02` (top panel) and the dataset with id `na14` to the dataset with id `na01` (bottom panel) of the NIREP repository. The data is rigidly aligned. For each panel, we show the following: The top row shows the coronal view, the middle row the axial view, and the bottom row the sagittal view of the 3D volume. The columns are (from left to right) (i) the template image m_0, (ii) the reference image m_1, (iii) the residual differences between the reference image and the template image (before registration; large differences are colored in black and no residual difference are colored in white), (iv) the residual differences between the deformed template image and the reference image (after registration), (v) an illustration of the velocity field (color represents orientation; see boundary), (vi) visualization of the determinant of the deformation gradient (color bar on top), and (vii) an illustration of the projection of the computed deformation map onto the corresponding plane

(standard deviation: 11.323 seconds), (ii) the computed deformation maps are diffeomorphic (up to numerical accuracy), and (iii) we overall obtain high-quality registration results with precise control on the determinant of the deformation gradient.

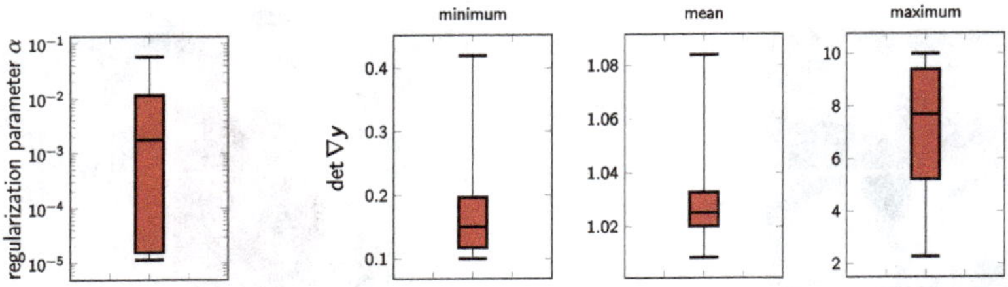

Fig. 8 Statistics for the estimation of the regularization parameter α across 240 registrations between all NIREP datasets. The data has been rigidly registered. We report the estimated regularization parameters α (right) and the statistics for the minimum, mean, and maximum of the determinant of the deformation gradient across each individual registration. The estimated regularization parameter α is 7.525*e*–3 with a standard deviation of 1.098*e*–2, a median of 1.773*e*–3, a minimal value of 1.141*e*–5, and a maximal value of 5.641*e*–2 across all 240 registrations. The 25th percentile is 1.563*e*–5 and the 75th percentile is 1.141*e*–2. The minimal, mean, and maximal values are 1.668*e*–1 (standard deviation: 6.119*e*–2), 1.028*e*0 (standard deviation: 1.183*e*–2), and 7.150*e*0 (standard deviation: 2.324*e*0), respectively

Table 2 Workload for estimating the regularization parameter α. We consider a squared L^2-distance measure. All reported numbers are computed across all 240 registrations. We report the number of outer iterations, the number of Hessian matvecs, the number of PDE solves, the relative mismatch after registration, the relative change of the norm of the gradient, and the runtime (in seconds). These numbers are for solving the inverse problem multiple times; we search for an optimal regularization parameter using a binary search

	Mean	Stdev	Min	Max	Median	1st QT	3rd QT
Iterations	20.538	2.268	17	28	20	19	22
Matvecs	166.971	124.749	39	388	92	64	326
PDE solves	447.975	318.877	134	1008	252	190	867
Mismatch	4.508*e*–2	3.993*e*–2	4.422*e*–3	1.756*e*–1	3.183*e*–2	1.000*e*–2	7.095*e*–2
Gradient	1.394*e*–2	5.733*e*–3	3.166*e*–3	3.350*e*–2	1.338*e*–2	8.995*e*–3	1.836*e*–2
Runtime	15.135	11.323	3.974	37.867	8.125	5.877	29.232

3.4 Registration Accuracy

In this section we assess the registration accuracy. In particular, we report the Dice values for the parameter search described in the former section. Aside from considering a squared L^2-distance, we also report registration accuracy for normalized cross correlation as a similarity measure (see appendix for details). In Fig. 9 we report the Dice score for the individual labels. We report the statistics for the 240 registration runs in Table 3. Here, we compute the union of all 33 labels and report the global Dice score. We report additional results in the appendix.

The most important observations are as follows: (i) CLAIRE yields an excellent agreement for the overall Dice with an increase from 0.551 (standard deviation:

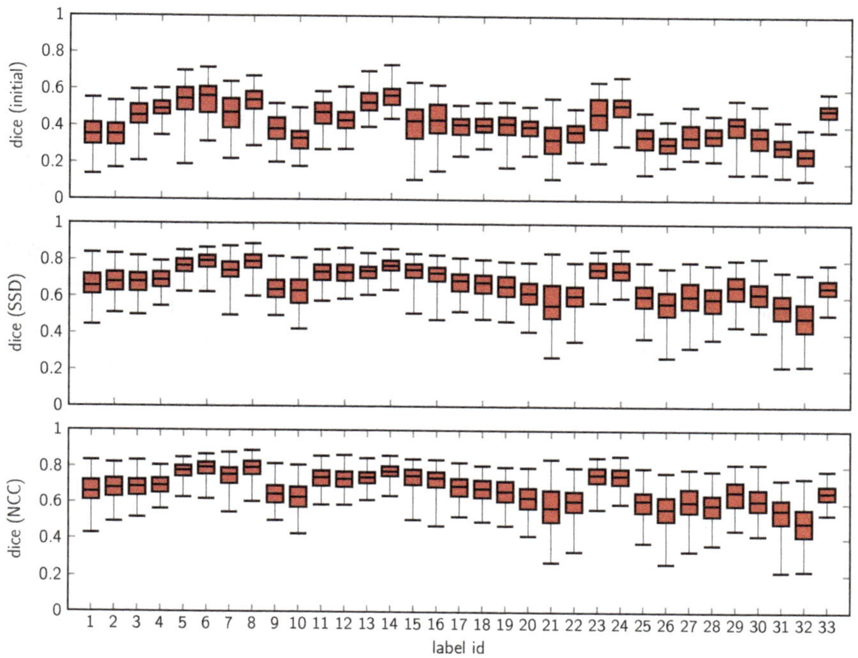

Fig. 9 Average Dice score for individual labels. We show box plots for the Dice score for each individual label. The statistics are computed for all 240 registrations. The top row corresponds to the Dice values before registration. The middle row shows values for the Dice score after registration using a squared L^2-distance as a similarity measure. The bottom row shows the results obtained for normalized cross correlation. We report statistics for the union of these labels in Table 3

Table 3 Average Dice values. We report the mean, min, max, and median value as well as the 1st quantile and the 3rd quantile. These values are computed for the union of all labels. We report the initial values in the first row. The values after diffeomorphic registration based on the squared L^2-distance and normalized cross correlation are reported in the second and third rows, respectively. We report the scores for the individual labels in Fig. 9

	Mean	Stdev	Min	Max	Median	1st quantile	3rd quantile
Initial	0.551	0.041	0.421	0.625	0.555	0.527	0.583
SSD	0.831	0.059	0.697	0.922	0.842	0.785	0.884
NCC	0.835	0.060	0.699	0.923	0.844	0.784	0.889

0.041) before registration to 0.831 (standard deviation: 0.059) for the squared L^2-distance and 0.835 (standard deviation: 0.060) for normalized cross correlation. (ii) The performance for the squared L^2-distance and normalized cross correlation are en par for our current implementation.

3.5 Convergence and Runtime

In the former section, we have seen how CLAIRE performs when searching for an optimal regularization parameter for each individual volume. In the current section, we fix the regularization parameter to the mean optimal value of $\alpha = 1.773e{-}3$ determined in the former section and focus on computational performance. We plot the residual versus the number of outer iterations in Fig. 10. Here, we solve the inverse problem for a fixed $\alpha = 1.773e{-}3$ without performing any scale, grid, or parameter continuation. We report the runtime for our parameter continuation scheme for a target regularization parameter $\alpha = 1.773e{-}3$ in Table 4.

The most important observations are as follows: (i) We can solve the inverse problem in under 4 seconds (on average, the runtime is 3.839 seconds; standard deviation: 0.648 seconds), with a minimum runtime of under 3 seconds and a maximum runtime of slightly above 6 seconds. (ii) We converge in about 12 iterations

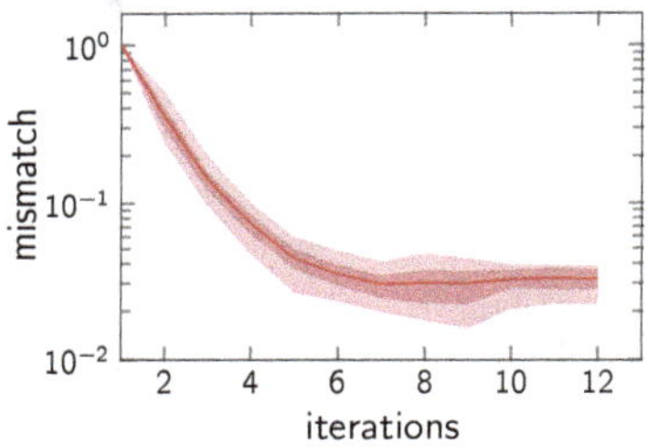

Fig. 10 Convergence behavior. We plot the relative reduction of the mismatch versus the number of iterations. The plot is generated for 240 registrations. We execute the algorithm for a regularization parameter value of $\alpha = 1.773e{-}3$. We solve the problem without using any continuation scheme. The average runtime is 3.615 seconds. The solid line represents the mean convergence for the data mismatch. We also show the envelopes for the 25th to 75th quantile and the 5th to the 95th quantile for the values of the mismatch. The average number of iterations is 10.554. We show the trend until iteration 12

Table 4 Workload for estimating the regularization parameter α. We consider a squared L^2-distance measure. All reported numbers are computed across all 240 registration. We report the number of outer iterations, the number of Hessian matvecs, the number of PDE solves, the relative mismatch after registration, the relative change of the norm of the gradient, and the runtime (in seconds). These numbers are for solving the inverse problem multiple times; we search for an optimal regularization parameter using a binary search

	Mean	Stdev	Min	Max	Median	1st QT	3rd QT
Iterations	10.554	1.246	9	15	10	10	11
Matvecs	24.525	5.659	14	47	24	21	28
PDE solves	82.158	13.664	58	136	82	74	89
Mismatch	$4.399e{-}2$	$1.037e{-}2$	$1.906e{-}2$	$8.087e{-}2$	$4.292e{-}2$	$3.691e{-}2$	$5.008e{-}2$
Gradient	$3.940e{-}2$	$6.485e{-}3$	$2.135e{-}2$	$4.973e{-}2$	$4.012e{-}2$	$3.623e{-}2$	$4.409e{-}2$
Runtime	3.839	0.648	2.704	6.399	3.779	3.451	4.159

to a stable solution of our problem (the mismatch stagnates), where a majority of the runs we have executed converge after only 10.554 (standard deviation: 1.246). (iii) Once we have determined an adequate regularization parameter for a particular application, we can solve the problem quickly with an accuracy that is equivalent to the more expensive parameter search considered in the section above as judged by the relative reduction of the mismatch.

4 Conclusions

We have reviewed our past work on scalable algorithms for diffeomorphic image registration. Several issues remain.

Our implementation currently only supports the registration of images acquired with the same imaging modality. Developing an effective solver for other distance measures remains subject to future work. We have worked on several numerical schemes for preconditioning the reduced-space Hessian. The spectral preconditioner is extremely efficient to apply but its performance deteriorates as we reduce the regularization parameter. This is true for all other schemes we have implemented to precondition the reduced Hessian. Although they are more effective than the simple spectral preconditioner, developing a scheme that has a rate of convergence that is mesh-independent and at the same time independent of the choice of the regularization (parameter) remains subject to future work.

Our 3D GPU implementation currently only supports stationary velocities. These velocities do not define a proper metric in the Riemannian space of diffeomorphic flows. While we have implemented a MATLAB prototype version of a solver that supports time-varying velocities, this implementation has not yet been ported to the C++ implementation of CLAIRE.

Another challenge in diffeomorphic image registration is how to handle data that underwent topological changes (e.g., the emergence of a tumor or tissue being removed due to clinical intervention). One possibility to handle this is to introduce additional biophysical constraints [93, 173–182]. On the downside, this makes the problem much more challenging to solve since we not only invert for a deformation map but also for the parameters of the model. More generic approaches to deal with changes in topology are described in [183–187].

Acknowledgments This work was in part supported by the National Science Foundation (NSF) through the grants DMS-2012825 and DMS-2145845. Any opinions, findings, and conclusions or recommendations expressed herein are those of the author and do not necessarily reflect the views of the NSF. This work was completed in part with resources provided by the Research Computing Data Core at the University of Houston. The author would like to thank George Biros, Malte Brunn, Amir Gholami, Naveen Himthani, Jae Youn Kim, and Miriam Schulte for their numerous contributions to this work.

Appendix

PDE Constraints

Below, we will revisit some of the problem formulations we have considered in our past work. These are extensions to the formulation considered in Sect. 2.2. We limit the description of our methodology to the most basic formulation for simplicity. The default formulation implemented in our current GPU version in CLAIRE is different [90, 91].

Nonstationary Velocities

In [30], we consider stationary and nonstationary velocities. For nonstationary velocities, the reduced gradient in (9) is given by

$$\boldsymbol{g}(\boldsymbol{v}) := \alpha\mathcal{L}\boldsymbol{v} + \lambda\nabla m.$$

(Near-)Incompressible Diffeomorphisms

In [30], we augment the formulation in (7) by introducing the incompressibility constraint $\nabla \cdot \boldsymbol{v} = 0$. A similar formulation has been considered in [40]. For the primal-dual optimal variables $(m^\star, \boldsymbol{v}^\star, \lambda^\star, \rho^\star)$, the associated KKT conditions are given by

$$\partial_t m^\star + \boldsymbol{v}^\star \cdot \nabla m^\star = 0 \qquad \text{in } (0, 1] \times \Omega, \tag{17a}$$

$$m^\star = m_0 \qquad \text{in } \{0\} \times \Omega, \tag{17b}$$

$$-\partial_t \lambda^\star + \nabla \cdot \lambda^\star \boldsymbol{v}^\star = 0 \qquad \text{in } [0, 1) \times \Omega, \tag{17c}$$

$$\lambda^\star = -(m^\star - m_1) \qquad \text{in } \{1\} \times \Omega, \tag{17d}$$

$$\nabla \cdot \boldsymbol{v}^\star = 0 \qquad \text{in } \Omega, \tag{17e}$$

$$\alpha\mathcal{L}\boldsymbol{v}^\star + \nabla\rho^\star + \int_0^1 \lambda^\star \nabla m^\star \,\mathrm{d}t = \boldsymbol{0} \qquad \text{in } \Omega. \tag{17f}$$

We eliminate the incompressibility constraint (17e) and the dual variable ρ from the optimality system stated above to obtain the expression

$$\alpha\mathcal{L}\boldsymbol{v} + \int_0^1 \lambda\nabla m \mathrm{d}t - \nabla\Delta^{-1}\nabla \cdot \int_0^1 \lambda\nabla m \,\mathrm{d}t$$

for the reduced gradient. The remaining PDE operators in (17) for m and λ in the associated KKT system are identical.

In [33], we relaxed the incompressiblity constraint by introducing an additional control variable w to obtain $\nabla \cdot \boldsymbol{v} = w$. This allows us to model near-incompressible deformations. After eliminating the constraint $\nabla \cdot \boldsymbol{v} = w$ and the associated dual variable ρ from the KKT system, we obtain the reduced gradient:

$$\alpha \mathcal{L} \boldsymbol{v}^\star + \int_0^1 \lambda^\star \nabla m^\star \mathrm{d}t - \nabla(\alpha(\beta(-\Delta^{-1} + \mathrm{id}))^{-1} + \mathrm{id})^{-1} \Delta^{-1} \nabla \cdot \int_0^1 \lambda^\star \nabla m^\star \, \mathrm{d}t,$$

Here, $\beta > 0$ denotes the regularization parameter of the regularizer for the second control variable w. We consider an H^1-norm. We refer to [33] for additional details. This represents the default model implemented in the hardware-accelerated implementation of CLAIRE [1, 90, 91]. The results reported in this study also consider this formulation. The regularization model for the velocity field is an H^1-seminorm.

Aside from this, we have also explored a model of incompressible flows that promotes shear [33]. To do so, we introduce a nonlinear regularization model. In particular, we replaced the regularization model for $\boldsymbol{v}$ by

$$|\boldsymbol{v}|_{H^1(\Omega)}^{(1+\nu)/2\nu} = \frac{2\nu}{\nu + 1} \int_\Omega (\mathcal{E}[\boldsymbol{v}] \colon \mathcal{E}[\boldsymbol{v}])^{(1+\nu)/2\nu} \, \mathrm{d}\boldsymbol{x},$$

where

$$\mathcal{E}[\boldsymbol{v}] := \frac{1}{2}\left((\nabla_d \boldsymbol{v}) + (\nabla_d \boldsymbol{v})^\mathsf{T}\right), \quad \nabla_d \boldsymbol{v} := \begin{pmatrix} (\nabla v_1)^\mathsf{T} \\ \vdots \\ (\nabla v_d)^\mathsf{T} \end{pmatrix} \in \mathbb{R}^{d,d},$$

denotes the strain tensor, and $\nu > 0$ controls the nonlinearity. With this regularization model in conjunction with the incompressibility constraint $\nabla \cdot \boldsymbol{v} = 0$, we obtain a Stokes-like optimality system with a viscosity that depends on the strain rate. The reduced gradient is given by

$$-\operatorname{div}\left(2 \operatorname{tr}(\mathcal{E}[\boldsymbol{v}]\mathcal{E}[\boldsymbol{v}])^{(1-\nu)/2\nu} \mathcal{E}[\boldsymbol{v}]\right) + \nabla \rho + \int_0^1 \lambda \nabla m \mathrm{d}t,$$

where $\operatorname{div}(\boldsymbol{A}) = (\nabla \cdot \boldsymbol{a}_1, \ldots, \nabla \cdot \boldsymbol{a}_d) \in \mathbb{R}^d$ for an arbitrary $d \times d$ matrix $\boldsymbol{A}$ with columns $\boldsymbol{a}_i \in \mathbb{R}^d$, $i = 1, \ldots, d$. In the limit $\nu \to \infty$ this model behaves like total variation regularization. For $\nu \in (0, 1)$ we obtain a shear thickening and for $\nu > 1$ a shear thinning fluid. Likewise to the linear case, we can eliminate the incompressiblity constraint and the associated dual variable ρ from the optimality system. We refer to [33] for additional details.

Optimal Transport

In our past work, we have not only introduced new hard or soft constraints for $\boldsymbol{v}$ but also considered a different forward model for transporting m. In particular, we use the continuity equation

$$\partial_t m + \nabla \cdot m\boldsymbol{v} = 0 \quad \text{in } [0, 1) \times \Omega,$$

with initial condition $m = m_0$ in $\{0\} \times \Omega$ to model the transport of the intensities of the template image m_0. In this model, mass is conserved. This establishes a connection to optimal transport [47, 188, 189]. We refer to [48] for additional details.

Deformation Gradient

In the context of image registration, the determinant of the deformation gradient $\det \nabla \boldsymbol{y}$ is often used to assess invertibility of $\boldsymbol{y}$ as well as a measure of local volume change in the context of morphometry and shape analysis. In the framework of continuum mechanics, we can obtain this information from the deformation tensor field $\boldsymbol{f} : [0, 1] \times \bar{\Omega} \to \mathbb{R}^{d,d}$, where $\boldsymbol{f}$ is related to $\boldsymbol{v}$ by

$$\partial_t \boldsymbol{f} + (\boldsymbol{v} \cdot \nabla_d)\boldsymbol{f} = (\nabla_d \boldsymbol{v})\boldsymbol{f} \;\text{ in }\; \Omega \times (0, 1], \qquad \boldsymbol{f} = \boldsymbol{I}_d \;\text{ in }\; \Omega \times \{0\}, \tag{18}$$

with periodic boundary conditions on $\partial\Omega$. Here, $\boldsymbol{I}_d = \operatorname{diag}(1, \ldots, 1) \in \mathbb{R}^{d,d}$. In our implementation we use $\det \boldsymbol{f}_1$ with $\boldsymbol{f}_1 := \boldsymbol{f}(t = 1, \cdot)$ as a surrogate for $\det \nabla_d \boldsymbol{y}$,

Normalized Cross Correlation

Aside from using the squared L^2-distance, we also consider normalized cross correlation as a distance measure. We note that we have not presented results for normalized cross correlation elsewhere. The choice of the similarity measure in general only affects the final condition of the dual variable. The normalized cross correlation distance measure is given by

$$\text{dist}_{\text{NCC}}(m(1), m_1) = 1 - \frac{\langle m(1), m_1\rangle^2_{L^2(\Omega)}}{\langle m_1, m_1\rangle_{L^2(\Omega)}\langle m(1), m(1)\rangle_{L^2(\Omega)}}, \tag{19}$$

where

$$\langle u, w\rangle_{L^2(\Omega)} = \int_\Omega u(\boldsymbol{x})w(\boldsymbol{x})\,\mathrm{d}\boldsymbol{x}$$

denotes the standard L^2-inner product on $\Omega \subset \mathbb{R}^d$ for arbitrary functions $u : \bar{\Omega} \to \mathbb{R}$, $w : \bar{\Omega} \to \mathbb{R}$. Using this distance, the final condition for the adjoint equation is given by

$$\lambda(1, \boldsymbol{x}) = -2\frac{\langle m_1, m(1)\rangle_{L^2(\Omega)}}{\|m(1)\|^2_{L^2(\Omega)}\|m_R\|^2_{L^2(\Omega)}}\left(\frac{\langle m_1, m(1)\rangle_{L^2(\Omega)}}{\|m(1)\|^2_{L^2(\Omega)}}m(1, \boldsymbol{x}) - m_1(\boldsymbol{x})\right).$$

Similarly, the expression for the final condition of the incremental dual variable $\tilde{\lambda}$ is given by

$$\tilde{\lambda}(1, \boldsymbol{x}) = \frac{2(q_1 m_1(\boldsymbol{x}) + q_2 m(1, \boldsymbol{x}) - q_3 \tilde{m}(1, \boldsymbol{x}))}{\|m_1\|^2_{L^2(\Omega)}},$$

where

$$q_1 = 2\frac{\langle m_1, m(1)\rangle_{L^2(\Omega)}\langle m(1), \tilde{m}(1)\rangle_{L^2(\Omega)}}{\|m(1)\|^4_{L^2(\Omega)}} - \frac{\langle m_R, \tilde{m}(1)\rangle_{L^2(\Omega)}}{\|m(1)\|^2_{L^2(\Omega)}},$$

$$q_2 = 4\frac{\langle m_R, m(1)\rangle^2_{L^2(\Omega)}\langle m(1), \tilde{m}(1)\rangle_{L^2(\Omega)}}{\|m(1)\|^6_{L^2(\Omega)}} - 2\frac{\langle m_R, m(1)\rangle_{L^2(\Omega)}\langle m_R, \tilde{m}(1)\rangle_{L^2(\Omega)}}{\|m(1)\|^4_{L^2(\Omega)}},$$

$$q_3 = \frac{\langle m_R, m(1)\rangle^2_{L^2(\Omega)}}{\|m(1)\|^4_{L^2(\Omega)}}.$$

Newton–Krylov Algorithm

We summarize our Newton–Krylov algorithm here. The outer iterations are given in Algorithm 1. The inner iterations (i.e., the computation of the search direction) are given in Algorithm 2. We describe this algorithm in some detail in Sect. 2.5.5.

Hardware

We execute CLAIRE on the Sabine Cluster of the Research Computing Data Core at the University of Houston. Sabine hosts a total of 5704 CPU cores in 169 compute and 12 GPU nodes. We limit the experiments to our GPU implementation. The associated nodes are equipped with an Intel Xeon E5-2680v4 CPUs (2 sockets with 28 cores) with 256 GB of memory. Each node is also equipped with 8 NVIDIA V100 GPUs with a total of 40,960 cores and 128 GB of memory.

Algorithm 1 Inexact Newton–Krylov method (outer iterations). We use the relative norm of the reduced gradient with tolerance $\epsilon_{\text{opt}} > 0$ as stopping criterion

$k \leftarrow 0$
initial guess $\mathbf{v}^{(k)} \leftarrow \mathbf{0}$
$\mathbf{m}^{(k)} \leftarrow$ solve state equation in (1b) forward in time given $\mathbf{v}^{(k)}$
$j^{(k)} \leftarrow$ evaluate objective functional (1a) given $\mathbf{m}^{(k)}$ and $\mathbf{v}^{(k)}$
$\boldsymbol{\lambda}^{(k)} \leftarrow$ solve adjoint equation (10) backward in time given $\mathbf{v}^{(k)}$ and $\mathbf{m}^{(k)}$
$\mathbf{g}^{(k)} \leftarrow$ evaluate reduced gradient (9) given $\mathbf{m}^{(k)}$, $\boldsymbol{\lambda}^{(k)}$ and $\mathbf{v}^{(k)}$
while $\|\mathbf{g}^{(k)}\|_\infty > \|\mathbf{g}^{(0)}\|_\infty \epsilon_{\text{opt}}$ **do**
    $\tilde{\mathbf{v}}^{(k)} \leftarrow$ solve $\mathbf{H}^{(k)}\tilde{\mathbf{v}}^{(k)} = -\mathbf{g}^{(k)}$ given $\mathbf{m}^{(k)}$, $\boldsymbol{\lambda}^{(k)}$, $\mathbf{v}^{(k)}$, and $\mathbf{g}^{(k)}$ (see Algorithm 2)
    $\gamma^{(k)} \leftarrow$ perform line search on $\tilde{\mathbf{v}}^{(k)}$ subject to Armijo condition
    $\mathbf{v}^{(k+1)} \leftarrow \mathbf{v}^{(k)} + \gamma^{(k)}\tilde{\mathbf{v}}^{(k)}$
    $\mathbf{m}^{(k+1)} \leftarrow$ solve state equation (1b) forward in time given $\mathbf{v}^{(k+1)}$
    $j^{(k+1)} \leftarrow$ evaluate (1a) given $\mathbf{m}^{(k+1)}$ and $\mathbf{v}^{(k+1)}$
    $\boldsymbol{\lambda}^{(k+1)} \leftarrow$ solve adjoint equation (10) backward in time given $\mathbf{v}^{(k+1)}$ and $\mathbf{m}^{(k+1)}$
    $\mathbf{g}^{(k+1)} \leftarrow$ evaluate (9) given $\mathbf{m}^{(k+1)}$, $\boldsymbol{\lambda}^{(k+1)}$ and $\mathbf{v}^{(k+1)}$
    $k \leftarrow k + 1$
end while

Algorithm 2 Newton step (inner iterations). We illustrate the solution of the reduced KKT system (15) using a PCG method at a given outer iteration $k \in \mathbb{N}$. We use a superlinear forcing sequence to compute the tolerance $\eta^{(k)}$ for the PCG method (inexact solve)

input: $\mathbf{m}^{(k)}$, $\boldsymbol{\lambda}^{(k)}$, $\mathbf{v}^{(k)}$, $\mathbf{g}^{(k)}$, $\mathbf{g}^{(0)}$
$l \leftarrow 0$
set $\epsilon_H \leftarrow \min\left(0.5, \|\mathbf{g}^{(k)}\|_\infty^{1/2}\right)$, $\tilde{\mathbf{v}}^{(l)} \leftarrow \mathbf{0}$, $\mathbf{r}^{(l)} \leftarrow -\mathbf{g}^{(k)}$
$\mathbf{z}^{(l)} \leftarrow$ apply preconditioner $\mathbf{M}^{-1}$ to $\mathbf{r}^{(l)}$
$\mathbf{s}^{(l)} \leftarrow \mathbf{z}^{(l)}$
while $l < n$ **do**
    $\tilde{\mathbf{m}}^{(l)} \leftarrow$ solve (12) forward in time given $\mathbf{m}^{(k)}$, $\mathbf{v}^{(k)}$ and $\tilde{\mathbf{v}}^{(l)}$
    $\tilde{\boldsymbol{\lambda}}^{(l)} \leftarrow$ solve (13) backward in time given $\boldsymbol{\lambda}^{(k)}$, $\mathbf{v}^{(k)}$, $\tilde{\mathbf{m}}^{(l)}$ and $\tilde{\mathbf{v}}^{(l)}$
    $\tilde{\mathbf{s}}^{(l)} \leftarrow$ apply $\mathbf{H}^{(l)}$ to $\mathbf{s}^{(l)}$ given $\boldsymbol{\lambda}^{(k)}$, $\mathbf{m}^{(k)}$, $\tilde{\mathbf{m}}^{(l)}$ and $\tilde{\boldsymbol{\lambda}}^{(l)}$ (see (11))
    $\kappa^{(l)} \leftarrow \langle \mathbf{r}^{(l)}, \mathbf{z}^{(l)} \rangle / \langle \mathbf{s}^{(l)}, \tilde{\mathbf{s}}^{(l)} \rangle$
    $\tilde{\mathbf{v}}^{(l+1)} \leftarrow \tilde{\mathbf{v}}^{(l)} + \kappa^{(l)}\mathbf{s}^{(l)}$
    $\mathbf{r}^{(l+1)} \leftarrow \mathbf{r}^{(l)} - \kappa^{(l)}\tilde{\mathbf{s}}^{(l)}$
    if $\|\mathbf{r}^{(l+1)}\|_2 < \epsilon_H$ **break**
    $\mathbf{z}^{(l+1)} \leftarrow$ apply preconditioner $\mathbf{M}^{-1}$ to $\mathbf{r}^{(l+1)}$
    $\mu^{(l)} \leftarrow \langle \mathbf{z}^{(l+1)}, \mathbf{r}^{(l+1)} \rangle / \langle \mathbf{z}^{(l)}, \mathbf{r}^{(l)} \rangle$
    $\mathbf{s}^{(l+1)} \leftarrow \mathbf{z}^{(l+1)} + \mu^{(l)}\mathbf{s}^{(l)}$
    $l \leftarrow l + 1$
end while
output: $\tilde{\mathbf{v}}^{(k)} \leftarrow \tilde{\mathbf{v}}^{(l+1)}$

Additional Results

We report more detailed results for the registration accuracy of CLAIRE in this section. The statistics for the Dice for the squared L^2-distance with respect to each individual label is reported in Table 5. The associated results for normalized cross correlation are reported in Table 6. These results are for the parameter search for 240 registration (all-to-all) of the NIREP dataset.

Table 5 Average Dice values. We report the mean, min, max, and median value as well as the 1st quantile and the 3rd quantile. These values are computed for each individual label for all 240 registrations. These results are obtained for the squared L^2-distance measure

Label id	Mean	Stdev	Min	Max	Median	1st quantile	3rd quantile
1	0.661	0.082	0.445	0.838	0.658	0.613	0.719
2	0.678	0.070	0.510	0.833	0.681	0.629	0.732
3	0.678	0.067	0.500	0.822	0.681	0.628	0.724
4	0.689	0.054	0.548	0.794	0.691	0.650	0.731
5	0.765	0.046	0.626	0.852	0.769	0.733	0.803
6	0.789	0.042	0.624	0.867	0.796	0.759	0.822
7	0.738	0.070	0.498	0.876	0.744	0.707	0.787
8	0.786	0.054	0.602	0.888	0.793	0.756	0.826
9	0.642	0.066	0.496	0.820	0.639	0.596	0.690
10	0.629	0.081	0.423	0.811	0.634	0.565	0.691
11	0.732	0.055	0.577	0.858	0.736	0.691	0.774
12	0.729	0.058	0.588	0.865	0.733	0.688	0.774
13	0.734	0.046	0.611	0.839	0.742	0.702	0.763
14	0.768	0.042	0.637	0.861	0.771	0.744	0.798
15	0.736	0.057	0.510	0.835	0.748	0.706	0.779
16	0.721	0.061	0.475	0.828	0.731	0.687	0.763
17	0.680	0.062	0.520	0.813	0.690	0.634	0.727
18	0.673	0.069	0.480	0.804	0.681	0.625	0.721
19	0.655	0.073	0.467	0.794	0.660	0.607	0.715
20	0.621	0.079	0.412	0.789	0.623	0.566	0.678
21	0.569	0.122	0.272	0.841	0.556	0.488	0.669
22	0.602	0.081	0.359	0.790	0.609	0.558	0.660
23	0.750	0.054	0.571	0.851	0.754	0.715	0.792
24	0.746	0.053	0.596	0.861	0.747	0.703	0.792
25	0.604	0.082	0.375	0.792	0.608	0.549	0.663
26	0.562	0.091	0.271	0.758	0.569	0.501	0.629
27	0.600	0.098	0.324	0.792	0.607	0.540	0.681
28	0.587	0.084	0.370	0.777	0.594	0.524	0.652
29	0.650	0.078	0.440	0.805	0.660	0.592	0.714
30	0.618	0.083	0.412	0.814	0.621	0.564	0.675
31	0.547	0.093	0.220	0.741	0.555	0.485	0.611
32	0.486	0.103	0.224	0.726	0.490	0.415	0.566
33	0.656	0.057	0.507	0.780	0.657	0.620	0.697

Table 6 Average Dice values. We report the mean, min, max, and median value as well as the 1st quantile and the 3rd quantile. These values are computed for each individual label for all 240 registrations. The results are obtained for the normalized cross correlation distance measure

Label id	Mean	Stdev	Min	Max	Median	1st quantile	3rd quantile
1	0.665	0.082	0.427	0.835	0.660	0.614	0.724
2	0.681	0.070	0.491	0.822	0.683	0.632	0.735
3	0.683	0.066	0.514	0.834	0.690	0.639	0.727
4	0.694	0.052	0.562	0.808	0.697	0.655	0.734
5	0.772	0.043	0.628	0.850	0.780	0.747	0.804
6	0.790	0.042	0.620	0.867	0.798	0.759	0.824
7	0.743	0.067	0.543	0.877	0.758	0.707	0.791
8	0.787	0.053	0.605	0.888	0.796	0.756	0.827
9	0.648	0.066	0.499	0.816	0.649	0.597	0.696
10	0.632	0.080	0.425	0.810	0.633	0.573	0.688
11	0.736	0.055	0.588	0.858	0.743	0.696	0.778
12	0.731	0.059	0.587	0.864	0.734	0.690	0.773
13	0.740	0.044	0.615	0.847	0.743	0.710	0.771
14	0.773	0.042	0.639	0.860	0.775	0.749	0.803
15	0.740	0.059	0.504	0.842	0.751	0.703	0.786
16	0.725	0.063	0.469	0.837	0.739	0.689	0.772
17	0.686	0.063	0.522	0.822	0.696	0.642	0.735
18	0.678	0.068	0.493	0.806	0.681	0.632	0.727
19	0.660	0.073	0.471	0.799	0.665	0.611	0.722
20	0.625	0.080	0.414	0.795	0.631	0.572	0.684
21	0.576	0.120	0.271	0.841	0.574	0.495	0.672
22	0.604	0.083	0.328	0.795	0.613	0.553	0.667
23	0.754	0.056	0.566	0.853	0.758	0.716	0.796
24	0.749	0.054	0.597	0.861	0.752	0.707	0.791
25	0.609	0.081	0.377	0.794	0.617	0.550	0.661
26	0.565	0.092	0.263	0.768	0.570	0.501	0.636
27	0.603	0.095	0.331	0.786	0.610	0.550	0.682
28	0.589	0.083	0.364	0.778	0.593	0.530	0.646
29	0.653	0.080	0.448	0.817	0.667	0.589	0.719
30	0.618	0.083	0.418	0.815	0.617	0.563	0.678
31	0.553	0.093	0.219	0.731	0.566	0.488	0.622
32	0.488	0.102	0.224	0.740	0.492	0.415	0.570
33	0.662	0.053	0.538	0.780	0.662	0.625	0.699

References

1. A. Mang et al. "CLAIRE: A distributed-memory solver for constrained large deformation diffeomorphic image registration". In: *SIAM Journal on Scientific Computing* 41.5 (2019), pp. C548–C584. (cit. on pp. 167, 171, 172, 177, 178, 180, 182, 183, 184, 185, 186, 187, 189, 191, 192, 201)

2. A. Mang. *CLAIRE: Constrained Large Deformation Diffeomorphic Image Registration.* https://andreasmang.github.io/claire. 2019. (cit. on p. 171)
3. M. Brunn et al. "CLAIRE: Constrained large deformation diffeomorphic image registration on parallel architectures". In: *The Journal of Open Source Software* 6.61 (2021), p. 3038. (cit. on pp. 167, 172)
4. B. Fischer and J. Modersitzki. "Ill-posed medicine—an introduction to image registration". In: *Inverse Problems* 24.3 (2008), pp. 1–16. (cit. on pp. 167, 168, 170)
5. J. Modersitzki. *Numerical methods for image registration.* New York: Oxford University Press, 2004.
6. J. Modersitzki. *FAIR: Flexible algorithms for image registration.* Philadelphia, Pennsylvania, US: SIAM, 2009. (cit. on p. 168)
7. L. Younes. *Shapes and diffeomorphisms.* 2nd ed. Vol. 171. Springer Verlag Berlin Heidelberg, 2019. (cit. on pp. 168, 170, 173, 174, 175)
8. M. F. Beg et al. "Computing large deformation metric mappings via geodesic flows of diffeomorphisms". In: *International Journal of Computer Vision* 61.2 (2005), pp. 139–157. (cit. on pp. 170, 171, 176, 177, 181)
9. A. Trouveé. *A infinite dimensional group approach for physics based models in pattern recognition.* Tech. rep. Laboratoire d'Analyse Numerique CNRS URA, Universiteé Paris, 1995. (cit. on p. 175)
10. A. Trouveé. "Diffeomorphism groups and pattern matching in image analysis". In: *International Journal of Computer Vision* 28.3 (1998), pp. 213–221. (cit. on pp. 170, 173)
11. P. Dupuis, U. Gernander, and M. I. Miller. "Variational problems on flows of diffeomorphisms for image matching". In: *Quarterly of Applied Mathematics* 56.3 (1998), pp. 587–600. (cit. on pp. 168, 170, 175, 176)
12. A. Borzi and V. Schulz. *Computational optimization of systems governed by partial differential equations.* Philadelphia, Pennsylvania, US: SIAM, 2012. (cit. on pp. 168, 170)
13. M. Hinze et al. *Optimization with PDE constraints.* Berlin, DE: Springer, 2009.
14. H. Antil et al. *Frontiers in PDE-constrained optimization.* Vol. 163. Springer, 2018. (cit. on p. 170)
15. A. Mang et al. "PDE-constrained optimization in medical image analysis". In: *Optimization and Engineering* 19.3 (2018), pp. 765–812. (cit. on pp. 168, 171, 172, 187)
16. L. T. Biegler et al. *Large-scale PDE-constrained optimization.* Springer, 2003. (cit. on pp. 170, 171)
17. M. D. Gunzburger. *Perspectives in flow control and optimization.* Philadelphia, Pennsylvania, US: SIAM, 2003. (cit. on p. 178)
18. J. L. Lions. *Optimal control of systems governed by partial differential equations.* Springer, 1971. (cit. on p. 170)
19. J. V. Hajnal, D. L. G. Hill, and D. J. Hawkes, eds. *Medical Image Registration.* Boca Raton, Florida, US: CRC Press, 2001. (cit. on p. 170)
20. A. Sotiras, C. Davatzikos, and N. Paragios. "Deformable medical image registration: A survey". In: *Medical Imaging, IEEE Transactions on* 32.7 (2013), pp. 1153–1190. (cit. on p. 170)
21. M. Burger, J. Modersitzki, and L. Ruthotto. "A hyperelastic regularization energy for image registration". In: *SIAM Journal on Scientific Computing* 35.1 (2013), B132–B148. (cit. on p. 170)
22. E. Haber and J. Modersitzki. "Image registration with guaranteed displacement regularity". In: *International Journal of Computer Vision* 71.3 (2007), pp. 361–372.
23. T. Rohlfing et al. "Volume-preserving nonrigid registration of MR breast images using free-form deformation with an incompressibility constraint". In: *Medical Imaging, IEEE Transactions on* 22 (2003), pp. 730–741.
24. M. Sdika. "A fast nonrigid image registration with constraints on the Jacobian using large scale constrained optimization". In: *Medical Imaging, IEEE Transactions on* 27.2 (2008), pp. 271–281. (cit. on p. 170)

25. L. Younes, B. Gris, and A. Trouveé. "Sub–Riemannian methods in shape analysis". In: *Handbook of Variational Methods for Nonlinear Geometric Data* (2020), pp. 463–495. (cit. on pp. 170, 175)
26. M. I. Miller and L. Younes. "Group actions, homeomorphism, and matching: A general framework". In: *International Journal of Computer Vision* 41.1/2 (2001), pp. 61–81. (cit. on p. 170)
27. T. Vercauteren et al. "Diffeomorphic demons: Efficient non-parametric image registration". In: *NeuroImage* 45.1 (2009), S61–S72. (cit. on pp. 171, 177)
28. G. E. Christensen, R. D. Rabbitt, and M. I. Miller. "Deformable templates using large deformation kinematics". In: *IEEE Transactions on Image Processing* 5.10 (1996), pp. 1435–1447.
29. L. Younes. "Jacobi fields in groups of diffeomorphisms and applications". In: *Quarterly of Applied Mathematics* 650.1 (2007), pp. 113–134. (cit. on p. 170)
30. A. Mang and G. Biros. "An inexact Newton–Krylov algorithm for constrained diffeomorphic image registration". In: *SIAM Journal on Imaging Sciences* 8.2 (2015), pp. 1030–1069. (cit. on pp. 170, 171, 172, 177, 178, 179, 180, 182, 185, 186, 187, 191, 200)
31. G. L. Hart, C. Zach, and M. Niethammer. "An optimal control approach for deformable registration". In: *Proc IEEE Conference on Computer Vision and Pattern Recognition*. 2009, pp. 9–16. (cit. on pp. 170, 171, 178)
32. A. Borzi, K. Ito, and K. Kunisch. "Optimal control formulation for determining optical flow". In: *SIAM Journal on Scientific Computing* 24.3 (2002), pp. 818–847. (cit. on pp. 170, 171, 178)
33. A. Mang and G. Biros. "Constrained H^1-regularization schemes for diffeomorphic image registration". In: *SIAM Journal on Imaging Sciences* 9.3 (2016), pp. 1154–1194. (cit. on pp. 170, 171, 172, 177, 178, 184, 187, 192, 201)
34. K. Chen and D. A. Lorenz. "Image sequence interpolation using optimal control". In: *Journal of Mathematical Imaging and Vision* 41 (2011), pp. 222–238. (cit. on pp. 170, 171)
35. J. Hinkle et al. "4D MAP image reconstruction incorporating organ motion". In: *Proc Information Processing in Medical Imaging*. LNCS 5636. 2009, pp. 676–687.
36. T. Mansi et al. "iLogDemons: A demons-based registration algorithm for tracking incompressible elastic biological tissues". In: *International Journal of Computer Vision* 92.1 (2011), pp. 92–111.
37. P. Ruhnau and C. Schnoörr. "Optical Stokes flow estimation: An imaging-based control approach". In: *Experiments in Fluids* 42 (2007), pp. 61–78. (cit. on p. 171)
38. K. A. Saddi, C. Chefd'hotel, and F. Cheriet. "Large deformation registration of contrast-enhanced images with volume-preserving constraint". In: *Proc SPIE Medical Imaging*. Vol. 6512. 2008, pp. 651203-1–651203-10. (cit. on p. 170)
39. A. Borzi, K. Ito, and K. Kunisch. "An optimal control approach to optical flow computation". In: *International Journal for Numerical Methods in Fluids* 40.1–2 (2002), pp. 231–240. (cit. on pp. 170, 178)
40. K. Chen and D. A. Lorenz. "Image sequence interpolation based on optical flow, segmentation and optimal control". In: *Image Processing, IEEE Transactions on* 21.3 (2012), pp. 1020–1030. (cit. on pp. 178, 200)
41. E. Lee and M. Gunzburger. "An optimal control formulation of an image registration problem". In: *Journal of Mathematical Imaging and Vision* 36.1 (2010), pp. 69–80.
42. E. Lee and M. Gunzburger. "Analysis of finite element discretization of an optimal control formulation of the image registration problem". In: *SIAM Journal on Numerical Analysis* 49.4 (2011), pp. 1321–1349.
43. F.-X. Vialard et al. "Diffeomorphic 3D image registration via geodesic shooting using an efficient adjoint calculation". In: *International Journal of Computer Vision* 97 (2012), pp. 229–241. (cit. on p. 171)
44. R. Herzog, J. W. Pearson, and M. Stoll. "Fast iterative solvers for an optimal transport problem". In: *Advances in Computational Mathematics* 45 (2019), pp. 495–517. (cit. on pp. 170, 171)

45. M. Benzi, E. Haber, and L. Taralli. "A preconditioning technique for a class of PDE-constrained optimization problems". In: *Advances in Computational Mathematics* 35.2-4 (2011), pp. 149–173. (cit. on p. 171)
46. E. Haber and R. Horesh. "A multilevel method for the solution of time dependent optimal transport". In: *Numerical Mathematics: Theory, Methods and Applications* 8.1 (2015), pp. 97–111.
47. T. ur Rehman et al. "3D nonrigid registration via optimal mass transport on the GPU". In: *Medical Image Analysis* 13.6 (2009), pp. 931–940. (cit. on pp. 172, 202)
48. A. Mang and L. Ruthotto. "A Lagrangian Gauss–Newton–Krylov solver for mass-and intensity-preserving diffeomorphic image registration". In: *SIAM Journal on Scientific Computing* 39.5 (2017), B860–B885. (cit. on pp. 170, 171, 172, 178, 182, 186, 187, 189, 202)
49. B. K. P. Horn and B. G. Shunck. "Determining optical flow". In: *Artificial Intelligence* 17.1-3 (1981), pp. 185–203. (cit. on p. 171)
50. E. M. Kalmoun, L. Garrido, and V. Caselles. "Line search multilevel optimization as computational methods for dense optical flow". In: *SIAM Journal on Imaging Sciences* 4.2 (2011), pp. 695–722. (cit. on p. 171)
51. R. Andreev, O. Scherzer, and W. Zulehner. "Simultaneous optical flow and source estimation: Space–time discretization and preconditioning". In: *Applied Numerical Mathematics* 96 (2015), pp. 72–81. (cit. on p. 171)
52. V. Barbu and G. Marinoschi. "An optimal control approach to the optical flow problem". In: *Systems & Control Letters* 87 (2016), pp. 1–9. (cit. on p. 171)
53. T. Vercauteren et al. "Symmetric log-domain diffeomorphic registration: A demons-based approach". In: *Proc Medical Image Computing and Computer-Assisted Intervention*. Vol. LNCS 5241. 5241. 2008, pp. 754–761. (cit. on p. 171)
54. B. B. Avants et al. "A reproducible evaluation of ANTs similarity metric performance in brain image registration". In: *NeuroImage* 54 (2011), pp. 2033–2044. (cit. on p. 171)
55. B. B. Avants et al. "Symmetric diffeomorphic image registration with cross-correlation: Evaluating automated labeling of elderly and neurodegenerative brain". In: *Medical Image Analysis* 12.1 (2008), pp. 26–41. (cit. on p. 171)
56. A. Bône et al. "Deformetrica 4: An open-source software for statistical shape analysis". In: *International Workshop on Shape in Medical Imaging*. Springer. 2018, pp. 3–13. (cit. on pp. 171, 172)
57. J. Fishbaugh et al. "Geodesic shape regression with multiple geometries and sparse parameters". In: *Medical Image Analysis* 39 (2017), pp. 1–17. (cit. on p. 171)
58. J. Ashburner. "A fast diffeomorphic image registration algorithm". In: *NeuroImage* 38.1 (2007), pp. 95–113. (cit. on pp. 171, 177)
59. Y. Cao et al. "Large deformation diffeomorphic metric mapping of vector fields". In: *Medical Imaging, IEEE Transactions on* 24.9 (2005), pp. 1216–1230. (cit. on p. 171)
60. D.-N. Hsieh et al. "Mechanistic modeling of longitudinal shape changes: Equations of motion and inverse problems". In: *SIAM Journal on Applied Dynamical Systems* 21.1 (2022), pp. 80–101.
61. T. Polzin et al. "Memory efficient LDDMM for lung CT". In: *Proc Medical Image Computing and Computer-Assisted Intervention*. Vol. LNCS 9902. 2016, pp. 28–36. (cit. on p. 171)
62. T. Polzin et al. "A discretize–optimize approach for LDDMM registration". In: *Riemannian Geometric Statistics in Medical Image Analysis*. Elsevier, 2020, pp. 479–532. (cit. on pp. 171, 176, 178)
63. M. Niethammer, G. L. Hart, and C. Zach. "An optimal control approach for the registration of image time-series". In: *Proceedings of the 48h IEEE Conference on Decision and Control*. IEEE. 2009, pp. 2427–2434.
64. S. Arguillere, M. I. Miller, and L. Younes. "Diffeomorphic surface registration with atrophy constraints". In: *SIAM Journal on Imaging Sciences* 9.3 (2016), pp. 975–1003. (cit. on pp. 171, 176)

65. M. Zhang and P. T. Fletcher. "Finite-dimensional Lie algebras for fast diffeomorphic image registration". In: *Proc Information Processing in Medical Imaging*. Vol. 24. 2015, pp. 249–259.
66. J. Ashburner and K. J. Friston. "Diffeomorphic registration using geodesic shooting and Gauss-Newton optimisation". In: *NeuroImage* 55.3 (2011), pp. 954–967. (cit. on p. 171)
67. M. I. Miller, A. Trouveé, and L. Younes. "Geodesic shooting for computational anatomy". In: *Journal of Mathematical Imaging and Vision* 24 (2006), pp. 209–228. (cit. on p. 171)
68. R. Azencott et al. "Diffeomorphic matching and dynamic deformable surfaces in 3D medical imaging". In: *Computational Methods in Applied Mathematics* 10.3 (2010), pp. 235–274. (cit. on pp. 171, 176, 191)
69. P. Zhang et al. "Diffeomorphic shape matching by operator splitting in 3D cardiology imaging". In: *Journal of Optimization Theory and Applications* 188 (2021), pp. 143–168. (cit. on p. 171)
70. A. Mang, J. He, and R. Azencott. "An operator-splitting approach for variational optimal control formulations for diffeomorphic shape matching". In: *Journal of Computational Physics* (2023). (cit. on pp. 171, 176)
71. A. Mang, A. Gholami, and G. Biros. "Distributed-memory large-deformation diffeomorphic 3D image registration". In: *Proceedings of the International Conference on High Performance Computing, Networking, Storage and Analysis*. 2016, pp. 842–853. (cit. on pp. 171, 172, 180, 181, 182, 183, 187)
72. A. Mang and G. Biros. "A semi-Lagrangian two-level preconditioned Newton–Krylov solver for constrained diffeomorphic image registration". In: *SIAM Journal on Scientific Computing* 39.6 (2017), B1064–B1101. (cit. on pp. 171, 172, 180, 181, 182)
73. M. Hernandez. "Gauss-Newton inspired preconditioned optimization in large deformation diffeomorphic metric mapping". In: *Physics in Medicine and Biology* 59.20 (2014), pp. 6085–6115. (cit. on p. 171)
74. V. Simoncini. "Reduced order solution of structured linear systems arising in certain PDE-constrained optimization problems". In: *Computational Optimization and Applications* 53.2 (2012), pp. 591–617. (cit. on p. 171)
75. A. Thorley et al. "Nesterov accelerated ADMM for fast diffeomorphic image registration". In: *Medical Image Computing and Computer Assisted Intervention*. 2021, pp. 150–160. (cit. on p. 171)
76. Y. T. Lee, K. C. Lam, and L. M. Lui. "Landmark-matching transformation with large deformation via n-dimensional quasi-conformal maps". In: *Journal of Scientific Computing* 67 (2016), pp. 926–954. (cit. on p. 171)
77. H.-W. Hsieh and N. Charon. "Diffeomorphic registration with density changes for the analysis of imbalanced shapes". In: *International Conference on Information Processing in Medical Imaging*. Springer. 2021, pp. 31–42. (cit. on p. 171)
78. A. François, P. Gori, and J. Glaune's. "Metamorphic image registration using a semi-Lagrangian scheme". In: *International Conference on Geometric Science of Information*. Springer. 2021, pp. 781–788. (cit. on p. 181)
79. E. Hartman et al. "Elastic shape analysis of surfaces with second-order Sobolev metrics: A comprehensive numerical framework". In: *International Journal of Computer Vision* (2023), pp. 1–27. (cit. on p. 176)
80. A. Bo^ne et al. "Learning the spatiotemporal variability in longitudinal shape data sets". In: *International Journal of Computer Vision* 128.12 (2020), pp. 2873–2896. (cit. on p. 171)
81. Z. Shen et al. "Accurate point cloud registration with robust optimal transport". In: *Advances in Neural Information Processing Systems* 34 (2021), pp. 5373–5389.
82. L. Tian et al. In: (2023), pp. 18084–18094.
83. B. B. Amor, S. Arguillère, and L. Shao. "ResNet-LDDMM: Advancing the LDDMM framework using deep residual networks". In: *arXiv preprint arXiv:2102.07951* (2021).
84. J. Krebs et al. "Learning a probabilistic model for diffeomorphic registration". In: *IEEE Transactions on Medical Imaging* 38.9 (2019), pp. 2165–2176.

85. S. Sun et al. "Topology-preserving shape reconstruction and registration via neural diffeomorphic flow". In: *Proceedings of the IEEE/CVF Conference on Computer Vision and Pattern Recognition*. 2022, pp. 20845–20855.
86. X. Yang et al. "Quicksilver: Fast predictive image registration—A deep learning approach". In: *NeuroImage* 158 (2017), pp. 378–396.
87. N. Wu and M. Zhang. "NeurEPDiff: Neural Operators to Predict Geodesics in Deformation Spaces". In: *International Conference on Information Processing in Medical Imaging*. Springer. 2023, pp. 588–600.
88. S. Bharati et al. "Deep learning for medical image registration: A comprehensive review". In: *arXiv preprint arXiv:2204.11341* (2022).
89. Y. Wu et al. "NODEO: A neural ordinary differential equation based optimization framework for deformable image registration". In: *Proceedings of the IEEE/CVF Conference on Computer Vision and Pattern Recognition*. 2022, pp. 20804–20813. (cit. on p. 171)
90. M. Brunn et al. "Multi-node multi-GPU diffeomorphic image registration for large-scale imaging problems". In: *Proc ACM/IEEE Conference on Supercomputing*. 2020, pp. 523–539. (cit. on pp. 171, 172, 177, 178, 180, 181, 182, 183, 184, 185, 189, 192, 200, 201)
91. M. Brunn et al. "Fast GPU 3D diffeomorphic image registration". In: *Journal of Parallel and Distributed Computing* 149 (2021), pp. 149–162. (cit. on pp. 171, 172, 177, 180, 181, 183, 184, 185, 200, 201)
92. M. Benzi, G. H. Golub, and J. Liesen. "Numerical solution of saddle point problems". In: *Acta Numerica* 14 (2005), pp. 1–137. (cit. on p. 171)
93. A. Gholami et al. "A framework for scalable biophysics-based image analysis". In: *Proceedings of the International Conference on High Performance Computing, Networking, Storage and Analysis*. 19. 2017, 19:1–19:13. (cit. on pp. 171, 172, 180, 183, 184, 192, 199)
94. V. Akcelik, G. Biros, and O. Ghattas. "Parallel multiscale Gauss-Newton-Krylov methods for inverse wave propagation". In: *Proc ACM/IEEE Conference on Supercomputing*. 2002, pp. 1–15. (cit. on p. 171)
95. V. Akcelik et al. "Parallel algorithms for PDE constrained optimization". In: ed. by M. A. Heroux, P. Raghavan, and H. D. Simon. Vol. 20. Parallel Processing for Scientific Computing. Philadelphia, Pennsylvania, US: SIAM, 2006. Chap. 16, pp. 291–322.
96. G. Biros and O. Ghattas. "Parallel Newton-Krylov methods for PDE-constrained optimization". In: *Proc ACM/IEEE Conference on Supercomputing*. 1999, pp. 28–40.
97. G. Biros and O. Ghattas. "Parallel Lagrange-Newton-Krylov-Schur methods for PDE-constrained optimization—Part I: The Krylov-Schur solver". In: *SIAM Journal on Scientific Computing* 27.2 (2005), pp. 687–713.
98. G. Biros and O. Ghattas. "Parallel Lagrange-Newton-Krylov-Schur methods for PDE-constrained optimization—Part II: The Lagrange-Newton solver and its application to optimal control of steady viscous flows". In: *SIAM Journal on Scientific Computing* 27.2 (2005), pp. 714–739.
99. L. T. Biegler et al. *Real-time PDE-constrained optimization*. SIAM, 2007.
100. O. Shenk et al. "Parallel scalable PDE-constrained optimization: Antenna identification in hyperthermia cancer treatment planning". In: *Computer Science—Research and Development* 23.3–4 (2009), pp. 177–183. (cit. on p. 171)
101. A. Eklund et al. "Medical image processing on the GPU–past, present and future". In: *Medical Image Analysis* 17.8 (2013), pp. 1073–1094. (cit. on p. 172)
102. O. Fluck et al. "A survey of medical image registration on graphics hardware". In: *Computer Methods and Programs in Biomedicine* 104.3 (2011), e45–e57.
103. J. Shackleford, N. Kandasamy, and G. Sharp. *High performance deformable image registration algorithms for manycore processors*. Waltham, Massachusetts, US: Morgan Kaufmann, 2013.
104. R. Shams et al. "A survey of medical image registration on multicore and the GPU". In: *Signal Processing Magazine, IEEE* 27.2 (2010), pp. 50–60. (cit. on p. 172)

105. J. Shackleford, N. Kandasamy, and G. Sharp. "On developing B-spline registration algorithms for multi-core processors". In: *Physics in Medicine and Biology* 55.21 (2010), pp. 6329–6351. (cit. on p. 172)
106. M. Modat et al. "Fast free-form deformation using graphics processing units". In: *Computer Methods and Programs in Biomedicine* 98.3 (2010), pp. 278–284.
107. D. P. Shamonin et al. "Fast parallel image registration on CPU and GPU for diagnostic classification of Alzheimer's disease". In: *Frontiers in Neuroinformatics* 7.50 (2014), pp. 1–15. (cit. on p. 172)
108. L. K. Ha et al. "Fast parallel unbiased diffeomorphic atlas construction on multi-graphics processing units". In: *Proc Eurographics Conference on Parallel Graphics and Visualization*. 2009, pp. 41–48. (cit. on p. 172)
109. L. Ha et al. "Multiscale unbiased diffeomorphic atlas construction on multiGPUs". In: *CPU Computing Gems Emerald Edition*. Elsevier Inc, 2011. Chap. 48, pp. 771–791.
110. S. Sommer. "Accelerating multi-scale flows for LDDKBM diffeomorphic registration". In: *Proc IEEE International Conference on Computer Visions Workshops*. 2011, pp. 499–505.
111. P. Valero-Lara. "Multi-GPU acceleration of DARTEL (early detection of Alzheimer)". In: *Proc IEEE International Conference on Cluster Computing*. 2014, pp. 346–354. (cit. on p. 172)
112. N. Himthani et al. "CLAIRE: Parallelized diffeomorphic image registration for large-scale biomedical imaging applications". In: *Journal of Imaging* 8.9 (2022), p. 251. (cit. on pp. 172, 177, 191)
113. T. Munson et al. *TAO 3.7 users manual*. Argonne National Laboratory, Mathematics and Computer Science Division. 2017. (cit. on p. 172)
114. R. T. Mills et al. "Toward performance-portable PETSc for GPU-based exascale systems". In: *Parallel Computing* 108 (2021), p. 102831. issn: 0167-8191.
115. S. Balay et al. *PETSc Web page*. 2023. url: https://petsc.org/. (cit. on p. 186)
116. S. Balay et al. *PETSc/TAO Users Manual*. Tech. rep. ANL-21/39 - Revision 3.20. Argonne National Laboratory, 2023. (cit. on pp. 172, 186)
117. J. Glaune's, A. Trouveé, and L. Younes. "Diffeomorphic matching of distributions: A new approach for unlabelled point-sets and sub-manifolds matching". In: *Proc IEEE Conference on Computer Vision and Pattern Recognition*. Vol. 2. 2004, pp. 712–718. (cit. on p. 173)
118. J. Glaunès et al. "Large deformation diffeomorphic metric curve mapping". In: *International Journal of Computer Vision* 80.3 (2008), pp. 317–336. (cit. on pp. 173, 175)
119. V. I. Arnold. "Sur la geéomeétrie diffeérentielle des groupes de Lie de dimension infinie et ses applications a l'hydrodynamique des fluides parfaits". In: *Annales de l'Institut Fourier* 16 (1966), pp. 319–361. (cit. on p. 173)
120. V. I. Arnold. *Les meéthodes matheématiques de la meéchanique classique*. MIR, Moscow, 1976.
121. D. G. Ebin and J. Marsden. "Groups of diffeomorphisms and the motion of an incompressible fluid". In: *Annals of Mathematics* 92.1 (1970), pp. 102–163. (cit. on p. 173)
122. W. P. Ziemer. *Weakly differentiable functions: Sobolev spaces and functions of bounded variation*. Vol. 120. Springer Science & Business Media, 1989. (cit. on p. 174)
123. M. I. Miller, A. Trouveé, and L. Younes. "On the metrics and Euler–Lagrange equations of computational anatomy". In: *Annual Review of Biomedical Engineering* 4.1 (2002), pp. 375–405. (cit. on p. 175)
124. M. Bauer, M. Bruveris, and P. W. Michor. "Constructing reparametrization invariant metrics on spaces of plane curves". In: *Differential Geometry and its applications* 34 (2014), pp. 139–165. (cit. on p. 175)
125. U. Grenander and M. I. Miller. "Computational anatomy: An emerging discipline". In: *Quarterly of Applied Mathematics* 56.4 (1998), pp. 617–694. (cit. on p. 175)
126. M. I. Miller. "Computational anatomy: Shape, growth and atrophy comparison via diffeomorphisms". In: *NeuroImage* 23.1 (2004), S19–S33.
127. L. Younes, F. Arrate, and M. I. Miller. "Evolutions equations in computational anatomy". In: *NeuroImage* 45 (2009), S40–S50.

128. M. I. Miller, A. Trouveé, and L. Younes. "Hamiltonian systems and optimal control in computational anatomy: 100 years since D'Arcy Thompson". In: *Annual Review of Biomedical Engineering* 17.447–509 (2015). (cit. on p. 175)
129. J. Glaunès, M. Vailland, and M. I. Miller. "Landmark matching via large deformation diffeomorphisms on the sphere". In: *Journal of Mathematical Imaging and Vision* 20 (2004), pp. 179–200. (cit. on p. 176)
130. S. Joshi and M. I. Miller. "Landmark matching via large deformation diffeomorphisms". In: *IEEE Transactions on Image Processing* 9.8 (2000), pp. 1357–1370. (cit. on p. 176)
131. S. Durrleman. "Statistical models of currents for measuring the variability of anatomical curves, surfaces and their evolution". PhD thesis. Universiteé Nice-Sophia Antipolis, France, 2010. (cit. on p. 176)
132. S. Durrleman et al. "A forward model to build unbiased atlases from curves and surfaces". In: *Proc Medical Image Computing and Computer-Assisted Intervention*. 2008, pp. 68–79. (cit. on p. 176)
133. J. A. Glaune's and S. Joshi. "Template estimation from unlabeled point set data and surfaces for computational anatomy". In: *Proc International Workshop on the Mathematical Foundations of Computational Anatomy*. 2006, pp. 29–39. (cit. on p. 176)
134. S. Kurtek et al. "Elastic geodesic paths in shape space of parameterized surfaces". In: *Pattern Analysis and Machine Intelligence, IEEE Transactions on* 34.9 (2012), pp. 1717–1730. (cit. on p. 176)
135. Y. Cao et al. "Diffeomorphic matching of diffusion tensor images". In: *2006 Conference on Computer Vision and Pattern Recognition Workshop*. IEEE. 2006, pp. 67–67. (cit. on p. 176)
136. V. Arsigny et al. "A Log-Euclidean framework for statistics on diffeomorphisms". In: *Proc Medical Image Computing and Computer-Assisted Intervention*. Vol. LNCS 4190. 2006, pp. 924–931. (cit. on p. 177)
137. M. Hernandez, M. N. Bossa, and S. Olmos. "Registration of anatomical images using paths of diffeomorphisms parameterized with stationary vector field flows". In: *International Journal of Computer Vision* 85.3 (2009), pp. 291–306.
138. M. Lorenzi and X. Pennec. "Geodesics, parallel transport and one-parameter subgroups for diffeomorphic image registration". In: *International Journal of Computer Vision* 105.2 (2013), pp. 111–127.
139. M. Lorenzi et al. "LCC-Demons: a robust and accurate symmetric diffeomorphic registration algorithm". In: *NeuroImage* 81 (2013), pp. 470–483. (cit. on p. 177)
140. A. Staniforth and J. Côteé. "Semi-Lagrangian integration schemes for atmospheric models—A review". In: *Monthly Weather Review* 119.9 (1991), pp. 2206–2223. (cit. on p. 181)
141. C. Sigg and M. Hadwiger. "Fast third-order texture filtering". In: vol. 2. GPU Gems. 2005, pp. 313–329. (cit. on p. 184)
142. D. Ruijters, B. M. ter Haar Romeny, and P. Suetens. "Efficient GPU-based texture interpolation using uniform B-splines". In: *Journal of Graphics Tools* 13.4 (2008), pp. 61–69. (cit. on p. 184)
143. D. Ruijters and P. Thévenaz. "GPU prefilter for accurate cubic B-spline interpolation". In: *The Computer Journal* 55.1 (2012), pp. 15–20. (cit. on p. 184)
144. F. Champagnat and Y. Le Sant. "Efficient cubic B-spline image interpolation on a GPU". In: *Journal of Graphics Tools* 16.4 (2012), pp. 218–232. (cit. on p. 184)
145. *Thrust: The C + + Parallel Algorithms Library*. https://nvidia.github.io/thrust. 2023. (cit. on p. 184)
146. J. Nocedal and S. J. Wright. *Numerical Optimization*. New York, New York, US: Springer, 2006. (cit. on pp. 185, 186)
147. S. Boyd and L. Vandenberghe. *Convex Optimization*. Cambridge University Press, 2004. (cit. on p. 185)
148. I. S. Duff, A. M. Erisman, and J. K. Reid. *Direct methods for sparse matrices*. Oxford University Press, 2017. (cit. on p. 185)
149. T. A. Davis. *Direct methods for sparse linear systems*. SIAM, 2006. (cit. on p. 185)

150. M. R. Hestenes and E. Stiefel. "Methods of conjugate gradients for solving linear systems". In: *Journal of Research of the National Bureau of Standards* 49.6 (1952), pp. 409–436. (cit. on p. 185)
151. R. S. Dembo and T. Steihaug. "Truncated-Newton algorithms for large-scale unconstrained optimization". In: *Mathematical Programming* 26.2 (1983), pp. 190–212. (cit. on p. 186)
152. S. C. Eisentat and H. F. Walker. "Choosing the forcing terms in an inexact Newton method". In: *SIAM Journal on Scientific Computing* 17.1 (1996), pp. 16–32. (cit. on p. 186)
153. T. Bui-Thanh et al. "Extreme-scale UQ for Bayesian inverse problems governed by PDEs". In: *Proceedings of the International Conference on High Performance Computing, Networking, Storage and Analysis*. IEEE. 2012, pp. 1–11. (cit. on p. 187)
154. T. Bui-Thanh et al. "A computational framework for infinite-dimensional Bayesian inverse problems Part I: The linearized case, with application to global seismic inversion". In: *SIAM Journal on Scientific Computing* 35.6 (2013), A2494–A2523.
155. A. Alexanderian et al. "A fast and scalable method for A-optimal design of experiments for infinite-dimensional Bayesian nonlinear inverse problems". In: *SIAM Journal on Scientific Computing* 38.1 (2016), A243–A272. (cit. on p. 187)
156. S. S. Adavani and G. Biros. "Multigrid algorithms for inverse problems with linear parabolic PDE constraints". In: *SIAM Journal on Scientific Computing* 31.1 (2008), pp. 369–397. (cit. on p. 187)
157. G. Biros and G. Doğan. "A multilevel algorithm for inverse problems with elliptic PDE constraints". In: *Inverse Problems* 24.1–18 (2008).
158. L. Giraud, D. Ruiz, and A. Touhami. "A comparative study of iterative solvers exploiting spectral information for SPD systems". In: *SIAM Journal on Scientific Computing* 27.5 (2006), pp. 1760–1786.
159. B. Kaltenbacher. "V-cycle convergence of some multigrid methods for illposed problems". In: *Mathematics of Computation* 72.244 (2003), pp. 1711–1730.
160. B. Kaltenbacher. "On the regularizing properties of a full multigrid method for ill-posed problems". In: *Inverse Problems* 17.4 (2001), pp. 767–788.
161. J. T. King. "On the construction of preconditioners by subspace decomposition". In: *Journal of Computational and Applied Mathematics* 29 (1990), pp. 195–205. (cit. on p. 187)
162. W. Griggs, V. E. Henson, and S. F. McCormick. *A multigrid tutorial*. SIAM, 2000. (cit. on p. 189)
163. O. Axelsson and P. S. Vassilevski. "A black box generalized conjugate gradient solver with inner iterations and variable step preconditioning". In: *SIAM Journal on Matrix Analysis and its Applications* 12.4 (1991), pp. 625–644. (cit. on p. 189)
164. Y. Notay. "Flexible conjugate gradients". In: *SIAM Journal on Scientific Computing* 22.4 (2000), pp. 1444–1460. (cit. on p. 189)
165. M. Gutknecht and S. Röllin. "The Chebisyev iteration revisited". In: *Parallel Computing* 28.2 (2002), pp. 263–283. (cit. on p. 189)
166. G. H. Golub and R. S. Varga. "Chebyshev semi-iterative methods, successive overrelaxation iterative methods, and second order Richardson iterative methods". In: *Numerische Mathematik* 3.1 (1961), pp. 147–156. (cit. on p. 189)
167. C. R. Vogel. *Computational methods for inverse problems*. Philadelphia, Pennsylvania, US: SIAM, 2002. (cit. on p. 190)
168. E. Haber and J. Modersitzki. "A multilevel method for image registration". In: *SIAM Journal on Scientific Computing* 27.5 (2006), pp. 1594–1607. (cit. on pp. 190, 191)
169. E. Haber, U. M. Ascher, and D. Oldenburg. "On optimization techniques for solving nonlinear inverse problems". In: *Inverse Problems* 16 (2000), pp. 1263–1280. (cit. on p. 191)
170. J. Wang and M. Zhang. "Deep Learning for Regularization Prediction in Diffeomorphic Image Registration". In: *Journal of Machine Learning for Biomedical Imaging* 17 (2021), pp. 1–20. (cit. on p. 191)
171. E. Al Safadi and X. Song. "Learning-based image registration with meta-regularization". In: *Proceedings of the IEEE/CVF Conference on Computer Vision and Pattern Recognition*. 2021, pp. 10928–10937. (cit. on p. 191)

172. G. E. Christensen et al. "Introduction to the non-rigid image registration evaluation project". In: *Proc Biomedical Image Registration*. Vol. LNCS 4057. 2006, pp. 128–135. (cit. on p. 192)
173. A. Gholami, A. Mang, and G. Biros. "An inverse problem formulation for parameter estimation of a reaction-diffusion model of low grade gliomas". In: *Journal of Mathematical Biology* 72.1 (2016), pp. 409–433. (cit. on p. 199)
174. K. Scheufele et al. "Image-driven biophysical tumor growth model calibration". In: *SIAM Journal on Scientific Computing* 42.3 (2020), B549–B580.
175. K. Scheufele et al. "Coupling brain-tumor biophysical models and diffeomorphic image registration". In: *Computer Methods in Applied Mechanics and Engineering* 347 (2019), pp. 533–567.
176. A. Mang et al. "SIBIA-GIS: Scalable biophysics-based image analysis for glioma segmentation". In: *Proc BraTS 2017 Workshop (MICCAI)*. 2017, pp. 197–204.
177. A. Gooya et al. "GLISTR: Glioma image segmentation and registration". In: *Medical Imaging, IEEE Transactions on* 31.10 (2013), pp. 1941–1954.
178. A. Mang et al. "Integrated biophysical modeling and image analysis: Application to neuro-oncology". In: *Annual Review of Biomedical Engineering* 22 (2020), pp. 309–341.
179. C. Hogea, C. Davatzikos, and G. Biros. "Brain-tumor interaction biophysical models for medical image registration". In: *SIAM Journal on Imaging Sciences* 30.6 (2008), pp. 3050–3072.
180. E. I. Zacharaki et al. "A comparative study of biomechanical simulators in deformable registration of brain tumor images". In: *Biomedical Engineering, IEEE Transactions on* 55.3 (2008), pp. 1233–1236.
181. E. I. Zacharaki et al. "Non-diffeomorphic registration of brain tumor images by simulating tissue loss and tumor growth". In: *NeuroImage* 46.3 (2009), pp. 762–774.
182. C. Hogea, C. Davatzikos, and G. Biros. "An image-driven parameter estimation problem for a reaction-diffusion glioma growth model with mass effects". In: *Journal of Mathematical Biology* 56.6 (2008), pp. 793–825. (cit. on p. 199)
183. H.-W. Hsieh and N. Charon. "Weight metamorphosis of varifolds and the LDDMM–Fisher–Rao metric". In: *Calculus of Variations and Partial Differential Equations* 61.5 (2022), p. 165. (cit. on p. 199)
184. X. Li et al. "Registration of images with varying topology using embedded maps". In: *Medical Imaging, IEEE Transactions on* 31.3 (2012), pp. 749–765 (cit. on p. 32)
185. A. François et al. "Weighted Metamorphosis for registration of images with different topologies". In: *International Workshop on Biomedical Image Registration*. Springer. 2022, pp. 8–17.
186. P.-L. Antonsanti et al. "Partial matching in the space of varifolds". In: *International Conference on Information Processing in Medical Imaging*. Springer. 2021, pp. 123–135.
187. Y. Sukurdeep, M. Bauer, and N. Charon. "A new variational model for shape graph registration with partial matching constraints". In: *SIAM Journal on Imaging Sciences* 15.1 (2022), pp. 261–292. (cit. on p. 199)
188. S. Angenent, S. Haker, and A. Tannenbaum. "Minimizing flows for the Monge–Kantrovich problem". In: *SIAM Journal on Mathematical Analysis* 35.1 (2003), pp. 61–97. (cit. on p. 202)
189. Y. Chen et al. "An efficient algorithm for matrix-valued and vector-valued optimal mass transport". In: *Journal of Scientific Computing* 77 (2018), pp. 79–100. (cit. on p. 202)

A Genomic Tree-Based Sparse Solver

Timothy A. Davis and Srinivas Subramanian

Abstract This article is concerned with improving an existing application of compressive sensing to metagenomics—the Quikr method, where nonnegative sparse recovery is performed using the Lawson Hanson algorithm. To enhance the computational speed of this algorithm, we offer GETS: a GEnomic Tree based Sparse solver. We exploit the inherent structure of the genomic problem to uncover an evolutionary family tree type relationship between the species. This genomic tree enables us to obtain a sparse representation of the problem which is created in the offline stage of GETS, a one time computation. This allows for reduced storage and asymptotic speed ups for our solver via sparse matrix computations. We conclude the article with the results of computational experiments performed with genomic datasets. These experiments illustrate the significant speed ups obtained by GETS over MATLAB's implementation of the Lawson Hanson algorithm.

1 Introduction

Metagenomics is the study of microbial communities from the content of their sequenced DNA or RNA. Compressive sensing techniques for metagenomics were developed in [6] and [7]. The Quikr method of [6] is a k-mer-based approach that solves the problem of reconstruction of concentrations of bacterial communities from an environmental sample. Prior methods worked by classifying samples in a read-by-read fashion which is time consuming for large datasets. Compressive sensing techniques enable a quick classification of an entire dataset simultaneously, thus achieving an order of magnitude reduction in the execution time. Essentially, Quikr measures the frequency of k-mers in a database of 16S rRNA genes for

T. A. Davis · S. Subramanian (✉)
Texas A&M University, College Station, TX, USA
e-mail: davis@tamu.edu; srini2092@tamu.edu

S. Foucart, S. Wojtowytsch (eds.), *Explorations in the Mathematics of Data Science*, Applied and Numerical Harmonic Analysis,
https://doi.org/10.1007/978-3-031-66497-7_9

known bacteria, calculates the frequency of k-mers in the given sample, and then reconstructs the concentrations of the bacteria in the sample by solving an underdetermined system of linear equations under a sparsity assumption. This method was extended in the WGSQuikr method of [7] to enable the accurate analysis of very large whole-genome shotgun datasets. Both methods utilize a novel technique for sparse recovery by solving an ℓ_1-squared nonnegative regularization problem. The article [4] showed that this optimization problem can be recast as a conventional nonnegative least squares problem, which can be solved using the Lawson-Hanson algorithm [8, Chapter 23], an iterative active set algorithm which is well suited for a fast implementation.

Improvements to the Lawson-Hanson algorithm were proposed specifically for use in chemometric applications in [1] and [9]. The ideas are mainly applicable to problems involving matrices which have a much larger number of rows than columns. This is in contrast to the compressive sensing setting of the genomic problem that we consider here, which involves a measurement matrix with the number of columns being much larger than the number of rows. For instance, [1] involves precomputing and storing the product of the transpose of a matrix with itself, for use in each iteration of the algorithm. This would be prohibitively expensive for the dimensions involved in the genomic problem. Likewise, in [9], the idea of grouping and computing common pseudoinverses is only applicable for problems involving a large number of measurement vectors for a given measurement matrix and utilizes a combinatorial approach that might only be suitable for matrices with a few columns.

We introduce GETS, a genomic tree-based sparse solver, which improves the computational speed of the Lawson-Hanson algorithm by exploiting the inherent structure of the genomic problem. This is primarily done by establishing an evolutionary family-tree-type relationship between the columns of the k-mer matrix of DNA sequences that allows for a terse representation of the data. This representation is created in the offline stage of GETS, a one-time computation which enables reduced storage and allows for large asymptotic speedups whenever the solver is used. The problem setting, the optimization formulations, and the algorithm are discussed in Sect. 2. An overview of the ideas to speed up the algorithm are provided in Sect. 3, while the details are covered in Sects. 4 and 5. GETS is primarily written in C, with an interface for use in MATLAB through its MEX functions. The computational experiments described in Sect. 6 illustrate the significant speed improvements obtained by GETS over MATLAB's implementation of the Lawson-Hanson algorithm. These experiments may be reproduced by running the MATLAB scripts present in the GETS package.[1]

[1] GETS is available at https://github.com/SrinivasSubra/GETS.

2 Solving the Problem via the Lawson-Hanson Algorithm

2.1 Problem Setting

Consider a database of 16S rRNA gene sequences for a number of N known bacterial organisms. For a fixed integer k, the number of occurrences of DNA subwords of length k in each gene sequence is measured to obtain a matrix $\mathbf{A}$ of k-mer counts. Its columns are indexed by organisms in the database and its rows are indexed by all 4^k possible DNA subwords of length k. The (i, j)th entry of $\mathbf{A}$ counts the number of occurrence of the ith subword in the DNA sequence of the jth organism. For instance, in one of our datasets, we have an integer matrix of 6-mer counts for a database of 188,326 genomes. With $k = 6$, the integer matrix $\mathbf{A} \in \mathbb{Z}^{m \times N}$ has $m = 4^k = 4096$ rows, and $N = 188{,}326$ columns. By normalizing the columns of $\mathbf{A}$, we obtain a frequency matrix $\mathbf{C}$ whose columns sum to one. It has nonnegative entries $\mathbf{C}_{i,j} \geq 0$ with $\sum_{i=1}^{m} \mathbf{C}_{i,j} = 1$, for all $j = 1, \ldots, m$. Given a sample of 16S rRNA reads, the frequency of all k-mers is calculated to obtain a vector $\mathbf{y}$, the sample k-mer frequency vector.

The problem of reconstructing the concentrations of bacterial communities from a given sample is expressed in the form of a compressive sensing problem. The k-mer frequency matrix of DNA sequences is the measurement matrix $\mathbf{C} \in \mathbb{R}^{m \times N}$. The sample k-mer frequency vector obtained from the DNA reads is the measurement vector $\mathbf{y} \in \mathbb{R}^m$. The composition of the bacterial sample is represented by a nonnegative vector $\overline{\mathbf{x}} \in \mathbb{R}^N$, where $\overline{\mathbf{x}}_j$ is the concentration of jth organism in the database. Here $\overline{\mathbf{x}}$ is assumed to be sparse, since there are expected to be relatively few different organisms in a given sample. The number of measurements $m = 4^k$ is much smaller than N, the number of organisms in the database. Therefore, the problem entails the recovery of sparse nonnegative $\overline{\mathbf{x}}$ from the underdetermined system $\mathbf{y} = \mathbf{C}\overline{\mathbf{x}}$.

2.2 Nonnegative Least Squares (NNLS)

Since $\mathbf{C}$ is a frequency matrix and $\overline{\mathbf{x}}$ is nonnegative, recovery can be performed by solving the nonnegative least squares problem (see [4]):

$$\min_{\mathbf{v} \in \mathbb{R}^N} \|\mathbf{y} - \mathbf{C}\mathbf{v}\|_2 \quad \text{subject to } \mathbf{v} \geq \mathbf{0} \tag{NNLS}$$

via the active set algorithm of Lawson and Hanson. Throughout this article, we use the following notation:

- $\mathbf{z}_{\mathcal{P}}$ is the restriction of a vector $\mathbf{z}$ to an index set $\mathcal{P}$. In MATLAB notation, $\mathbf{z}_{\mathcal{P}} = \mathbf{z}(\mathcal{P})$.
- $\mathbf{C}_{\mathcal{P}}$ is the column submatrix of a matrix $\mathbf{C}$ consisting of the columns indexed by $\mathcal{P}$. In MATLAB notation, $\mathbf{C}_{\mathcal{P}} = \mathbf{C}(:, \mathcal{P})$.

Algorithm 1 Lawson-Hanson NNLS($\mathbf{C}, \mathbf{y}, \epsilon$)

1: **Initialize:** $\mathcal{P} := \emptyset$, $\mathcal{R} := \{1, 2, \ldots, n\}$, $\mathbf{x} := \mathbf{0}$, $\mathbf{z} := \mathbf{0}$, $\mathbf{w} := \mathbf{C}^{\mathsf{T}}\mathbf{y}$.
2: **while** $\mathcal{R} \neq \emptyset$ and $\max(\mathbf{w}_{\mathcal{R}}) > \epsilon$ **do**
3: Find an index $t \in \mathcal{R}$ such that $\mathbf{w}_t = \max(\mathbf{w}_{\mathcal{R}})$
4: Move t from $\mathcal{R}$ to $\mathcal{P}$
5: Compute $\mathbf{z}_{\mathcal{P}}$ as the solution to the least squares problem $\mathbf{C}_{\mathcal{P}}\mathbf{z}_{\mathcal{P}} \cong \mathbf{y}$
6: Set $\mathbf{z}_{\mathcal{R}} := \mathbf{0}$
7: **while** $\min(\mathbf{z}_{\mathcal{P}}) \leq 0$ **do**
8: Set $\alpha := \min\left\{ \dfrac{x_i}{x_i - z_i} : z_i \leq 0, i \in \mathcal{P} \right\}$
9: Set $\mathbf{x} = \mathbf{x} + \alpha(\mathbf{z} - \mathbf{x})$
10: Move from $\mathcal{P}$ to $\mathcal{R}$ all indices $j \in \mathcal{P}$ for which $x_j < \epsilon$
11: Compute $\mathbf{z}_{\mathcal{P}}$ as solution to the least squares problem $\mathbf{C}_{\mathcal{P}}\mathbf{z}_{\mathcal{P}} \cong \mathbf{y}$
12: Set $\mathbf{z}_{\mathcal{R}} := \mathbf{0}$
13: **end while**
14: Set $\mathbf{x} := \mathbf{z}$
15: Set $\mathbf{w} := \mathbf{C}^{\mathsf{T}}(\mathbf{y} - \mathbf{C}\mathbf{x})$
16: **end while**

- $\mathbf{C}_{\mathcal{P}}\mathbf{z}_{\mathcal{P}} \cong \mathbf{y}$ denotes the least squares problem $\min \|\mathbf{y} - \mathbf{C}_{\mathcal{P}}\mathbf{u}\|_2$ with the solution at $\mathbf{u} = \mathbf{z}_{\mathcal{P}}$.

The Lawson-Hanson algorithm (Algorithm 1) takes as input a matrix $\mathbf{C} \in \mathbb{R}^{m \times N}$, a vector $\mathbf{y} \in \mathbb{R}^m$, and a tolerance $\epsilon \in \mathbb{R}$ for the stopping criterion.

Index sets $\mathcal{P}$ and $\mathcal{R}$ are modified in the course of the algorithm. The set $\mathcal{P}$ will be the support of vectors $\mathbf{x}$ and $\mathbf{z}$ at the end of each iteration of the main loop. The vectors $\mathbf{w}$ and $\mathbf{z}$ provide working space. On termination, $\mathbf{x}$ will be the solution vector and $\mathbf{w}$ will be the dual vector satisfying the Karush–Kuhn–Tucker (KKT) optimality conditions for (NNLS), namely,

$$\mathbf{x}_{\mathcal{P}} > \mathbf{0}, \quad \mathbf{x}_{\mathcal{R}} = \mathbf{0}, \quad \mathbf{w}_{\mathcal{P}} = \mathbf{0}, \quad \mathbf{w}_{\mathcal{R}} \leq \mathbf{0}.$$

The inner loop at Step 7 is entered if any of the components of $\mathbf{z}_{\mathcal{P}}$ become negative after the least squares problem at Step 5. It serves as a course correction by updating $\mathbf{x}$ using a convex combination of the current and previous iterate, until the least squares problem at Step 11 yields a nonnegative solution. In our implementation, the vectors $\mathbf{x}$ and $\mathbf{z}$ are stored as sparse vectors. The index sets $\mathcal{P}$ and $\mathcal{R}$, which are complements of one another, are stored in a binary indication vector of length N. The set $\mathcal{P}$ is also implicitly available as the support of $\mathbf{x}$.

2.3 *ℓ_1-Squared Nonnegative Regularization (NNREG)*

In practice, we need to account for error in the measurement because the acquisition of the vector $\overline{\mathbf{x}}$ is not perfect. The measurement vector $\mathbf{y} = \mathbf{C}\overline{\mathbf{x}} + \mathbf{e}$ now contains a noise term $\mathbf{e} \in \mathbb{R}^m$. There are several strategies possible for the approximate recovery of $\bar{\mathbf{x}}$, typically involving an optimization problem with a nonnegative constraint. As demonstrated in [4], a novel technique is to solve an ℓ_1-squared nonnegative regularization problem, namely,

$$\min_{\mathbf{v}\in\mathbb{R}^N} \|\mathbf{v}\|_1^2 + \lambda^2\|\mathbf{y} - \mathbf{C}\mathbf{v}\|_2^2 \quad \text{subject to } \mathbf{v} \geq \mathbf{0} \tag{NNREG}$$

for some large $\lambda > 0$, a regularization parameter. The squared ℓ_1-norm along with the nonnegative constraint enables a recasting of this problem into a nonnegative least squares problem. Indeed, with $\mathbf{v} \geq 0$, we have $\|\mathbf{v}\|_1 = \sum_{j=1}^N v_j$. Following [4], this observation allows (NNREG) to be written as

$$\min_{\mathbf{v}\in\mathbb{R}^N} \|\widetilde{\mathbf{y}} - \widetilde{\mathbf{C}}\mathbf{v}\|_2^2 \quad \text{subject to } \mathbf{v} \geq \mathbf{0} \tag{$\widetilde{\text{NNLS}}$}$$

where $\widetilde{\mathbf{y}} \in \mathbb{R}^{m+1}$ and $\widetilde{\mathbf{C}} \in \mathbb{R}^{(m+1)\times N}$ are defined as

$$\widetilde{\mathbf{y}} = \left[\begin{array}{c}\lambda\mathbf{y}\\ \hline 0\end{array}\right] \quad \text{and} \quad \widetilde{\mathbf{C}} = \left[\begin{array}{c}\lambda\mathbf{C}\\ \hline 1\ \dots\ 1\end{array}\right].$$

Therefore, we can use the Lawson-Hanson algorithm to solve (NNREG), since ($\widetilde{\text{NNLS}}$) is equivalent to (NNLS) with ($\mathbf{C}$, $\mathbf{y}$) replaced with ($\widetilde{\mathbf{C}}$, $\widetilde{\mathbf{y}}$). For our implementation in GETS, it is useful to reformulate the algorithm computations involving the matrix $\widetilde{\mathbf{C}}$ in terms of computations involving the matrix $\mathbf{C}$. With vectors $\mathbf{x} \in \mathbb{R}^N$, $\widetilde{\mathbf{r}} \in \mathbb{R}^{m+1}$, and defining $\mathbf{r} = \widetilde{\mathbf{r}}(1 : m)$, we observe that

$$\widetilde{\mathbf{C}}\mathbf{x} = \left[\begin{array}{c}\lambda\mathbf{C}\mathbf{x}\\ \hline \sum_{j=1}^N x_j\end{array}\right] \quad \text{and} \quad \widetilde{\mathbf{C}}^{\mathsf{T}}\widetilde{\mathbf{r}} = \left[\begin{array}{c|c}\lambda\mathbf{C}^{\mathsf{T}} & \begin{array}{c}1\\ \vdots\\ 1\end{array}\end{array}\right]\left[\begin{array}{c}\mathbf{r}\\ \hline \widetilde{r}_{m+1}\end{array}\right] = \lambda\mathbf{C}^{\mathsf{T}}\mathbf{r} + \widetilde{r}_{m+1}\begin{bmatrix}1\\ \vdots\\ 1\end{bmatrix}.$$

Then, setting $\widetilde{\mathbf{r}} = \widetilde{\mathbf{y}} - \widetilde{\mathbf{C}}\mathbf{x}$ allows us to compute $\mathbf{w} = \widetilde{\mathbf{C}}^{\mathsf{T}}(\widetilde{\mathbf{y}} - \widetilde{\mathbf{C}}\mathbf{x})$, since

$$\widetilde{\mathbf{r}} = \widetilde{\mathbf{y}} - \widetilde{\mathbf{C}}\mathbf{x} = \left[\begin{array}{c}\lambda(\mathbf{y} - \mathbf{C}\mathbf{x})\\ \hline -\sum_{j=1}^N x_j\end{array}\right] \quad \text{yields} \quad \mathbf{w} = \widetilde{\mathbf{C}}^{\mathsf{T}}\widetilde{\mathbf{r}} = \lambda^2\mathbf{C}^{\mathsf{T}}(\mathbf{y} - \mathbf{C}\mathbf{x}) - \sum_{j=1}^N x_j\begin{bmatrix}1\\ \vdots\\ 1\end{bmatrix}.$$

For the least squares step, we solve the problem:

$$\widetilde{\mathbf{C}}_{\mathcal{P}}\mathbf{z}_{\mathcal{P}} = \left[\begin{array}{c}\lambda\mathbf{C}_{\mathcal{P}}\\ \hline 1\ \dots\ 1\end{array}\right]\mathbf{z}_{\mathcal{P}} \cong \left[\begin{array}{c}\lambda\mathbf{y}\\ \hline 0\end{array}\right] = \widetilde{\mathbf{y}}$$

by forming the submatrix $\widetilde{\mathbf{C}}_{\mathcal{P}}$ on the fly. Therefore, we do not need to store or work directly with the matrix $\widetilde{\mathbf{C}}$ (in fact, in the subsequent sections we will see that we do not actually need to store the matrix **C** either). Since the implementation of (NNREG) is a straightforward extension of (NNLS), in the rest of this article, we will describe the conceptual details with reference to (NNLS) only. The computational experiments for solving (NNREG) problems are described in Sect. 6.

3 Exploiting the Structure of the Problem

In MATLAB, the Lawson-Hanson algorithm is implemented as the function `lsqnonneg`. However, this implementation is not efficient for our purpose, since it does not exploit the structure of the problem. In our implementation in GETS, we exploit the algorithm structure along with the genomics of the problem. In terms of time consumed during each iteration of the algorithm, the following steps are significant:

- Computing $\mathbf{Cx}$
- Solving $\mathbf{C}_{\mathcal{P}}\mathbf{z}_{\mathcal{P}} \cong \mathbf{y}$
- Computing $\mathbf{C}^{\mathsf{T}}\mathbf{r}$, where the residual $\mathbf{r} = \mathbf{y} - \mathbf{Cx}$

In this section, we provide an overview of tackling these aspects of the problem. We begin with the basic observation that at each iteration of the algorithm, the variable vector $\mathbf{x}$ is sparse. Then, the computation $\mathbf{Cx}$ is quite simple, since $\mathbf{Cx} = \mathbf{C}_{\mathcal{P}}\mathbf{x}_{\mathcal{P}}$. This is essentially equivalent to a dense matrix-vector multiplication problem involving small dimensions, since the submatrix $\mathbf{C}_{\mathcal{P}}$ is of size $m \times s$, where s denotes the sparsity of $\mathbf{x}$ in the current iteration. Thus, the complexity is just $O(ms)$ flops instead of $O(mN)$ flops which is required in a full dense matrix-vector product.

Next, we consider the least squares problem $\mathbf{C}_{\mathcal{P}}\mathbf{z}_{\mathcal{P}} \cong \mathbf{y}$ in Step 5. This is an overdetermined least squares problem which normally requires $O(ms^2)$ flops. The Lawson-Hanson algorithm is an active set algorithm where each iteration of the outer loop involves updating the support $\mathcal{P}$ by adding an index to it. The submatrix $\mathbf{C}_{\mathcal{P}}$ is thus updated by the addition of a column. We exploit this fact to utilize a QR update method which speeds up the least squares computations by bringing down the complexity to $O(ms)$ flops (Note: If the inner loop is entered, which does not happen often in practice, one or more columns of $\mathbf{C}_{\mathcal{P}}$ are removed in Step 10. Since we keep track of the columns of $\mathbf{C}_{\mathcal{P}}$, a QR update is always possible in the outer loop. See Sect. 5).

The main focus of this article is in tackling the major computational bottleneck—computing $\mathbf{C}^{\mathsf{T}}\mathbf{r}$, a dense matrix-vector product of complexity $O(mN)$ flops, since the residual $\mathbf{r}$ is dense. We consider the following question:

How can we exploit the genomics of the problem to speed up the computations?

The matrix **C** was originally obtained after column normalizing an integer matrix **A**, whose (i, j)th entry counts the number of occurrence of the ith subword in the DNA sequence of the jth organism. This suggests that columns corresponding to genomically similar organisms would not be very different. We establish an evolutionary family-tree-type relationship between the columns of the integer matrix $\mathbf{A}$ that allows for a sparse representation of the data. This representation is created in the offline stage, which is a one-time computation. This enables a compact storage of the data and allows for large asymptotic speedups in the computation of $\mathbf{C}^\mathsf{T}\mathbf{r}$ during each iteration of the algorithm.

Thus, for the given data of a k-mer matrix $\mathbf{A}$ of DNA sequences, we obtain a significant overall speedup of the Lawson-Hanson algorithm every time the solver is called to recover any sparse vector $\overline{\mathbf{x}}$ from the measurement vector $\mathbf{y} = \mathbf{C}\overline{\mathbf{x}}$ as input. The ideas discussed above are explored in the following sections.

4 Genomic Tree-Based Computations

4.1 *Offline Stage*

Let us consider a similarity matrix $\mathbf{\Delta} \in \mathbb{Z}^{N\times N}$ to quantify the differences in the columns of $\mathbf{A}$. Define $\mathbf{\Delta}(i, j) := \mathrm{nnz}\left(\mathbf{A}(:, i) - \mathbf{A}(:, j)\right)$, where we use the MATLAB notation $\mathbf{A}(:, i)$ to denote the ith column of $\mathbf{A}$ and $\mathrm{nnz}(\mathbf{M})$ to denote the number of nonzeros in a matrix $\mathbf{M}$. We compute the minimum spanning tree (MST) on the graph of this matrix with edge weights $\mathbf{\Delta}(i, j)$. Since the graph of $\mathbf{\Delta}$ would be dense, we use Prim's algorithm to compute the MST. We use the adjacency matrix representation of the graph for the implementation. We do not actually store $\mathbf{\Delta}$, a large matrix requiring significant memory because N can be large. Instead, we compute the edge weights $\mathbf{\Delta}(i, j)$ implicitly, on the fly, during the course of Prim's algorithm. Computing an edge weight $\mathbf{\Delta}(i, j)$ takes $O(m)$ flops; therefore, the entire computational complexity for the MST would be $O(N^2m)$ flops.

The MST gives us a tree of relationships between the organisms. The tree can be represented by a length N `parent` array. If $\texttt{parent}(i) = j$ in the tree, we would expect $\mathbf{\Delta}(i, j)$ to be small due to the genomic similarities between the organisms. Since the MST relies on genomic similarities, we can consider it to be a genomic tree somewhat reminiscent to a phylogenetic tree. In this article, we are only concerned with the computational benefits offered by this tree. The MST enables the creation of a sparse difference matrix which is formed by taking the differences between the columns of a child and its parent in the tree. Let $\mathbf{D} \in \mathbb{Z}^{m\times N}$ denote the difference matrix. Then $\mathbf{D}$ is defined as $\mathbf{D}(:, i) = \mathbf{A}(:, i) - \mathbf{A}(:, j)$, where $j = \texttt{parent}(i)$ for all nodes i (except for the root node which has no parent). Thus, $\mathbf{D}$ would be a sparse matrix as the parent and child columns would differ only in few places. The sparseness of $\mathbf{D}$ was numerically confirmed for our genomic datasets (see Sect. 4.3). Before delving into the remaining offline computations, we first look at how this $\mathbf{D}$ matrix can be used to compute $\mathbf{C}^\mathsf{T}\mathbf{r}$ in $O(\mathrm{nnz}(\mathbf{D}))$ flops.

4.2 *Fast Computation of* $\mathbf{C}^{\mathsf{T}}\mathbf{r}$

We make use of the difference matrix $\mathbf{D}$ in computing $\mathbf{C}^{\mathsf{T}}\mathbf{r}$, where $\mathbf{r}$ is the residual. Let **diag** denote a diagonal matrix consisting of the reciprocals of the column sums of the matrix $\mathbf{A}$. As the matrix $\mathbf{C}$ was obtained after column normalizing $\mathbf{A}$, we have $\mathbf{C} = \mathbf{A} * \mathbf{diag}$ and $\mathbf{C}^{\mathsf{T}}\mathbf{r} = (\mathbf{A} * \mathbf{diag})^{\mathsf{T}}\mathbf{r} = \mathbf{diag} * (\mathbf{A}^{\mathsf{T}}\mathbf{r})$. Therefore, we only require the integer matrix $\mathbf{A}$ for the computations. We will first compute $\mathbf{w} = \mathbf{A}^{\mathsf{T}}\mathbf{r}$ and then scale $\mathbf{w}$ using **diag** afterwards. For each i, we need to compute $\mathbf{w}(i) = \mathbf{A}(:, i)^{\mathsf{T}}\mathbf{r}$. Since $\mathbf{D}(:, i) = \mathbf{A}(:, i) - \mathbf{A}(:, j)$ and $j = \mathtt{parent}(i)$, we have

$$\begin{aligned} \mathbf{w}(i) &= \big(\mathbf{D}(:, i) + \mathbf{A}(:, j)\big)^{\mathsf{T}}\mathbf{r}, \\ &= \mathbf{A}(:, j)^{\mathsf{T}}\mathbf{r} + \mathbf{D}(:, i)^{\mathsf{T}}\mathbf{r}, \\ &= \mathbf{w}(j) + \mathbf{D}(:, i)^{\mathsf{T}}\mathbf{r}. \end{aligned} \quad (\mathtt{TREEMULT})$$

As the columns of $\mathbf{D}$ are sparse, we can compute $\mathbf{D}(:, i)^{\mathsf{T}}\mathbf{r}$ very quickly in $\mathrm{nnz}(\mathbf{D}(:, i))$ flops. This allows us to obtain an asymptotic speedup in computing $\mathbf{C}^{\mathsf{T}}\mathbf{r}$—the complexity reduces from $O(mN)$ to $O(\mathrm{nnz}(\mathbf{D}))$ flops. The speedup is proportional to the sparsity of the difference matrix $\mathbf{D}$. Any spanning tree would give a difference matrix that can be used to compute $\mathbf{C}^{\mathsf{T}}\mathbf{r}$, but a MST would yield the sparsest difference matrix and hence the largest speedup.

In (TREEMULT), we need to compute the parent value $\mathbf{w}(j)$ prior to computing the child value $\mathbf{w}(i)$. In order to access a parent node and compute its $\mathbf{w}$ value before reaching its child node in an efficient manner, we make use of a topological ordering of the MST. A topological ordering of the tree ensures that the parent node j occurs before node i in the ordering. We obtain a topological ordering in the offline stage of our implementation, by performing a reverse post ordering of the MST via depth-first search. The reverse post ordering is represented by a `post` array. One can apply `post` on a new node of the reordered tree to obtain the old node. To do the reverse, one can apply `invpost`, the inverse of the `post` array, on the old node. The parent-child relationships remain the same, so we can think of the reordered tree as the old tree with the nodes renumbered by applying `invpost`. In this reordered tree, the new "parent" array `ipt` is given by `ipt = invpost(parent(post))`. We work with the reordered tree throughout the Lawson-Hanson algorithm. For instance, $\mathbf{A}$ is now replaced by $\mathbf{A}(:, \mathtt{post})$, its column reordered version. Likewise, we work with the reordered versions of $\mathbf{D}$ and **diag**. Computing $\mathbf{w}(i)$ in (TREEMULT) then becomes $\mathbf{w}(i) = \mathbf{w}(\mathtt{ipt}(i)) + \mathbf{D}(:, i)^{\mathsf{T}}\mathbf{r}$. The index i can simply be varied from 2 to N in a loop without issue, since the topological ordering ensures that $\mathtt{ipt}(i) < i$ (the root node at $i = 1$ has no parent). All the reordering-related computations are performed in the offline stage. If the original indices of the solution are needed, one can apply the `post` array on the support of the solution. That is, with $\mathcal{P}$ as the support of the solution vector $\mathbf{x}$ upon termination, the original support is given by $\mathtt{post}(\mathcal{P})$.

4.3 Compact Data Representation

In the C programming language, the size of the datatype double is 64 bits, short is 16 bits, and char is 8 bits. The normalized matrix **C**, which is stored in double format, is expensive for memory access in comparison with storing only an integer matrix **A**. In our large dataset, we have a k-mer matrix $\mathbf{A}_{\text{large}} \in \mathbb{Z}^{m \times n}$ of dimensions $m = 4096$ and $n = 188{,}326$. We observed that $\mathbf{A}_{\text{large}}$ has entries no bigger than 50; hence, we store this as a char matrix. Storing **C** requires 8 times more memory space than storing **A**. Therefore, we never work directly with the matrix **C** in the entire algorithm. Instead, we only store the matrix **A** and scale on the fly. For instance, for the computation **Cx**, we compute $\mathbf{A}(\mathbf{diag} * \mathbf{x})$. We just need to scale the entries of the sparse vector **x** with **diag** before performing a dense matrix times sparse vector operation. The computation of $\mathbf{C}^\mathsf{T}\mathbf{r}$ and the least squares problem using only **A** and **diag**, without **C**, are noted in the corresponding sections.

The expected sparseness of the difference matrix **D** corresponding to the large dataset was confirmed numerically. We found that it comprised approximately 93% zeros, 6.5% binary nonzeros with $+1$s and -1s, and 0.5% nonbinary values ranging between -40 and $+40$. We exploit this distribution of values by splitting **D** into a sum of three sparse matrices, $\mathbf{D} = \mathbf{D}_{+1} + \mathbf{D}_{-1} + \mathbf{D}_{\text{nb}}$. The nonzero values of **D** are split into $+1$s in $\mathbf{D}_{+1}$, -1s in $\mathbf{D}_{-1}$, and nonbinary values in $\mathbf{D}_{\text{nb}}$. The benefit of this lies in even more compact storage—for the sparse matrices $\mathbf{D}_{+1}$ and $\mathbf{D}_{-1}$ we do not need to store the numerical values; we only need to store the location of the binary entries as opposed to both location and values in a *compressed column form* [2]. The dot products $\mathbf{D}_{+1}^\mathsf{T}\mathbf{r}$ and $\mathbf{D}_{-1}^\mathsf{T}\mathbf{r}$ require only addition operations. We store the entries of the very sparse $\mathbf{D}_{\text{nb}}$ matrix as a char array. We create a data structure treedata, a C struct, for this purpose that encapsulates all the data required for the solver—the post ordering of the tree, parent array of the reordered tree, reordered **A**, **diag**, and the sparse components of **D**. The offline stage of GETS involves taking the original k-mer matrix **A** as input and performing all the required one time computations to generate this compact data representation. The output treedata thus generated is saved so that it is always available to be used as an input for the solver. This enables the fast computations previously discussed thereby providing significant speedups to the Lawson-Hanson algorithm whenever the solver is used.

The C struct declaration of treedata is provided below to illustrate the discussion here.

```
typedef struct gets_treedata_struct
{
 csi m; /* number of rows */
 csi n; /* number of columns */
 csi *bposp; /* column pointers for binary +1 entries of D */
 csi *bnegp; /* column pointers for binary -1 entries of D */
 short *bposi; /* row indices for binary +1 entries of D */
 short *bnegi; /* row indices for binary -1 entries of D */
 csi *xp; /* column pointers for nonbinary entries of D */
```

```
  short *xi; /* row indices for nonbinary entries of D */
  char *xx; /* numerical values for nonbinary entries of D */
  char *A; /* reordered unnormalized integer matrix A */
  csi *diag; /* column sums of reordered A */
  csi *ipt; /* parent array in the reordered tree */
  csi *post; /* post ordering of the MST */
} treedata ;
```

GETS uses the CSparse package [2] for implementing sparse vectors and utilizing certain functions. The `csi` datatype from CSparse is an integer datatype of size 64 bits. In `treedata`, the integers `m` and `n` are the dimensions of the integer k-mer matrix $\mathbf{A}$. This m-by-n dense matrix is represented by the `char` array `A` (after its columns are reordered). The `csi` arrays `diag`, `ipt`, and `post` are all of size n. The array `diag` stores the column sums of $\mathbf{A}$, which are used for the previously mentioned scaling operations done by **diag**. The arrays `ipt` and `post` are as defined in Sect. 4.2. The sparse m-by-n matrix $\mathbf{D}_{\text{nb}}$ is essentially represented in *compressed column form* by the arrays `xp`, `xi`, and `xx`. The array `xi`, of type `short`, contains the row indices of nonzero entries. Its size therefore equals nnz($\mathbf{D}_{\text{nb}}$), the number of nonzero entries in $\mathbf{D}_{\text{nb}}$, and its values will be less than or equal to m. The array `xx`, of the same size and of type `char`, contains the numerical values of the nonzero entries in $\mathbf{D}_{\text{nb}}$. The array `xp`, of size n+1, contains the column pointers for the *compressed column form* such that $\texttt{xp}(j) + 1$ is the location in the `xi` and `xx` arrays of the first nonzero entry of $\mathbf{D}_{\text{nb}}$ in column j. Row indices of entries in column j are stored in $\texttt{xi}(\texttt{xp}(j) + 1)$ through $\texttt{xi}(\texttt{xp}(j + 1))$ for $j = 1 : n$, and the corresponding numerical values are stored in the same location in `xx`. The first entry $\texttt{xp}(1)$ is 0, while the last entry $\texttt{xp}(n{+}1)$ is nnz($\mathbf{D}_{\text{nb}}$). Likewise, `bposp` and `bposi` are the arrays of column pointers and row indices used to represent the sparse matrix $\mathbf{D}_{+1}$, while the arrays `bnegp` and `bnegi` correspond to the sparse matrix $\mathbf{D}_{-1}$. As mentioned before, we do not require any arrays to store the numerical values of $\mathbf{D}_{+1}$ and $\mathbf{D}_{-1}$. (Note: The C programming language actually uses *zero-based* indexing, but the above descriptions are *one-based* for the sake of consistency).

To solve larger-sized problems, we might need to use bigger-sized datatypes than discussed above. The values in the arrays of row indices `bposi`, `bnegi`, and `xi` can range from 1 to $m = 4^k$. They are declared as `short` type in `treedata`, since we assume that $m < 2^{15}$. This would work for k-mers in the range $k = 1, \ldots, 7$. For 8-mers or larger, a datatype of size bigger than `short` would need to be chosen accordingly. For instance, declaring the arrays of row indices as 32-bit integers would work for k-mers up to $k = 15$. Further, `treedata` assumes that the values in the matrices $\mathbf{A}$ and $\mathbf{D}_{\text{nb}}$ are in the range -2^7 to $2^7 - 1$, since the arrays `A` and `xx` are declared as `char`. As discussed earlier, this was indeed the case for our datasets. If this is not the case for some given dataset, the datatype can be changed to `short` or a bigger-sized integer datatype in the offline stage, according to the range of values in $\mathbf{A}$ and $\mathbf{D}_{\text{nb}}$.

5 Solving Least Squares Using QR Updates

We use Householder-based QR updates for speeding up the least squares problem $\mathbf{C}_{\mathcal{P}}\mathbf{z}_{\mathcal{P}} \cong \mathbf{y}$ in Step 5. At each outer iteration of the Lawson-Hanson algorithm, the $m \times s$ submatrix $\mathbf{C}_{\mathcal{P}}$ is augmented with a new column. We thus use a left-looking method to update our QR factorization to solve the new problem with $s+1$ columns by exploiting the QR factorization of the prior $\mathbf{C}_{\mathcal{P}}$ matrix with s columns. Golub and Van Loan [5, Algorithm 5.2.1], a Householder-based QR factorization method, is similar to this left-looking approach, but since it is a right-looking method, it is not suitable for QR updates in Step 5 as it requires all columns of the matrix being factorized to be available at once. Throughout the algorithm, the submatrix $\mathbf{C}_{\mathcal{P}}$ is maintained in the form of a "VR matrix," an economical representation of its QR factorization. Let $\mathbf{H}_1, \ldots, \mathbf{H}_s$ be the Householder matrices that have previously been computed (implicitly), such that if $\mathbf{Q} = \mathbf{H}_1 \ldots \mathbf{H}_s$, then $\mathbf{Q}^{\mathsf{T}}\mathbf{C}_{\mathcal{P}} = \mathbf{R}$ is upper triangular. The matrix $\mathbf{Q}$ and the Householder matrices $\mathbf{H}_j = \mathbf{I} - \beta_j \mathbf{v}_j \mathbf{v}_j^{\mathsf{T}}$ are never explicitly formed, but are represented by the Householder vectors $\mathbf{v}_j$ and the corresponding scalars β_j.

The "VR matrix" is a factored form representation of the QR factorization that stores the matrix $\mathbf{R}$ in its the upper triangular part and the essential parts of the s Householder vectors in its lower triangular part. Let the array C denote the "VR matrix", of the same dimensions as the submatrix $\mathbf{C}_{\mathcal{P}}$. Then its upper triangular part $C(1:s, 1:s) = \mathbf{R}$, while $C(j+1:m, j)$ houses the essential part of the jth Householder vector $\mathbf{v}_j$, for $j = 1:s$. This is enabled by the use of [5, Algorithm 5.1.1] which creates a Householder vector v that is normalized so that $v(1) = 1$ does not need to be stored—we only need to store the remaining part of v, its essential part. Therefore, in order to apply the jth Householder vector $\mathbf{v}_j$ on some vector $\mathbf{c}$, we make use $C(j+1:m, j)$ and β_j to overwrite $\mathbf{c}(j:m)$ with $\mathbf{H}_j\mathbf{c}(j:m)$.

With the vector $\mathbf{b} = \mathbf{Q}^{\mathsf{T}}\mathbf{y}$, the solution to the problem $\mathbf{C}_{\mathcal{P}}\mathbf{z}_{\mathcal{P}} \cong \mathbf{y}$ is obtained by solving the upper triangular system $\mathbf{R}\mathbf{z}_{\mathcal{P}} = \mathbf{b}(1:s)$. When a new index is added to $\mathcal{P}$, we need to update the submatrix by a new column. Since we never actually store the normalized $\mathbf{C}$ matrix, we access the corresponding column of $\mathbf{A}$ and normalize it on the fly using the corresponding **diag** value to obtain $C(:, s+1) = \mathbf{C}_{\mathcal{P}}(:, s+1)$. The following QR update algorithm is then used to update the QR factorization as well as the current right hand side vector $\mathbf{b}$. It takes as input the integer s, the "VR matrix" represented by array C which contains the previous s Householder vectors $\mathbf{v}_j$, the corresponding scalars β_j, and the vector $\mathbf{b}$.

We perform the QR updates by first applying all the previous Householder vectors $\mathbf{v}_1, \ldots, \mathbf{v}_s$ to the new column $C(:, s+1)$. Next, we construct the new Householder vector $\mathbf{v}_{s+1}$ and β_{s+1} from $C(s+1:m, s+1)$. Subsequently, we set $C(s+1, s+1) = \|C(s+1:m, s+1)\|_2$ so that $C(1:s+1, s+1) = \mathbf{R}(1:s+1, s+1)$ and overwrite $C(s+2:m, s+1)$ with the essential part of $\mathbf{v}_{s+1}$. Lastly, we update

Algorithm 2 QR update

for $j = 1 : s$ **do**
    $C(j : m, s + 1) = (\mathbf{I} - \beta_j \mathbf{v}_j \mathbf{v}_j^\mathsf{T}) C(j : m, s + 1)$
end for
Construct Householder vector $\mathbf{v}_{s+1}$ and β_{s+1} from $C(s+1 : m, s+1)$ using [5, Algorithm 5.1.1]
$C(s + 1, s + 1) = \|C(s + 1 : m, s + 1)\|_2$
$C(s + 2 : m, s + 1) = \mathbf{v}_{s+1}(2 : m - s)$
$\mathbf{b}(s + 1 : m) = (\mathbf{I} - \beta_{s+1} \mathbf{v}_{s+1} \mathbf{v}_{s+1}^\mathsf{T}) \mathbf{b}(s + 1 : m)$

the current right-hand side vector $\mathbf{b}$, by applying the Householder vector $\mathbf{v}_{s+1}$ on $\mathbf{b}(s + 1 : m)$. With this, the QR update steps for the new index are complete and the "VR matrix" form is thus maintained for subsequent use. This method allows us to solve the least squares problem in Step 5 with reduced complexity in $O(ms)$ flops. Without QR updates, the problem would require a full QR factorization of complexity $O(ms^2)$ flops during each iteration.

If the inner loop is entered in Step 7, one or more indices in $\mathcal{P}$ will be removed prior to solving the least squares problem in Step 11. In this case, we do a full QR factorization from scratch by using successive QR updates. Thus, the "VR matrix" form of $\mathbf{C}_\mathcal{P}$ is always maintained, and regardless of whether the inner loop is entered, Step 5 in the outer loop is always solved using QR updates. In our experiments (see Sect. 6), we observed that the inner loop is infrequently entered and typically exited within one or two iterations. Moreover, in each problem, the total time spent in the inner loop is small. For instance, for our large dataset, in 86 % of the problems the total time spent in the inner loop was less than 1% of the problem runtime. Only in 4% of the problems did the total time spent in the inner loop exceed 5% of the problem runtime (the largest was 8.5%). In addition, revising the QR factorization would require an irregular downdate, because one or more columns are being removed from arbitrary locations in the $\mathbf{C}_\mathcal{P}$ matrix. Thus, the approach of performing the factorization from scratch when the inner loop is entered does not impose a significant cost on the algorithm performance. In situations where many columns are removed relative to the number of columns of the submatrix $\mathbf{C}_\mathcal{P}$, revising the QR factorization would require a large-rank irregular downdate. Prior work [3] suggests that large-rank update/downdates (relative to the size of the matrix) can be slower than simply factorizing the matrix from scratch.

Remark 1 The outer loop condition of Step 2 ensures that a new column added to $\mathbf{C}_\mathcal{P}$ will never be redundant. Indeed, once the condition of Step 2 is satisfied, the outer loop continues, and the new index t is chosen in Step 3. Therefore, the new column satisfies $\mathbf{C}(:, t)^\mathsf{T}(\mathbf{y} - \mathbf{Cx}) > \epsilon$. Moreover, after the least squares step in the previous iteration, the residual $\mathbf{y} - \mathbf{Cx}$ will be orthogonal to the column space of $\mathbf{C}_\mathcal{P}$. This implies that $\mathbf{C}(:, t)$ cannot lie in the column space of $\mathbf{C}_\mathcal{P}$. Thus, $\mathbf{C}_\mathcal{P}$ will always have full rank and the least squares problem will never be rank deficient.

6 Computational Experiments

The datasets used for our numerical experiments pertain to two different databases of 16S rRNA gene sequences for known bacteria. We were provided with integer k-mer matrices (for a fixed $k \sim 6$) with $m = 4^k$ rows and N columns and several sample k-mer frequency vectors of size m. Specifically, the dataset consists of:

- $\mathbf{A}_{\text{small}}$, an integer matrix of 6-mer counts for a database of 10,046 genomes with dimensions $m = 4096$ and $N = 10{,}046$. This corresponds to the "small database" used in the experiments of Quikr [6].
- $\mathbf{A}_{\text{large}}$, an integer matrix of 6-mer counts for a database of 188,326 genomes with dimensions $m = 4096$ and $N = 188{,}326$. This was derived from the "large database" used in the experiments of Quikr.
- A total of 211 different sample 6-mer frequency vectors $\mathbf{y}_i$, which are the measurement vectors of size $m = 4096$.

6.1 Offline Stage

The terse representations of the small and large datasets, $\mathbf{D}_{\text{small}}$ and $\mathbf{D}_{\text{large}}$, were created in the offline stage. Given an integer k-mer matrix $\mathbf{A}$, GETS performs the required one time computations to generate the compact representation for the problem in the form of the `treedata` data structure. For the small dataset, with $\mathbf{A}_{\text{small}}$ as input, GETS generated $\mathbf{D}_{\text{small}}$ in a runtime of about 2 minutes. Correspondingly, for the large dataset, with $\mathbf{A}_{\text{large}}$ as input, $\mathbf{D}_{\text{large}}$ was generated in about 8 hours. They were saved to be available as inputs for the GETS solver for the experiments described below.

6.2 Speed Comparison of Solvers

To test how fast our solver performs, we compared GETS with MATLAB's `lsqnonneg`. We found that GETS is about 15 times faster than `lsqnonneg` for the large dataset and about 6 times faster for the small dataset. The experiments were performed on laptop with 16 GB memory and a 2.6 GHz 6-Core Intel Core i7 processor.

In our experiments, we solved 211 problems each for the small and large datasets. In each problem, given the measurement vector $\mathbf{y}_i = \mathbf{C}\overline{\mathbf{x}_i} + \mathbf{e}_i$, we seek to approximately recover the vector $\overline{\mathbf{x}_i}$, where $i = 1 : 211$. The vector $\mathbf{e}_i$ indicates the unknown measurement error. The matrix $\mathbf{C} \in \mathbb{R}^{m \times N}$ is the normalized version of the integer matrix $\mathbf{A}$, with column sums equal to one. The experiments were performed for the small and large datasets with $\mathbf{A} = \mathbf{A}_{\text{small}}, \mathbf{A}_{\text{large}}$. For each problem, the solution was obtained by solving (NNREG) via (NNLS) as described in Sect. 2.3. For (NNREG), smaller the value of λ chosen, the sparser the reconstruction obtained. The larger it is, the more closely the k-mer counts will be fit. Our 16S rRNA datasets, $\mathbf{A}_{\text{small}}$ and $\mathbf{A}_{\text{large}}$, were obtained from the Quikr datasets of [6]. Therefore, we used their choice of $\lambda = 10{,}000$ as the regularization parameter, since it worked well for the Quikr experiments. Changing the dimensions of the problem, for instance, for problems involving k-mer matrices with larger values of k, will most likely require a change in the selection of λ. In that case, depending on the inputs, a procedure such as the one used in [7] may be considered—WGSQuikr uses an adaptive procedure which consists of running (NNREG) multiple times with different parameters.

Let $\mathbf{x}_G$ and $\mathbf{x}_M$ denote the solutions generated by GETS and MATLAB's `lsqnonneg`, respectively. We compared $\|\mathbf{x}_G - \mathbf{x}_M\|_2$, the error in ℓ_2 norm between the two solutions. The maximum ℓ_2 error was found to be 2.59158e-14 for the small dataset and 3.90415e-14 for the large dataset. The solutions $\mathbf{x}_G$ and $\mathbf{x}_M$ were found to have the same support in every experiment. As expected, both MATLAB and GETS generated the same solutions within tolerance.

For the small dataset, the average time taken by MATLAB to solve a problem was 2.2446 seconds, while GETS took 0.3667 seconds. Correspondingly, for the large dataset, the MATLAB average was 66.7561 seconds, while the GETS average was just 4.3911 seconds. To illustrate the speedup for a specific problem in the large dataset, consider the problem where both GETS and MATLAB took the maximum time to solve it. MATLAB solved the problem in 229.1092 seconds, while GETS took just 16.2336 seconds.

For each problem we considered the speedup ratio, which is defined as the ratio of the time taken by MATLAB divided by the time taken by GETS. The histogram plots of the speedup ratios for the problems are provided in Fig. 1. For the small dataset, the mean speedup ratio was 5.9647, while the median speedup ratio was 5.9125. For the large dataset, the mean speedup ratio was 15.2900, while the median speedup ratio was 15.3311. Thus, GETS provides a significant speed improvement in both cases, with the large dataset exhibiting a greater speedup. We

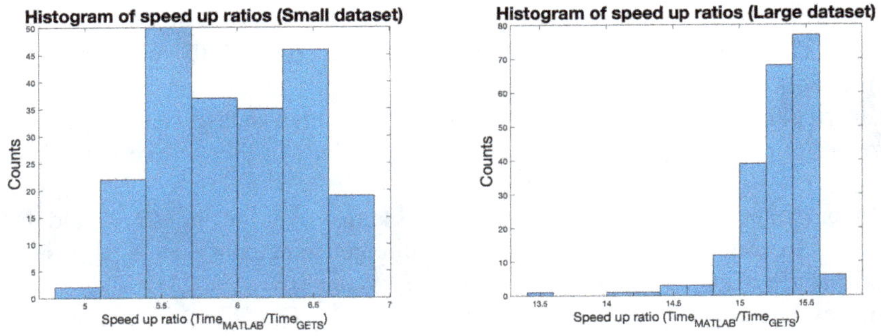

Fig. 1 Histogram of speedup ratios for the problems of small and large datasets. A total of 211 problems were solved in each case

believe that this is mainly due to the larger dimensions involved and the sparser representation in $\mathbf{D}_{\text{large}}$ (detailed in Sect. 4.3). As discussed in Sect. 4.2, the sparser the representation, the greater the asymptotic speedup obtained in computing $\mathbf{C}^{\mathsf{T}}\mathbf{r}$—the major computational bottleneck of the Lawson-Hanson algorithm.

Acknowledgments We thank Simon Foucart for introducing the problem and for valuable discussions and suggestions. We also thank David Koslicki for providing the genomic data that was used in the computational experiments.

References

1. Bro, R., De Jong, S.: A fast non-negativity-constrained least squares algorithm. Journal of Chemometrics **11**, 393–401 (1997)
2. Davis, T. A.: Direct methods for sparse linear systems. Society for Industrial and Applied Mathematics (2006)
3. Davis, T. A., Hager, W. W.: Dynamic supernodes in sparse Cholesky update/downdate and triangular solves. ACM Transactions on Mathematical Software **35**, 1–23 (2009)
4. Foucart, S., Koslicki, D.: Sparse recovery by means of nonnegative least squares. IEEE Signal Processing Letters **21**, 498–502 (2014)
5. Golub, G. H., Van Loan, C. F.: Matrix computations. John Hopkins University Press (2013)
6. Koslicki, D., Foucart, S., Rosen, G.: Quikr: a method for rapid reconstruction of bacterial communities via compressive sensing. Bioinformatics **29**, 2096–2102 (2013)
7. Koslicki, D., Foucart, S., Rosen, G.: WGSQuikr: fast whole-genome shotgun metagenomic classification. PLoS ONE **9**, e91784 (2014)
8. Lawson, C. L., Hanson, R. J.: Solving least squares problems. Prentice-Hall Series in Automatic Computation (1974)
9. Van Benthem, M. H., Keenan, M. R.: Fast algorithm for the solution of large-scale non-negativity-constrained least squares problems. Journal of Chemometrics **18**, 441–450 (2004)

A Qualitative Difference Between Gradient Flows of Convex Functions in Finite- and Infinite-Dimensional Hilbert Spaces

Jonathan W. Siegel and Stephan Wojtowytsch

Abstract We consider gradient flow/gradient descent and heavy-ball/ac-celerated gradient descent optimization for convex objective functions. In the gradient flow case, we prove the following:

1. If f does not have a minimizer, the convergence $f(x_t) \to \inf f$ can be arbitrarily slow.
2. If f does have a minimizer, the excess energy $f(x_t) - \inf f$ is integrable/summable in time. In particular, $f(x_t) - \inf f = o(1/t)$ as $t \to \infty$.
3. In infinite-dimensional Hilbert spaces, this is optimal: $f(x_t) - \inf f$ can decay to 0 as slowly as any given function which is monotone decreasing and integrable at ∞, even for a fixed quadratic objective.
4. In finite dimension (or more generally, for all gradient flow curves of finite length), this is *not* optimal: For instance, we show that every gradient flow x_t of a convex function f in finite dimension satisfies $\liminf_{t\to\infty} \big(t \cdot \log^2(t) \cdot \big\{f(x_t) - \inf f\big\}\big) = 0$.

This improves on the commonly reported $O(1/t)$ rate (at least for the lower limit) and provides a sharp characterization of the energy decay law. We also note that it is impossible to establish a rate $O(1/(t\phi(t)))$ for the full limit for any function ϕ which satisfies $\lim_{t\to\infty} \phi(t) = \infty$, even asymptotically.

Similar results are obtained in related settings for (1) discrete-time gradient descent, (2) stochastic gradient descent with multiplicative noise, and (3) the heavy-ball ODE. In the case of stochastic gradient descent, the summability of $\mathbb{E}[f(x_n) - \inf f]$ is used to prove that $f(x_n) \to \inf f$ almost surely—an improvement on the

---

J. W. Siegel
Department of Mathematics, Texas A&M University, College Station, TX, USA
e-mail: jwsiegel@tamu.edu

S. Wojtowytsch (✉)
Department of Mathematics, University of Pittsburgh, Pittsburgh, PA, USA
e-mail: s.woj@pitt.edu

S. Foucart, S. Wojtowytsch (eds.), *Explorations in the Mathematics of Data Science*, Applied and Numerical Harmonic Analysis,
https://doi.org/10.1007/978-3-031-66497-7_10

convergence almost surely up to a subsequence which follows from the $O(1/n)$ decay estimate.

1 Introduction

In this note, we discuss two approaches in gradient-based optimization for convex objective functions: steepest descent and the momentum method. They are described by the gradient flow ODE $\dot{x} = -\nabla f(x)$ and the heavy-ball ODE $\ddot{x} = -\alpha\dot{x} - \nabla f(x)$, respectively. The heavy-ball ODE is Newton's second law for a particle of mass $m = 1$ under the influence of a potential force $-\nabla f$ and a Stokes-type friction with coefficient α.

If the objective function f is merely convex, but not strongly convex, it is often favorable to let the coefficient of friction decay to zero, as the objective function can be very flat at its minimum. Constant friction in this setting dissipates kinetic energy too quickly, resulting in very slowly converging trajectories. Nesterov [35] and Su et al. [43] illustrate that the scaling $\alpha(t) := \frac{\alpha^*}{t}$ with $\alpha^* \geq 3$ balances the desirable effects of friction (extracting sufficient energy to dampen around the minimizer) against the undesirable (slowing down dynamics on the path to the minimizer).

In real applications, it is often not necessary to find a minimizer of f, but a point of low objective value. We therefore focus on studying the risk decay curves $g(t) = f(x(t)) - \inf f$ for gradient flows and heavy balls.

The rates which are commonly reported for convex optimization are $O(1/t)$ for gradient flows and $O(1/t^2)$ for the heavy-ball ODE. In discrete-time convex optimization, the rate $O(1/k^2)$ is optimal for any iterative method for which $x_{n+1} - x_n \in \text{span}\{\nabla f(x_0), \ldots, \nabla f(x_n)\}$ [34, Section 2.1.2], and it is achieved by Nesterov's accelerated gradient method. However, at any finite iteration k, a different (convex, quadratic) function on a space of dimension $\geq 2k$ is used to construct a lower bound. For a fixed convex function, Nesterov's method decreases the objective function as $o(1/k^2)$ [4], but generally not as $O(1/k^{2+\varepsilon})$ for any $\varepsilon > 0$. Su, Boyd, and Candes demonstrate in [43, Theorem 5] that

$$\int_0^\infty t\big(f(x(t)) - \inf f\big)\,\mathrm{d}t < +\infty. \tag{1}$$

In this work, we show that:

1. The characterization of the risk decay by (1) is essentially sharp even for a *fixed* quadratic function on a separable Hilbert space.
2. For gradient descent, the sharp characterization is given by the corresponding integrability condition that

$$\int_0^\infty \big(f(x(t)) - \inf f\big)\,\mathrm{d}t < +\infty.$$

3. In *finite-dimensional Hilbert* spaces, gradient flows decrease the excess objective value $f(x(t)) - \inf f$ faster in the sense that a stricter decay condition holds than mere integrability. In particular, we show that

$$\liminf_{t\to\infty} t \cdot \log^2(t) \cdot \big(f(x(t)) - \inf f\big) = 0.$$

However, we demonstrate that this statement cannot be strengthened by replacing the lower limit with an upper limit. The upper limit cannot be improved beyond the statement that $f(x(t)) - \inf f = o(1/t)$, even in dimension $d = 1$.

The improvement is based on the fact that gradient flow trajectories of convex objective functions have finite length in finite-dimensional Hilbert spaces. In the infinite-dimensional case, this may not be true.
4. We extend the integrability condition to gradient descent in discrete time both in the deterministic setting and the stochastic setting with noise which scales in a multiplicative fashion. Here we show that the SGD iterates satisfy

$$\sum_{n=0}^{\infty} \mathbb{E}\big[f(x_n) - \inf f\big] < +\infty$$

and deduce that $f(x_n) \to \inf f$ almost surely. Absent summability, the statement would remain true almost surely only along a subsequence.
5. All previous results are valid under the assumption that f is convex *and* that there exists a point x^* such that $f(x^*) = \inf f$. We show that without this assumption, the decay $f(x(t)) \to \inf f$ may be arbitrarily slow for both gradient flow and heavy-ball ODE.

The improvement to the bounds is qualitative and asymptotic in nature. While quantitative and non-asymptotic bounds have great benefits, the constants involved in the bounds are rarely available in practice. We therefore maintain that a sharp qualitative understanding of the convergence towards a minimum value is helpful.

Convex functions without minimizers are very common, for example, in applications in machine learning for classification if the cross-entropy loss function is used.

We believe that some of these results may be familiar to experts in the field, but we have been unable to find references for many of them. A principal goal of this work is to address this gap and provide a simple introduction to sharp results on gradient flows and accelerated gradient methods.

The article is structured as follows: Continuous-time gradient flows are discussed in Sect. 2 with special attention to the impact of finite dimension and the existence of minimizers. Corresponding results are obtained for gradient descent and stochastic gradient descent in Sect. 3. Momentum methods are only considered in continuous time in Sect. 4, but references to corresponding discrete-time results are provided in the appropriate places. A technical result concerning convex functions whose derivative is $L^{1/2}$-integrable on $(0, \infty)$ is postponed until the appendix.

1.1 Significance in Machine Learning

Our main motivation for the study of gradient-based optimizers is the recent popularity of simple first-order optimization algorithms in machine learning and specifically in deep learning. They have been used with great success to minimize high-dimensional functions like

$$L(w) = \mathbb{E}_{(x,y)\sim\mu}\big[\|h(w,x) - y\|^2\big] \quad \text{or} \quad L_n(w) = \frac{1}{n}\sum_{i=1}^{n} \big\|h(w,x_i) - y_i\big\|^2$$

as a discretized version. Here h denotes a parametrized function (e.g., a neural network) with parameters ("weights") $w \in \mathbb{R}^m$ and data $x \in \mathbb{R}^d$. The expression inside the expectation or sum can be more generally of the form $\ell\big(h(w,x), y\big)$—in classification, the cross-entropy loss

$$\ell_{ce}\big(h(w,x),\, y\big) := -\log\left(\frac{\exp(h(w,x)\cdot y)}{\sum_{j=1}^{k}\exp(h(w,x)\cdot e_j)}\right)$$

is more popular than the "mean squared error" (MSE)-loss = ℓ^2-loss [24]. Here $y = e_l$ is a vector corresponding to the label $l \in \{1,\dots,k\}$ of the point x. The "loss function" ℓ inside the expectation/empirical average $L(h) = \mathbb{E}[\ell(h(x), y)]$ is typically convex in the first argument, but not necessarily strictly convex. For instance, cross-entropy loss fails to be strictly convex since $\ell(h+\lambda(1,\dots,1),\, y) \equiv \ell(h,y)$ for all $\lambda \in \mathbb{R}$, i.e., the loss function is constant in one direction. Additionally, $\inf_h \ell_{ce}(h,y) = \lim_{\lambda\to\infty}\ell_{ce}(\lambda y, y) = 0$, but $\ell_{ce}(h,y) > 0$ for any h, y, so ℓ_{ce} does not admit minimizers.

If, for instance, $h(W,x) = Wx$ is a linear model with $W \in \mathbb{R}^{k\times d}$ and ℓ is convex, then L_n is convex. If additionally the dataset is linearly separable, i.e., there exists W^* such that all points are classified correctly in the sense that

$$h(W^*, x_i)\cdot y_i > \max_{e_j \neq y_i} h(W^*, x_i)\cdot e_j,$$

then there exists no minimizer of cross-entropy loss L_n since $L_n(W) > 0$ for all W but $\lim_{\lambda\to\infty} L_n(\lambda W^*) = 0$. The second point remains valid if the parametrized function class is not linear, but merely a cone (e.g., a class of neural networks).

We remark that cross-entropy has other favorable properties which distinguish it from the worst-case scenarios discussed above. For instance, ℓ_{ce} has exponential tails, i.e., it vanishes very quickly at infinity. Such properties can be captured, for example, in the language of Polyak-Łojasiewicz (PL) conditions, which have been used very successfully to study gradient flows (but not heavy-ball methods). Notably, mere convexity is not enough to study gradient-based optimization for objective functions without minimizers.

For parameterized functions $h(w, \cdot)$ which depend on their parameters in a nonlinear fashion (such as neural networks), the functional $L_n(w) := \frac{1}{n}\sum_{i=1}^n \ell\big(h(w, x_i), y_i\big)$ generally fails to be convex, even if the loss function ℓ is convex in the first argument:

- In the underparametrized regime of neural network learning, Safran and Shamir proved that the loss landscape generally contains many nonoptimal local minimizers [38].
- In the overparametrized regime, the set of minimizing parameters generically is a high-dimensional submanifold of the parameter space, at least for smoothly parametrized function classes [9]. Negative Hessian eigenvalues have been observed numerically close to the set of minimizers [1, 39, 40], and their existence has been justified theoretically in [47].

Even for simple neural networks with a single hidden layer and quadratic loss, the existence of minimizers is closely linked to the choice of activation function [13, 26]: For ReLU networks, minimizers often exist, while the energy infimum may only be attained asymptotically for activation functions like tanh with more complicated behavior at infinity.

Despite this non-convexity of the loss landscape, it has been observed that the weights of a neural network may remain in a "good region" and closely follow the trajectory of optimizing a linear model which is obtained by linearizing the neural network at the law of its initialization. This "neural tangent kernel" (NTK) was considered for gradient descent in [2, 14, 15, 17, 25] and for momentum-based optimization in [31]. The NTK is linear in its parameters, and thus, the loss landscape associated to the parameter optimization process is convex if ℓ is a convex function. This suggests that locally around a good initialization, the optimization landscape looks similar to that of a convex function. The crucial observation is that trajectories of gradient flows remain in this "good" region for all time under suitable conditions.

Even globally, for very wide networks there are no strict local minimizers which are not global minimizers [45, 46]. In this way, convex optimization informs the intuition of parameter optimization in deep learning, at least in a heavily overparametrized regime.

The empirical efficacy of momentum-based optimization algorithms further supports the idea that asymptotic convexity plays a role: While (stochastic) gradient descent is provably effective in more flexible function classes [27], acceleration is generally impossible under weaker assumptions such as the Polyak-Łojasiewicz condition [48]. Yet, in practice, momentum-based optimizers vastly outperform memory-less gradient descent.

Finding exact minimizers of the empirical loss function L_n is not always attractive in machine learning, where the true goal is to find minimizers of the (unknown) function L. The question whether a parameter w "generalizes" well (performs well on previously unseen data sampled from the same distribution which generated the training data (x_i, y_i), $i = 1, \ldots, n$) is generally considered more important than how close it is to the true optimal parameter w^*. At the optimal parameter for a given data sample, we may "overfit" to random noise in the training data with possibly disastrous implications for generalization.

Of course, we require good performance on the training set, but we may not train until we converge to a minimizer. This motivates us to primarily consider the rate of convergence of $f(x_t)$ rather than that of x_t.

1.2 Technical Tools

Most proofs in the following are elementary. We recall a statement which will be used frequently throughout the article.

Lemma 1 *Let* $f : (0, \infty) \to (0, \infty)$ *be a monotone decreasing function such that* $\int_0^\infty f(x)\,\mathrm{d}x < \infty$. *Then*

$$f(x) \leq \frac{2\int_0^\infty f(t)\,\mathrm{d}t}{x} \qquad \textit{and} \qquad \lim_{x\to\infty} x \cdot f(x) = 0.$$

Proof Since f is decreasing, we find that

$$\int_{x/2}^\infty f(t)\,\mathrm{d}t \geq \int_{x/2}^x f(t)\,\mathrm{d}t \geq \int_{x/2}^x f(x)\,\mathrm{d}t = \frac{x}{2}\cdot f(x) \geq 0.$$

On the other hand, since f is integrable, we find that

$$\lim_{x\to\infty}\int_{x/2}^\infty f(t)\,\mathrm{d}t = 0.$$

The result follows by the sandwich criterion. □

We briefly note that it is impossible to quantify the convergence in Lemma 1. A stronger version of Example 1 is given below in Example 2.

Example 1 Assume that $\phi : [0, \infty) \to \mathbb{R}$ is a monotone increasing function such that $\lim_{t\to\infty}\phi(t) = +\infty$. We aim to show that there exists a monotone decreasing integrable function $g : [0, \infty) \to \mathbb{R}$ such that

$$\limsup_{t\to\infty}\big(t\cdot\phi(t)\cdot g(t)\big) = +\infty.$$

In other words, we cannot guarantee that $g(t) \leq \frac{C}{t\,\phi(t)}$ for all large times for any given constant C. Of course, if $1/(t\cdot\phi(t))$ fails to be integrable (e.g., for $\phi(t) = \log t$), then $\liminf_{t\to\infty}\big(t\cdot\phi(t)\cdot g(t)\big) = 0$.

Let R_n be a monotone increasing sequence of positive numbers such that the series

$$\sum_{n=1}^{\infty} \frac{1}{\sqrt{\phi(R_n)}} \qquad \text{converges. Then} \qquad g(t) := \sum_{n=1}^{\infty} \frac{1}{R_n\sqrt{\phi(R_n)}}\, 1_{(0,R_n]}(t)$$

satisfies

$$\int_0^\infty g(t)\,\mathrm{d}t = \sum_{n=1}^{\infty} R_n \cdot \frac{1}{R_n\sqrt{\phi(R_n)}} = \sum_{n=1}^{\infty} \frac{1}{\sqrt{\phi(R_n)}} < +\infty$$

and

$$\begin{aligned}
\limsup_{t\to\infty}\big(t\cdot\phi(t)\cdot g(t)\big) &\geq \limsup_{n\to\infty}\big(R_n\cdot\phi(R_n)\cdot g(R_n)\big)\\
&\geq \limsup_{n\to\infty}\left(R_n\cdot\phi(R_n)\cdot\frac{1}{R_n\sqrt{\phi(R_n)}}1_{(0,R_n]}(R_n)\right)\\
&= \lim_{n\to\infty}\sqrt{\phi(R_n)} = +\infty.
\end{aligned}$$

Furthermore, g is a sum of decreasing functions and thus decreasing as well. It is easy to extend the example to functions which are continuous.

We will often refer to the derivative of a monotone function. All operations we require are well-defined when using the measure-valued derivative since monotone functions have bounded variation. If the reader prefers to avoid such technical distinctions, it suffices to assume additionally C^1-regularity whenever we refer to a monotone decreasing function g. See also [18, Sections 6.3, 6.4] for a detailed discussion of the regularity of convex functions.

2 Gradient Flows in Continuous Time

2.1 Existence and Uniqueness

We prove that the gradient flow equation has a unique solution in all situations where we will need it.

Lemma 2 *Let H be a separable Hilbert space and $F : H \to \mathbb{R}$ a convex function. Assume that F is (Frechet) differentiable and that the map $x \mapsto DF(x) \in H$ is continuous.*

Then, for every $x_0 \in H$*, there exists a unique continuous curve* $x : [0, \infty) \to H$ *such that* $x(0) = x_0$ *and* x *is* C^1*-smooth on* $(0, \infty)$ *with* $\dot{x}(t) = -DF(x)$.

Proof If H is finite-dimensional, the existence of a solution to the gradient flow ODE $\dot{x} = -\nabla F(x)$ with given initial condition follows, for instance, from Peano's existence theorem [44, Theorem 2.19] for all C^1-functions F, not just the convex ones, and indeed for any ODE of the form $x' = V(t, x)$ with a continuous vector field V.

For general F, gradient flows are not unique unless we assume that DF is a Lipschitz-continuous function (Picard-Lindelöf theorem, [44, Theorem 2.2]). If F is convex, the solutions are unique since the gradient flow is contracting. To see this, note that the first-order convexity condition

$$F(z) \geq F(x) + \langle DF(x), z - x\rangle \qquad \Leftrightarrow \qquad F(x) - F(z) \leq \langle DF(x), x - z\rangle$$

implies that the gradient of a convex function is a monotone operator in the sense that

$$\begin{aligned}\langle DF(x) - DF(z), x - z\rangle &= \langle DF(x), x - z\rangle + \langle DF(z), z - x\rangle \\ &\geq F(x) - F(z) + F(z) - F(x) = 0.\end{aligned}$$

Hence, if $x(t)$ and $z(t)$ satisfy

$$\begin{cases} \dot{x} = -\nabla f(x) & t > 0 \\ x = x_0 & t = 0 \end{cases} \qquad \text{and} \qquad \begin{cases} \dot{z} = -\nabla f(z) & t > 0 \\ z = z_0 & t = 0 \end{cases},$$

then

$$\frac{d}{dt}\|x - z\|^2 = 2\langle x - z, \dot{x} - \dot{z}\rangle = -2\langle x - z, DF(x) - DF(z)\rangle \leq 0.$$

In particular, if $x_0 = z_0$, then $x(t) = z(t)$ for all $t \geq 0$, i.e., the gradient flow is unique.

Peano's theorem does not hold in infinite-dimensional Hilbert spaces—see [20, 23] for counterexamples. The question of existence of gradient flows therefore has to be settled differently. The easiest alternative is strengthening the regularity assumption on the map $x \mapsto DF(x)$ since the Picard-Lindelöf theorem remains valid in infinite dimension [12, Chapter 1.1]. Due to the monotonicity of the gradient, this is not in fact necessary [12, Theorem 3.2]. In fact much more general existence theory is available under much weaker assumptions which apply also to the setting of partial differential equations (see, e.g., [19, Chapter 9.6]). □

2.2 Gradient Flows for Convex Functions: General Observations

We present gradient flows in the context of finite-dimensional Euclidean spaces, but most arguments carry over directly to Hilbert spaces. We generally only require f to be a convex C^1-function and $x(t)$ a C^1-solution of the gradient flow equation. Only rarely do we take second derivatives.

For the sake of avoiding technical complications, we avoid thinking about infinite-dimensional spaces except for Sect. 2.3, where they are needed for a counterexample. We note however that differences between gradient flows in finite-dimensional and infinite-dimensional Hilbert spaces are well-documented—[6] gives an example of a gradient flow of a convex function in an infinite-dimensional Hilbert space which does not converge in the norm topology. A comparable example for heavy-ball optimization is given in [3].

Let $f : \mathbb{R}^m \to \mathbb{R}$ be a convex C^1-function and assume that x solves the gradient-flow equation $\dot{x}_t = -\nabla f(x_t)$. Then by construction, the energy dissipation identity

$$\frac{d}{dt} f(x_t) = \nabla f(x_t) \cdot \dot{x}_t = -\|\nabla f(x_t)\|^2 \leq 0$$

holds. We review some additional well-known results.

Lemma 3 *Let f a convex C^1-function and $x^* \in \mathbb{R}^m$. Then the function $L : [0, \infty) \to \mathbb{R}$,*

$$L(t) = t\left(f(x_t) - f(x^*)\right) + \frac{1}{2}\|x_t - x^*\|^2$$

is nonincreasing.

We refer to L as the Lyapunov function associated with x^*.

Proof Due to the first-order convexity condition, we find that

$$\begin{aligned} L'(t) &= \left(f(x_t) - f(x^*)\right) + t\,\nabla f(x_t) \cdot \dot{x}_t + \langle x_t - x^*, \dot{x}_t\rangle \\ &= f(x_t) + \langle \nabla f(x_t), x^* - x_t\rangle - f(x^*) - t\,\|\nabla f(x_t)\|^2 \\ &\leq -t\,\|\nabla f(x_t)\|^2 \\ &\leq 0. \end{aligned}$$

□

As an immediate corollary, we find that gradient flows are consistent in convex optimization, irrespective of whether the minimum is attained, or even finite.

Corollary 1 *If f is convex, C^1, and x_t a gradient flow of f, then* $\lim_{t\to\infty} f(x_t) = \inf_{x\in\mathbb{R}^m} f(x)$.

Proof Since $\frac{d}{dt} f(x_t) = \langle \nabla f(x_t), \dot{x}_t\rangle = -\|\nabla f(x_t)\|^2$, we find that $f(x_t)$ is monotone decreasing in time. In particular, the limit $\lim_{t\to\infty} f(x_t)$ exists (but may be $-\infty$ if f is not bounded from below).

Assume for the sake of contradiction that $\lim_{t\to\infty} f(x_t) > \inf_{x\in\mathbb{R}^m} f(x)$. Choose $x^* \in \mathbb{R}^m$ such that $f(x^*) < \lim_{t\to\infty} f(x_t)$ and consider the associated function L. Then

$$L(0) \geq L(t) \geq t\left(f(x_t) - f(x^*)\right) \geq \frac{\lim_{s\to\infty} f(x_s) - f(x^*)}{2}\,t$$

for all sufficiently large t. As the term on the right grows uncontrollably as $t \to \infty$, we have reached a contradiction. □

2.3 Gradient Flows for Convex Functions with Minimizers: Hilbert Spaces

The assumption that f has a minimizer has profound impact. In particular, if $f(x^*) = \inf_{x\in\mathbb{R}^m} f(x)$, the first term in L is nonnegative. We conclude that

$$\left(f(x_t) - \inf_{x\in\mathbb{R}^m} f(x)\right) \leq \frac{L(t)}{t} \leq \frac{L(0)}{t} = \frac{\|x_0 - x^*\|^2}{2t},$$

i.e., the excess objective $f(x_t) - \inf f$ decays at least as fast as C/t for some $C > 0$. The question remains whether the rate of $1/t$ is optimal, and the fact that the energy dissipation $t\|\nabla f(x_t)\|^2$ of L is much larger than the decay rate $-\|\nabla f(x_t)\|^2$ of f suggests otherwise. We have so far only used the *sign* of the term. We see that this is indeed not the case—unlike the upper bound C/t, the excess objective value $f(x_t) - \inf f$ is in fact *integrable* at infinity.

Lemma 4 *Assume that f is a convex C^1-function which has a minimizer x^* and x is a gradient flow of f. Then*

$$\int_0^\infty \left(f(x_t) - \inf f\right) \mathrm{d}t \leq \frac{\|x_0 - x^*\|^2}{2}.$$

Proof Note that

$$\frac{d}{dt}\|x_t - x^*\|^2 = 2\,\langle x_t - x^*, \dot{x}_t\rangle = 2\,\langle \nabla f(x_t), x^* - x_t\rangle \leq 2\big\{f(x^*) - f(x_t)\big\} \leq 0.$$

In particular, the function $\|x_t - x^*\|^2$ is monotone decreasing and bounded from below and thus has a limit. We find that

$$\frac{\|x^* - x_0\|^2}{2} \geq \frac{\|x^* - x_0\|^2}{2} - \frac{\|x^* - x_T\|^2}{2} \geq \int_0^T f(x_t) - f(x^*)\,\mathrm{d}t$$

for all $T > 0$. The result follows by taking $T \to \infty$. □

Proof (Alternative Proof of Lemma 4) Again, let L be the Lyapunov function associated to a minimizer x^*. Then we compute that

$$\begin{aligned}\int_0^\infty \big(f(x_t) - \inf f\big)\,\mathrm{d}t &= -\int_0^\infty \int_t^\infty \frac{d}{ds} f(x_s)\,\mathrm{d}s\,\mathrm{d}t = \int_0^\infty \int_t^\infty \|\nabla f(x_s)\|^2\,\mathrm{d}s\,\mathrm{d}t\\ &= \int_0^\infty \int_0^s \|\nabla f(x_s)\|^2\,\mathrm{d}t\,\mathrm{d}s = \int_0^\infty s\,\|\nabla f(x_s)\|^2\,\mathrm{d}s\\ &\leq -\int_0^\infty L'(s)\,\mathrm{d}s \leq L(0).\end{aligned}$$

We use that $\lim_{t\to\infty} L(t) \geq 0$ since $L \geq 0$. It is possible to exchange the order of integration here due to a theorem of Tonelli; see, e.g., [30, Kapitel 8.5]. □

Since $f(x_t) - \inf f$ is additionally decreasing, Lemmas 1 and 4 imply the following.

Corollary 2 *Assume that f is a convex function which has a minimizer and x is a gradient flow of f. Then*

$$\lim_{t\to\infty} t \cdot \big(f(x_t) - \inf_{x\in\mathbb{R}^m} f(x)\big) = 0.$$

Additionally, we can compare $f(x_t)$ to a function which is non-integrable at infinity such as $1/(t\,\log t)$; we immediately obtain the qualitative statement that

$$\liminf_{t\to\infty} \big(t \cdot \log t \cdot \big(f(x_t) - \inf f\big)\big) = 0.$$

If this were not the case, then there would be $\varepsilon, T > 0$ such that $f(x_t) - \inf f \geq \frac{\varepsilon}{t\,\log t}$ for all $t > T$. Since the right-hand side integrates to infinity, so would the left, leading to a contradiction.

Stronger statements are available, but harder to formulate [36]. To illustrate that the improvement from integrability can be made quantitative, we obtain a non-asymptotic risk bound for the optimal iterate in a given range. While uncommon in convex optimization, such "optimal iterate" bounds are the norm in non-convex optimization.

Lemma 5 *Let f be a convex C^1-function. For any $t > 1$, we have*

$$\min_{t\leq s\leq t\log t}\left(s\cdot\left(f(x_s)-\inf f\right)\right)\leq\frac{\|x_0-x^*\|^2}{2\,\log(\log t)}.$$

Proof For simplicity, $\inf f = 0$. If we have $f(x_s) \geq \frac{\varepsilon}{s\,\log(\log t)}$ for some $\varepsilon > 0$ and $t \leq s \leq t\log t$, then

$$\frac{\|x_0-x^*\|^2}{2}\geq\int_0^\infty f(x_s)\,\mathrm{d}s\geq\frac{\varepsilon}{\log(\log t)}\int_t^{t\log t}\frac{1}{s}\,\mathrm{d}s=\varepsilon$$

where the first inequality is due to Lemma 4. □

In other words, $f(x_t) - \inf f$ decays slightly faster than $O(1/t)$ in a way that can be made precise. We now demonstrate that the characterization of Lemma 4 is sharp, at least in infinite-dimensional Hilbert spaces. The existence and regularity of a gradient flow curve in a Hilbert space follows by the Picard-Lindelöff theorem in the infinite-dimensional case just as it does in the finite-dimensional situation. Note that the (quadratic) objective function F below has a Lipschitz-continuous gradient, meaning that the result can be applied.

Note that by "gradient flows in Hilbert spaces" we refer to gradient flows of continuous convex functionals defined on a Hilbert space. This does *not* cover PDEs which arise as gradient flows of convex functionals such as the Dirichlet energy which are only defined on a dense subset. Much greater care must be taken in that context to prove existence and interpret gradients.

Lemma 6 *Let H be a separable Hilbert space of infinite dimension. Then there exists a strictly convex quadratic function $F : H \to [0,\infty)$ such that $F(0) = 0$ and the following is true: For any monotone decreasing integrable function $g : [0,\infty) \to \mathbb{R}$, there exists an initial condition $u_0 \in H$ such that the solution $u(t)$ of the gradient flow equation*

$$\begin{cases}\dot u=-\nabla F(u) & t>0\\ u=u_0 & t=0\end{cases}$$

satisfies

$$F(u(t))\geq g(t)\quad\forall\, t\geq 1.$$

Proof Note that a monotone decreasing integrable function g automatically satisfies $\lim_{t\to\infty} g(t) = 0$.

As the Lemma makes no claims for $t < 1$, we replace g by $g(t)1_{[1,\infty)}(t) + g(1)\,1_{(-\infty,1)}(t)$, which we also refer to as g by abuse of notation. Conveniently, the modified function g is bounded and defined on the whole real line. We replace the modified g by a weighted average $\tilde g$ over the interval $(t-1, t)$. Since g is monotone decreasing, the weighted average is, too, and it is larger than g:

$$\tilde{g}(t) = \frac{\int_{t-1}^{t} g(s)\,(t-s)(s-t-1)\,\mathrm{d}s}{\int_0^1 s(1-s)\,\mathrm{d}s} = 6\int_{t-1}^{t} g(s)\,(t-s)(s-t-1)\,\mathrm{d}s.$$

The benefit of $\tilde{g}$ over g is that it is continuously differentiable with derivative

$$\tilde{g}'(t) = 6\int_{t-1}^{t} g(s)\big((s-t-1)-(t-s)\big)\,\mathrm{d}s = 6\int_{t-1}^{t} g(s)\,\big(2(s-t)-1\big)\,\mathrm{d}s.$$

Again, we will refer to $\tilde{g}$ by g from now on.

Without loss of generality, we may consider $H = L^2(1,\infty)$ by isometry. Consider

$$F: H \to \mathbb{R}, \qquad F(u) = \frac{1}{2}\int_1^\infty \frac{u^2(s)}{s}\,\mathrm{d}s.$$

Since F is a continuous quadratic form, it is Fréchet differentiable with gradient $\big(\nabla F(u)\big)(s) = \frac{u(s)}{s}$. The gradient flow of F acts pointwise in s: $u(t,s) = e^{-t/s}u_0(s)$, so if $t \geq 1$, then

$$F(u(t)) = \frac{1}{2}\int_1^\infty \frac{u_0^2(s)}{s}\,e^{-2t/s}\,\mathrm{d}s \geq \frac{1}{2}\int_t^\infty \frac{u_0^2(s)}{s}\,e^{-2t/s}\,\mathrm{d}s \geq \frac{1}{2e^2}\int_t^\infty \frac{u_0^2(s)}{s}\,\mathrm{d}s.$$

In order to identify the lower bound, we require

$$g(t) = \frac{1}{2e^2}\int_t^\infty \frac{u_0^2(s)}{s}\,\mathrm{d}s \qquad \Leftrightarrow \qquad g'(t) = -\frac{u_0^2(t)}{2e^2\,t} \text{ and } \lim_{t\to\infty} g(t) = 0.$$

We therefore select $u_0(s) := \sqrt{-2e^2 s\, g'(s)}$ and verify that

$$\frac{1}{2e^2}\int_1^R u_0^2(s)\,\mathrm{d}s = -\int_1^R s\,g'(s)\,\mathrm{d}s = g(1) - Rg(R) + \int_1^R g(s)\,\mathrm{d}s.$$

Since g is integrable and monotone decreasing, we have $\limsup_{R\to\infty} Rg(R) = 0$ by Lemma 1. Thus,

$$\frac{1}{2e^2}\int_1^\infty u_0^2(s)\,\mathrm{d}s = g(1) + \int_1^\infty g(s)\,\mathrm{d}s < \infty.$$

In particular, $u_0 \in H$ is a valid initial condition. □

Notably, the convergence of $t\big(f(x_t) - \inf f\big)$ to zero can be arbitrarily slow, even for a fixed quadratic functional on an infinite-dimensional Hilbert space, depending only on the initial condition. This quadratic form is "infinitely flat" by its minimum: If $\phi : (0,\infty) \to (0,\infty)$ is any monotone increasing function such that $\lim_{r\to 0}\phi(r) = 0$, then there exists a sequence $u_n \in H$ such that

$$\lim_{n\to\infty} u_n = 0, \qquad \lim_{n\to\infty} \frac{F(u_n)}{\phi(\|u_n\|)} = 0.$$

In our example, such a sequence is given, for example, by

$$u_n = \frac{1}{n}\,1_{\{R_n < s < 1+R_n\}}, \qquad R_n = \frac{1}{\phi(1/n)} \qquad \text{since } F(u_n) \leq \frac{1}{n\,R_n}$$

and $\|u_n\|_{L^2} = 1/n$.

2.4 Gradient Flows for Convex Functions with Minimizers: Real Line

We can analyze the one-dimensional case more directly. Let x_t be a gradient flow curve for a C^2-smooth convex function $f : \mathbb{R}^m \to \mathbb{R}$. Define $g(t) = f(x_t)$. Then $g'(t) = -\left\|\nabla f(x_t)\right\|^2$ and

$$g''(t) = -\frac{d}{dt}\left\|\nabla f(x_t)\right\|^2 = -2\,\nabla f \cdot (D^2 f)\dot{x} = 2\,\nabla f \cdot (D^2 f)\nabla f \geq 0,$$

i.e., g is C^2-smooth, strictly monotone decreasing, and convex (or constant, if x_0 is a minimizer of f). Focusing on the one-dimensional case, the gradient flow has a limit since x_t is either monotone increasing or decreasing. To see this, note that $\dot{x}_t = -f'(x_t)$ cannot change sign along the gradient flow without passing through a minimizer, at which point the trajectory stops moving. More generally, the gradient flow curves of a convex function (with minimizers) on a finite-dimensional space have finite length [21, 32], so we find that

$$\int_0^\infty \sqrt{-g'(t)}\,\mathrm{d}t = \int_0^\infty \|\nabla f(x_t)\|\,\mathrm{d}t = \int_0^\infty \|\dot{x}_t\|\,\mathrm{d}t < \infty.$$

We see that this in fact characterizes the energy decay in gradient flows completely.

Lemma 7 *Let* $g : [0,\infty) \to [0,\infty)$ *be a monotone decreasing convex* C^2*-function such that*

$$|g'(0)| + \int_0^\infty \sqrt{-g'(t)}\,\mathrm{d}t < +\infty.$$

Then there exist:

1. *A convex C^1-function $\phi : \mathbb{R} \to [0, \infty)$ such that $\phi(x) = 0$ if and only if $x = 0$.*
2. *A gradient flow x_t of ϕ such that $\phi(x_t) = g(t)$ for all $t \in \mathbb{R}$.*

We note that $\lim_{t\to\infty} g(t)$ exists since g is monotone decreasing. Without loss of generality, we assume that $\lim_{t\to\infty} g(t) = 0$. We show that the conditions of Lemma 7 recover two previous characterizations, at least in part.

1. First, we note that for a convex decreasing function g tending to zero we have

$$g(t) = \int_t^\infty -g'(s)\,ds \le \sqrt{-g'(t)} \int_t^\infty \sqrt{-g'(s)}\,ds$$

$$\le \frac{1}{t}\int_0^t \sqrt{-g'(s)}\,ds \cdot \int_t^\infty \sqrt{-g'(s)}\,ds \le \frac{\left(\int_0^\infty \sqrt{-g'(s)}\,ds\right)^2}{4t}.$$

In the setting of gradient flows, $\int_0^\infty \sqrt{-g'(t)}\,dt = \int_0^\infty \|\nabla f(x_t)\|\,dt$ is the length of the gradient flow curve. In particular, this replaces the estimate

$$\big(f(x_t) - \inf f\big) \le \frac{\|x_0 - x^*\|^2}{2t}$$

with a better constant $1/4$ in place of $1/2$, but with the length of the trajectory rather than the Euclidean distance of its endpoints. This can vastly overestimate the true constant, but without access to the original geometry, it is a valid replacement. In one dimension, it is a strict improvement.

2. Next, we show that $\int_0^\infty g(t)\,dt < \infty$. Namely, since $\sqrt{-g'}$ is monotone decreasing, we have

$$0 \le g(t) = -\int_t^\infty g'(s)\,ds \le \sqrt{-g'(t)} \int_t^\infty \sqrt{-g'(s)}\,ds$$

$$\le \sqrt{-g'(t)}\left(\int_0^\infty \sqrt{-g'(s)}\,ds\right),$$

so $g \in L^1(0, \infty)$.

Proof (Proof of Lemma 7) **Setup** Since g is monotone, convex, and integrable, we note that $g'(t) < 0$ or $g(s) \equiv 0$ for all $s \ge t$. The second case is a simpler variation, so we may assume that $g'(t) \neq 0$ for any $t \in (0, \infty)$.

$$\Psi : [0, \infty) \to [0, \infty), \qquad \Psi(t) = \int_t^\infty \sqrt{-g'(s)}\,ds$$

is C^2-smooth and strictly monotone decreasing. In particular, we can define ϕ on the interval $(0, X]$ by

$$\phi\left(\int_t^\infty \sqrt{-g'(s)}\,\mathrm{d}s\right) = g(t) \qquad \text{where} \quad X = \int_0^\infty \sqrt{-g'(s)}\,\mathrm{d}s.$$

It is easy to extend ϕ in a C^1-fashion as $\phi(x) = x^2$ for $x \leq 0$ and $\phi(x) = \phi(X) + \phi'(X)(x - X)$ for $x > X$. For the finiteness of derivatives, see the next step.

Gradient Flow of ϕ. We can easily compute the derivatives as

$$g'(t) = -\phi'\left(\int_t^\infty \sqrt{-g'(s)}\,\mathrm{d}s\right)\sqrt{-g'(t)}.$$

Note in particular that $\phi'(X) = \sqrt{-g'(0)}$ and $\phi'(0) = \lim_{t\to\infty}\sqrt{-g'(t)} = 0$, since $\sqrt{-g'}$ is integrable and monotone decreasing (since g is convex). We compute further that

$$\begin{aligned} g''(t) &= \phi''\left(\int_t^\infty \sqrt{-g'(s)}\,\mathrm{d}s\right)\sqrt{-g'(t)}^2 - \phi'\left(\int_t^\infty \sqrt{-g'(s)}\,\mathrm{d}s\right)\frac{d}{dt}\sqrt{-g'(t)} \\ &= -\phi''(\Psi(t))\,g'(t) + \phi'(\Psi(t))\,\frac{g''(t)}{2\sqrt{-g'(t)}} \\ &= -\phi''\,g' - \sqrt{-g'}\,\frac{g''}{2\sqrt{-g'}} \\ &= -\phi''\,g' + g''/2, \end{aligned}$$

so

$$\phi''\left(\int_t^\infty \sqrt{-g'(s)}\,\mathrm{d}s\right) = -\frac{g''(t)}{2g'(t)} \geq 0.$$

If g is (strictly) monotone decreasing and convex, we see that ϕ is convex as well.

Gradient Flow of ϕ. Consider the gradient flow s_t of ϕ with initial condition X, i.e., the solution of the ODE:

$$\dot{s}_t = -\phi'(s_t) = -\sqrt{-g'(\Psi^{-1}(s_t))}.$$

Note that $S_t := \Psi(t)$ satisfies

$$\dot{S}_t = \Psi'(t) = -\sqrt{-g'(t)} = -\sqrt{-g'\big(\Psi^{-1}(\Psi(t))\big)} = -\sqrt{-g'\big(\Psi^{-1}(S_t)\big)},$$

i.e., s_t and S_t solve the same differential equation. We conclude that $s_t = \Psi(t)$. As usual, we have

$$\frac{d}{dt}\phi(s_t) = \phi'(s_t)\,\dot{s}_t = -\big(\phi'(s_t)\big)^2 = g'\big(\Psi^{-1}(s_t)\big) = g'(t).$$

Again, we conclude that $\phi(s_t) = g(t)$ since both functions approach 0 as $t \to \infty$ due to Corollary 2.

Second Case. In the case $g \equiv 0$ on $[t, \infty)$, the same arguments go through. We note, though, that ϕ cannot be $C^{1,1}$-smooth since the gradient flow reaches a global minimizer *in finite time*, i.e., it is not possible to solve the gradient flow backwards in time. If $\phi \in C^{1,1}$, this would be the case. □

Based on Lemmas 6 and 7, we demonstrate that there is a fundamental difference between the gradient flows of convex functions in finite-dimensional and infinite-dimensional Hilbert spaces. More precisely, while the integrability of $\sqrt{-g'}$ implies the integrability of g, the two are not equivalent:

Consider the function $g_\alpha(t) = \frac{1}{t\,(\log t)^\alpha}$ for $t \geq 2$, extended to $(-\infty, 2)$ as a linear C^1-function. Then

$$\int_2^\infty g_\alpha(t) = \frac{\log(t)^{1-\alpha}}{1-\alpha}\Bigg|_{t=2}^{t\to\infty} = \frac{(\log 2)^{1-\alpha}}{\alpha - 1} < +\infty,$$

for $\alpha > 1$ but

$$g_\alpha'(t) = -\left(\frac{\alpha}{t^2\,(\log t)^{1+\alpha}} + \frac{1}{t^2\,(\log t)^\alpha}\right) \leq -\frac{1}{t^2\,(\log t)^\alpha}.$$

In particular

$$\int_2^\infty \sqrt{-g_\alpha'(t)}\,\mathrm{d}t \geq \int_2^\infty \frac{1}{t\,(\log t)^{\alpha/2}}\,\mathrm{d}t = +\infty$$

if $\alpha \leq 2$ since

$$\int_2^\infty \frac{1}{t\,\log t}\,\mathrm{d}t = \log(\log t)\Bigg|_{t=2}^{t=\infty} = +\infty$$

and log is monotone increasing, $\alpha/2 > 0$. Thus, g_α is not the decay function for the gradient flow of a convex function in finite dimension for $\alpha \in (1, 2]$. We want to conclude that:

1. $\liminf_{t\to\infty}\big(t\,(\log t)^2 \cdot f(x_t)\big) = 0$ for the gradient flow of a convex function in finite dimension.
2. There is a qualitatively different condition on the decay rate in finite dimension compared to Hilbert spaces.

The question is: Is there a convex function $\tilde g_\alpha$ such that

$$\liminf_{t\to\infty} \frac{\tilde g_\alpha(t)}{g_\alpha(t)} \geq 1, \qquad \lim_{t\to\infty} \tilde g_\alpha(t) = 0 \qquad \text{and} \int_2^\infty \sqrt{-\tilde g_\alpha'(t)}\,\mathrm{d}t < \infty?$$

If such a function $\tilde g_\alpha$ exists, then g_α itself may not arise as the decay curve of a gradient flow, but it does not serve as a lower barrier, even asymptotically. We prove that this is not possible in Appendix "Appendix: A Comparison Principle for Convex Functions with Derivatives in $L^{1/2}$". Namely, we prove the following auxiliary statement.

Lemma 8 *Let $g, G : [0, \infty) \to [0, \infty)$ be decreasing, differentiable convex functions such that*

$$\lim_{t\to\infty} G(t) = \lim_{t\to\infty} g(t) = 0, \qquad \liminf_{t\to\infty} \frac{G(t)}{g(t)} > 0.$$

Then

$$\int_1^\infty \sqrt{-g'(t)}\,\mathrm{d}t = +\infty \qquad \Rightarrow \qquad \int_1^\infty \sqrt{-G'(t)}\,\mathrm{d}t = +\infty.$$

To understand Lemma 8, consider a simpler task first: Minimize $\int_0^\infty \sqrt{|G'(t)|}\,\mathrm{d}t$ in the class of functions G such that $G(0) = 1$ and $\lim_{t\to\infty} G(t) = 0$. This problem is not well-posed as the function $G_r(t) = \max\{1 - rt, 0\}$ achieves

$$\int_0^\infty \sqrt{|G_r'(t)|}\,\mathrm{d}t = \int_0^{1/r} \sqrt{r}\,\mathrm{d}t = \frac{1}{\sqrt{r}}, \qquad r > 0.$$

As $r \to \infty$, the integral approaches zero, i.e., the energy infimum is 0 and is not attained. This is due to the fact that for the square root of the derivative, short and steep segments are heavily discounted.

The function G_r is convex for all $r > 0$. An extension of the argument above could be used to construct $G \geq g$ such that $\|G'\|_{L^{1/2}(0,\infty)}$ is arbitrarily small by introducing many short, steep segments and keeping G mostly constant away from these fast transitions. However, in combination, the constraints that G must be convex and (a version of) $G \geq g$ induce a nontrivial competition: G should be as steep as possible, since large derivatives on short segments are heavily discounted. However, it cannot concentrate steep segments in many places since its derivative is a monotone function.

The proof of Lemma 8 is the most technically challenging part of the article. It primarily uses the concavity of the function $z \mapsto \sqrt{z}$ and the statement remains valid for more general concave functions. We postpone the proof to the Appendix in order to focus on the application to gradient flows for now.

In particular, we have shown the following.

Corollary 3 *Let $f : \mathbb{R} \to \mathbb{R}$ be a convex C^2-function and $x^* \in \mathbb{R}$ such that $f(x^*) = \inf f$. Then it is not possible that $f(x_t) - \inf f \geq \frac{\varepsilon}{t \log^2 t}$ for a fixed $\varepsilon > 0$ and all large t.*

We deduce that

$$\liminf_{t\to\infty} \left(t \log^2 t \cdot \left(f(x_t) - \inf f\right)\right) = 0. \tag{2}$$

We note however that a substantial improvement over the rate $O(1/t)$ is not possible, and in fact Lemma 7 shows that $f(x_t) - \inf f$ may satisfy

$$\limsup_{t\to\infty} \left(t \cdot \log^\alpha(t) \cdot \left(f(x_t) - \inf f\right)\right) = +\infty$$

for any $\alpha > 2$, even in one dimension.

It is tempting, but unfortunately incorrect to assume that the lower limit in (2) could be replaced by a proper limit for functions which satisfy the hypothesis that $\int_0^\infty \sqrt{-g'(t)}\,\mathrm{d}t < +\infty$. We extend Example 1 to this scenario and show that no stronger version of Lemma 1 can be achieved, even under the stronger condition of finite path length.

Example 2 Let $\phi : (0, \infty) \to (0, \infty)$ be a monotone increasing function such that $\lim_{t\to\infty} \phi(t) = +\infty$. We will show that there exists a convex nonincreasing function $g : (0, \infty) \to (0, \infty)$ such that

$$\lim_{t\to\infty} g(t) = 0, \qquad \int_0^\infty \sqrt{-g'(t)}\,\mathrm{d}t < +\infty, \qquad \limsup_{t\to\infty} \left(t \cdot \phi(t) \cdot g(t)\right) = +\infty.$$

Let R_n be a sequence such that $\sum_{n=1}^\infty \frac{1}{\sqrt[3]{\phi(R_n)}} < \infty$. Define

$$\sqrt{-g'(t)} = \sum_{n=1}^\infty \frac{1}{R_n \sqrt[3]{\phi(R_n)}}\, 1_{(0,2R_n]}(t).$$

Then

1. $\sqrt{-g'}$ is monotone decreasing, i.e., g' is monotone increasing, i.e., g is convex.
2. $\sqrt{-g'}$ is integrable by the same argument as in Example 1.

Ignoring cross-terms in the binomial formula for $-g' = \sqrt{-g'}^2$, we note that

$$g(t) = \int_t^\infty -g'(t)\,\mathrm{d}t \geq \int_t^\infty \sum_{n=1}^\infty \frac{1}{R_n^2 \left(\phi(R_n)\right)^{2/3}}\, 1_{(0,2R_n]}(t)\,\mathrm{d}t.$$

Then in particular

$$
\begin{aligned}
\limsup_{t\to\infty}\big(t\cdot\phi(t)\cdot g(t)\big) &\geq \limsup_{n\to\infty}\left(R_n\cdot\phi\left(R_n\right)\cdot g(R_n)\right)\\
&\geq \limsup_{n\to\infty} R_n\cdot\phi(R_n)\int_{R_n}^{2R_n}\frac{1}{R_n^2\left(\phi(R_n)\right)^{2/3}}\,1_{(0,2R_n]}(t)\,\mathrm{d}t\\
&= \limsup_{n\to\infty}\left(R_n\cdot\phi(R_n)\cdot\frac{1}{R_n\left(\phi(R_n)\right)^{2/3}}\right)\\
&= \lim_{n\to\infty}\phi(R_n)^{1/3} = +\infty.
\end{aligned}
$$

Again, it is easy to generalize the example to a version where g is infinitely smooth.

2.5 *Gradient Flows for Convex Functions with Minimizers: Finite Dimension*

In this section, we show that there is no difference between the decay rates which can be guaranteed for convex functions on finite-dimensional spaces and convex functions on the real line. We exploit two facts:

1. Only the geometry of the objective function along the gradient direction matters to the gradient flow, i.e., we can consider the objective function only along the curve traced by the gradient flow itself (in a suitable reparametrization).
2. A gradient flow curve in a finite-dimensional space always has finite length [32]. More generally, gradient flow curves satisfy the "self-contracting" property that

$$
t_1 < t_2 < t_3 \quad\Rightarrow\quad \|\gamma(t_2)-\gamma(t_3)\| \leq \|\gamma(t_1)-\gamma(t_3)\|
$$

   in the Euclidean norm [11, 16]. In fact, gradient-flow curves are completely characterized by the more restrictive strong self-contracting property [16]. Self-contracting curves were shown to have finite length in fairly general circumstances [42].

We note, however, that even gradient flows in two dimensions can be surprisingly complicated. In [11], the authors construct a convex function and a gradient flow trajectory which winds around the unique minimizer infinitely often. Such a construction is even possible for a function which is analytic except at the minimizer [10].

We note that these results do not hold in infinite-dimensional Hilbert spaces. In [6], the authors construct the gradient flow of a convex function on a Hilbert space which does not converge to a minimizer in the norm topology. In particular, as it does not converge to a limit, the gradient flow curve has infinite length.

Lemma 9 *Assume that H is a Hilbert space, $f : H \to \mathbb{R}$ is a convex C^2-function with a Lipschitz-continuous gradient, and x_t is a gradient flow of f. If x has finite length, then there exist:*

1. *A convex C^1-function $g : \mathbb{R} \to \mathbb{R}$ which has a minimizer and,*
2. *A gradient flow s_t of g such that*

$$f(x_t) = g(s_t) \qquad \forall\, t > 0.$$

In particular, the statement applies to all gradient flow lines in finite-dimensional Hilbert spaces. The lemma remains true without the finite length assumption, but becomes somewhat less instructive as g does not have a minimizer in this case.

Proof **Setup** Let us compare two curves: the solution of the gradient flow equation $\dot{x}_t = -\nabla f(x_t)$ or the time-normalized gradient flow $\dot{z}(t) = -\frac{\nabla f(z(t))}{\|\nabla f(z(t))\|}$. Then

$$x_t = z(\phi(t)) \qquad \text{where} \quad \phi(t) = \int_0^t \|\nabla f(x_s)\|\, \mathrm{d}s.$$

Since ∇f is Lipschitz-continuous, a gradient flow cannot reach a minimizer in finite time (since we can uniquely solve the ODE backwards in time). Thus, $\nabla f(x_s) \neq 0$ for all $s \in [0, \infty)$, meaning that ϕ is strictly monotone, hence invertible. To see this, consider $y(s) = x(\phi^{-1}(s))$ and note that

$$\begin{aligned} \frac{d}{ds} y(s) &= \dot{x}(\phi(s)) \left(\phi^{-1}\right)'(s) = -\nabla f\left(x(\phi^{-1}(s))\right) \frac{1}{\phi'(\phi^{-1}(s))} \\ &= \frac{-\nabla f(x(\phi^{-1}(s)))}{\|\nabla f(x(\phi^{-1}(s)))\|} = -\frac{\nabla f(y(s))}{\|\nabla f(y(s))\|}. \end{aligned}$$

Since y and z solve the same ODE and $\nabla f/\|\nabla f\|$ is locally Lipschitz-continuous on the set where $\nabla f \neq 0$, we find by the uniqueness assertion of the Picard-Lindelöff theorem that $y \equiv z$.

Recall that the assumption of finite length means that

$$\lim_{t\to\infty} \int_0^t \|\dot{x}_s\|\, \mathrm{d}s = \lim_{t\to\infty} \int_0^t \|\nabla f(x_s)\|\, \mathrm{d}s = \lim_{t\to\infty} \phi(t) < \infty.$$

We denote $R := \lim_{t\to\infty} \phi(t)$.
Step 1. In this step, we construct $g : [0, R] \to \mathbb{R}$ as $g(s) = f(z(s))$. Then

$$\begin{aligned} g'(s) &= \nabla f(z(s)) \cdot \dot{z}(s) = -\|\nabla f(z(s))\| \\ g''(s) &= -\frac{\nabla f(z(s))}{\|\nabla f(z(s))\|} \cdot D^2 f(z(s)) \cdot \dot{z}(s) = \frac{\nabla f(z)}{\|\nabla f(z)\|} \cdot D^2 f(z) \cdot \frac{\nabla f(z)}{\|\nabla f(z)\|} \geq 0 \end{aligned}$$

since $D^2 f$ is nonnegative semi-definite. The function g is therefore convex and C^1-smooth on its domain of definition.

We extend g to the entire real line by setting $g(s) = \inf f$ if $s > R$ and $g(s) = g(0) + g'(0)s$ if $s < 0$. This results in a C^1-extension, but we note that it could easily be made C^2-smooth, at least at $s = 0$.

Step 2. We consider a gradient flow curve s of $g : \mathbb{R} \to \mathbb{R}$ such that $s(0) = 0$. Then $g(s(0)) = f(x_0)$ by construction and

$$\dot{s}_t = -g'(s_t) = -\|\nabla f(z(s_t))\| \quad \Rightarrow \quad \frac{d}{dt} g(s_t) = -|g'(s_t)|^2 = -\|\nabla f(z(s_t))\|^2.$$

In particular $f(x_t) = g(s_t)$ for all t since their derivatives coincide and they take the same value at $t = 0$. □

2.6 Gradient Flows for Convex Functions Without Minimizers

For convex functions without minimizers, the decay of energy along a gradient flow can be arbitrarily slow, even in one dimension.

Lemma 10 *Assume that* $g : [0, \infty) \to [0, 1]$ *is a bounded measurable function such that* $\lim_{t\to\infty} g(t) = 0$*. Then there exists a convex function* $f : \mathbb{R} \to \mathbb{R}$ *and a gradient flow* x_t *of* g *such that* $\inf_{x\in\mathbb{R}} f(x) = 0$ *and* $f(x_t) \geq g(t)$ *for all* $t \geq 1$.

Proof **Step 1**. We make two adjustments.

1. Replace g by $\tilde{g}(t) = \max_{s\geq t} g(s)$ to ensure that the function is monotone nonincreasing.
2. Replace $g(t)$ by $\int_{t-1}^{t} g(s)\,\mathrm{d}s \geq g(t)$ (since g is nonincreasing).

Using the two modifications, we may assume that g is C^1-smooth and monotone nonincreasing.

Step 2. We construct a *convex* function $\phi \geq g$ such that $\lim_{t\to\infty} \phi(t) = 0$. Namely, set

$$\phi(t) = \int_t^\infty (s - t)\, \frac{-g'(s)}{s}\, \mathrm{d}s.$$

Then $\lim_{t\to\infty} \phi(t) = 0$ since g' is integrable and $\frac{s-t}{s} \leq 1$. Furthermore,

$$\phi'(t) = \int_t^\infty \frac{g'(s)}{s}\, \mathrm{d}s$$

is monotone increasing since the domain of integration is shrinking and $g' \leq 0$. Thus, g is convex. Finally, we note that

$$\phi(t) = \int_t^\infty (s-t)\,\frac{-g'(s)}{s}\,\mathrm{d}s \leq \int_t^\infty \big(-g'(s)\big)\,\mathrm{d}s = g(t)$$

for all $t > 0$ since $-g' \geq 0$ and $\frac{s-t}{s} \leq 1$. On the other hand,

$$\begin{aligned}\phi(t) &= \int_t^\infty \frac{s-t}{s}\big(-g'(s)\big)\,\mathrm{d}s \\ &\geq \int_{2t}^\infty (1-t/s)\big(-g'(s)\big)\,\mathrm{d}s \geq \frac{1}{2}\int_{2t}^\infty \big(-g'(s)\big)\,\mathrm{d}s = \frac{g(2t)}{2}.\end{aligned}$$

Rescaling g before, we can reach $\phi(t) \geq g(t)$ instead.

Step 3. By the same argument as in Lemma 7, we see that there exists a convex function

$$f:\mathbb{R}\to\mathbb{R}, \qquad f\left(\int_0^t \sqrt{-\phi'(s)}\,\mathrm{d}s\right) = \phi(t)$$

and a gradient flow x_t of f such that $f(x_t) = \phi(t) \geq g(t)$. The integrability of $\sqrt{\phi'}$ is not needed in this context as we do not insist that f has a minimizer. □

3 Gradient Descent in Discrete Time

In this section, we prove that the improved convergence result of Lemma 4 carries over to the explicit Euler time-stepping scheme for the gradient flow ODE, i.e., the gradient descent (GD) algorithm.

3.1 Deterministic Gradient Descent

We prove an analogue of Lemma 4 for gradient descent in discrete time.

Lemma 11 *Let $f : \mathbb{R}^m \to \mathbb{R}$ be a convex function, $x^* \in \mathbb{R}^m$ such that $f(x^*) = \inf_{x\in\mathbb{R}^m} f(x)$ and $x_1 \in \mathbb{R}^m$ an initial condition. Assume that ∇f is L-Lipschitz continuous with respect to the Euclidean norm. If the sequence $(x_n)_{n\in\mathbb{N}}$ follows the gradient descent law $x_{n+1} = x_n - \eta\nabla f(x_n)$ for a step-size $0 < \eta < 2/L$, then $f(x_n)$ is monotone decreasing and*

$$\eta\sum_{n=0}^\infty \big(f(x_n) - f(x^*)\big) \leq \frac{\|x_0 - x^*\|^2}{2} + \frac{\eta}{2(1-L\eta/2)}\big(f(x_0) - f(x^*)\big).$$

If $\eta = 1/L$, then this becomes

$$\sum_{n=0}^{\infty} \big(f(x_n) - f(x^*)\big) \leq \frac{L\,\|x_0 - x^*\|^2}{2} + \big(f(x_0) - f(x^*)\big).$$

Proof **Step 1**. Denote $g_n = \nabla f(x_n)$. The discrete-time analogue of the continuous-time energy dissipation identity $\frac{d}{dt} f(x_t) = -\|\nabla f(x_t)\|^2$ is

$$f(x_{n+1}) \leq f(x_n) - \left(1 - \frac{L\eta}{2}\right) \eta\,\|\nabla f(x_n)\|^2,$$

as proved, e.g., in [8, Lemma 3.4]. If $\eta < 2/L$, the factor $\gamma = \left(1 - \frac{L\eta}{2}\right)\eta$ is nonnegative.

Step 2. The sequence $L_n = \|x_n - x^*\|^2$ satisfies

$$\begin{aligned} L_{n+1} &= \|x_{n+1} - x_n + x_n - x^*\|^2 \\ &= \|\eta \nabla f(x_n)\|^2 + 2\langle -\eta\nabla f(x_n),\, x_n - x^*\rangle + \|x_n - x^*\|^2 \\ &\leq \eta^2\,\|\nabla f(x_n)\|^2 - 2\eta\big(f(x_n) - f(x^*)\big) + L_n \end{aligned}$$

by the first-order convexity condition. In particular, we have

$$\begin{aligned} \|x_0 - x^*\|^2 = L_0 &\geq \sum_{n=0}^{\infty} \big(L_n - L_{n+1}\big) \\ &\geq 2\eta \sum_{n=0}^{\infty} \big(f(x_n) - f(x^*)\big) - \frac{\eta^2}{\gamma} \sum_{n=0}^{\infty} \gamma\,\|\nabla f(x_n)\|^2 \\ &\geq 2\eta \sum_{n=0}^{\infty} \big(f(x_n) - f(x^*)\big) - \frac{\eta^2}{\gamma} \sum_{n=0}^{\infty} \big(f(x_n) - f(x_{n+1})\big), \end{aligned}$$

so

$$\eta \sum_{n=0}^{\infty} \big(f(x_n) - f(x^*)\big) \leq \frac{\|x_0 - x^*\|^2}{2} + \frac{\eta^2}{2\gamma}\big(f(x_0) - f(x^*)\big).$$ □

A discrete-time analogue of Lemma 1 with essentially the same proof states that if a_n is a monotone decreasing and summable sequence, then $\lim_{n\to\infty}(n \cdot a_n) = 0$. As a consequence, we note the following.

Corollary 4 *Let* $f : \mathbb{R}^m \to \mathbb{R}$ *be a convex function,* $x^* \in \mathbb{R}^m$*, such that* $f(x^*) = \inf_{x\in\mathbb{R}^m} f(x)$ *and* $x_1 \in \mathbb{R}^m$ *an initial condition. Assume that* ∇f *is L-Lipschitz* *continuous with respect to the Euclidean norm. If the sequence* $(x_n)_{n\in\mathbb{N}}$ *follows the gradient descent law* $x_{n+1} = x_n - \eta\nabla f(x_n)$ *for a step-size* $\eta < 2/L$, *then*

1. $\lim_{n\to\infty} n \cdot \big(f(x_n) - f(x^*)\big) = 0.$
2. $\liminf_{n\to\infty} n\,\log(n) \cdot \big(f(x_n) - f(x^*)\big) = 0.$

We conjecture that also in the discrete-time setting, a stronger result in the spirit of Corollary 3 can be obtained under the additional finite path length assumption, i.e., if

$$\sum_{n=1}^{\infty} \|x_n - x_{n+1}\| = \eta \sum_{n=1}^{\infty} \|\nabla f(x_n)\| < \infty.$$

The finite path-length assumption is true for gradient-descent trajectories in finite-dimensional spaces due to [21].

3.2 Stochastic Gradient Descent with Multiplicative Noise

We briefly consider a generalization of the gradient descent scheme discussed previously to the case where only stochastic gradient estimates are available. For background on conditional expectations, see, e.g., Chapter 8 in [28].

We assume that all random variables are defined on the same probability space, which remains implicit in our analysis and is only required to be large enough to support sufficiently many independent variables. We assume that a (random) initial condition X_0 is given and that we have access to a random variable g_0 such that $\mathbb{E}\big[g_0|\sigma(X_0)\big] = \nabla f(X_0)$ where $\sigma(X_0)$ denotes the σ-algebra generated by X_0. We can thus take a first step in the stochastic gradient procedure $X_1 = X_0 - \eta g_0$.

More generally, after taking n steps, we set $\mathcal{F}_n = \sigma(X_0, \dots X_n)$ and assume that we again have a gradient estimate g_n such that $\mathbb{E}[g_n|\mathcal{F}_n] = \nabla f(X_n)$, so that we can define $X_{n+1} = X_n - \eta g_n$.

An additional assumption is required in analyses of stochastic gradient descent to quantify the oscillations of g_n around its mean. We make the multiplicative noise scaling assumption:

$$\mathbb{E}\big[\|g_n - \nabla f(X_n)\|^2 \,\big|\, \mathcal{F}_n\big] \le \sigma^2\, \|\nabla f(X_n)\|^2.$$

This assumption has recently gained popularity to model the overparametrized regime in deep learning with mean squared error [7, 22, 47].

Lemma 12 *Assume that* f *is convex,* ∇f *is* L*-Lipschitz continuous with respect to the Euclidean norm, and consider the SGD trajectory with estimators* g_n *as described above and a step size* $\eta < \frac{2}{L(1+\sigma^2)}$.

Under these conditions, $\mathbb{E}\big[f(X_n)\big]$ *is monotone decreasing. If* $f(x^*) = \inf f$*, then*

$$\eta\sum_{n=0}^{\infty}\mathbb{E}\big[f(X_n)-\inf f\big] \leq \frac{\mathbb{E}\big[\|X_0-x^*\|^2\big]}{2}+\frac{\eta(1+\sigma^2)}{1-L(1+\sigma^2)\eta/2}\,\mathbb{E}\big[f(X_0)-\inf f\big].$$

In particular, with the choice $\eta = 1/(L(1+\sigma^2))$, which is optimal in terms of proving decay, this becomes

$$\sum_{n=0}^{\infty}\mathbb{E}\big[f(X_n)-\inf f\big] \leq \frac{L(1+\sigma^2)}{2}\,\mathbb{E}\big[\|X_0-x^*\|^2\big]+2(1+\sigma^2)\,\mathbb{E}\big[f(X_0)-\inf f\big].$$

The setup in the proof is well known in stochastic optimization—many details can be found, e.g., in [22, 47] and the references cited there.

We note that it would also be possible to consider a random variable X^* which is $\mathcal{F}_0$-measurable such that $f(X^*) \equiv 0$ almost surely. In convex functions which do not admit a unique minimizer, this allows us to select the unique closest point projection of X_0 onto the closed and convex set of minimizers. The σ-algebras do not change under this modification, but the constant may be reduced significantly.

Proof **Step 1: Energy dissipation** In the stochastic setting, the expected energy dissipation identity

$$\mathbb{E}[f(x_{n+1})] \leq \mathbb{E}[f(x_n)] - \eta\left(1-\frac{L(1+\sigma^2)\,\eta}{2}\right)\mathbb{E}[\|\nabla f(x_n)\|^2]$$

holds if the noise scales multiplicatively [22, Lemma 16].

Step 2: Summability Consider the sequence $L_n := \mathbb{E}\big[\|X_n - x^*\|^2\big]$. We find that

$$\begin{aligned}
L_{n+1} &= \mathbb{E}\big[\|X_{n+1}-X_n+X_n-x^*\|^2\big]\\
&= \mathbb{E}\big[\|X_{n+1}-X_n\|^2\big]+2\,\mathbb{E}\big[\langle X_{n+1}-X_n,\, X_n-x^*\rangle\big]+L_n\\
&= \eta^2\,\mathbb{E}\big[\|g_n\|^2\big]-2\eta\,\mathbb{E}\big[\langle g_n,\, X_n-x^*\rangle\big]+L_n\\
&\leq \eta^2(1+\sigma^2)\,\mathbb{E}\big[\|\nabla f(X_n)\|^2\big]-2\eta\,\mathbb{E}\big[\langle \nabla f(X_n),\, X_n-x^*\rangle\big]+L_n\\
&\leq \eta^2(1+\sigma^2)\,\mathbb{E}\big[\|\nabla f(X_n)\|^2\big]-2\eta\mathbb{E}\big[f(X_n)-f(x^*)\big]+L_n.
\end{aligned}$$

Since $0 \leq L_n$ for all n we see that

$$\begin{aligned}
L_0 \geq L_0 - L_{n+1} &= \sum_{k=0}^{n}(L_k-L_{k+1})\\
&\geq 2\eta\sum_{k=0}^{n}\mathbb{E}\big[f(X_n)-f(x^*)\big]-\eta^2(1+\sigma^2)\sum_{k=0}^{n}\mathbb{E}\big[\|\nabla f(X_n)\|^2\big]
\end{aligned}$$

and hence

$$\eta \sum_{k=0}^{n} \mathbb{E}\big[f(X_n) - \inf f\big] \le \frac{L_0}{2} + \frac{\eta^2(1+\sigma^2)}{\eta(1-\frac{L(1+\sigma^2)\eta}{2})}\mathbb{E}\big[f(X_0) - \inf f\big]$$

$$\le \frac{\mathbb{E}\big[\|X_0 - x^*\|^2\big]}{2} + \frac{\eta(1+\sigma^2)}{1 - L(1+\sigma^2)\eta/2}\mathbb{E}\big[f(X_0) - \inf f\big].$$

□

As a consequence of the summable decay, we immediately see that

$$\lim_{n\to\infty} n \cdot \mathbb{E}\big[f(X_n) - \inf f\big] = 0.$$

The more quantitative estimate

$$\mathbb{E}[f(x_n) - f(x^*)] \le (1+\sigma^2)\,\frac{\mathbb{E}[f(x_0) - f(x^*)] + \frac{L}{2}\,\mathbb{E}\big[\|x_0 - x^*\|^2\big]}{n+1+\sigma^2}.$$

is derived in [22, Lemma 7]. As an application, we prove that the random variables $f(X_n)$ converge to $\inf f$ in a stronger fashion.

Corollary 5 *Let X_n be stochastic gradient iterates for f. If $\mathbb{E}[f(X_0) + \|X_0\|^2] < \infty$, then $f(X_n)$ converges to $\inf f$ both almost surely and in L^1.*

Proof **Convergence in L^1.** As $\sum_{n=0}^{\infty} \mathbb{E}\big[f(X_n) - f(x^*)\big]$ converges, we find that $\lim_{n\to\infty} \mathbb{E}\big[f(X_n) - \inf f\big] = 0$. Since $f \ge \inf f$, this is the same as L^1-convergence

$$\lim_{n\to\infty} \|f_n - \inf f\|_{L^1} = \lim_{n\to\infty} \mathbb{E}\big[|f(X_n) - \inf f|\big] = \lim_{n\to\infty} \mathbb{E}\big[f(X_n) - \inf f\big] = 0.$$

Convergence Almost Surely The fact that the sequence $\|f(X_n) - \inf f\|_{L^1} = \mathbb{E}\big[f(X_n) - f(x^*)\big]$ is summable implies convergence almost surely by Klenke [28, Theorem 6.12]. □

Without exploiting summability, we only get convergence almost surely for a subsequence $f(X_{n_k})$ of $f(X_n)$—precisely such a sequence for which $\|f(X_{n_k}) - \inf f\|_{L^1}$ is summable.

4 Momentum Method

4.1 Continuous-Time Heavy-Ball ODE

In our analysis of accelerated gradient descent, let us consider the continuous-time limit of the Nesterov time-stepping scheme derived by Su et al. [43], which is

$$\dot{x}(t) = v(t), \qquad \dot{v}(t) = -\frac{\alpha}{t}\, v(t) - \nabla f(x_t). \tag{3}$$

Due to [43, Theorem 1], for each $x_0 \in \mathbb{R}^d$, (3) has a unique solution for $v_0 = 0$ if ∇f is Lipschitz-continuous. As α/t blows up at 0, this cannot be derived directly from common existence theorems, but suitable uniform estimates can be derived when replacing α/t by $\alpha/\max\{\varepsilon, t\}$ and sending $\varepsilon \to 0^+$.

We have the following result concerning the integrability of the objective error $f(x_t) - f(x^*)$.

Theorem 1 *Let H be a Hilbert space and suppose that $f : H \to \mathbb{R}$ is a convex differentiable function and $x^* \in \arg\min_{x\in\mathbb{R}^d} f(x)$. Then if x_t and $v(t)$ are defined via the differential equation* (3) *with $\alpha > 3$, we have*

$$\int_0^\infty t(f(x_t) - f(x^*))dt \le \frac{(\alpha-1)^2 \, \|x_0 - x^*\|^2}{2(\alpha-3)} < \infty. \tag{4}$$

The optimal bound is attained for $\alpha = 5$ where $\frac{(\alpha-1)^2}{2(\alpha-3)} = 4$.

This result was previously obtained in [43, Theorem 5]. We provide a brief since we will use many of the same concepts below. In Section 4.2 of [43], the authors demonstrate that $\alpha < 2$ does not generally lead to energy decay even for quadratic functions and that $\alpha < 3$ is generally inadmissible by considering $f(x) = |x|$.

In particular, "on average" $f(x_t)$ must decay faster than $O(1/t^2)$. A more quantitative statement does not follow immediately in this context since $f(x_t) - f(x^*)$ is not guaranteed to be monotone decreasing (and in many cases is not, see Sect. 4.2). However, it was observed in [4] that $f(x_t) - f(x^*) = o(1/t^2)$ also for the discrete-time Nesterov algorithm.

Proof The proof closely follows the argument from [43]. We consider the Lyapunov function:

$$L(t) = t^2(f(x_t) - f(x^*)) + \frac{1}{2}\|(\alpha-1)(x_t - x^*) + tv(t)\|^2 \tag{5}$$

Differentiating this, we obtain

$$\begin{aligned} L'(t) = 2t(f(x_t) - f(x^*)) &+ t^2\langle \nabla f(x_t), v(t)\rangle \\ &- \langle t\nabla f(x_t), (\alpha-1)(x_t - x^*) + tv(t)\rangle, \end{aligned} \tag{6}$$

since $\frac{d}{dt}[(\alpha-1)(x_t - x^*) + tv(t)] = -t\nabla f(x_t)$. Simplifying this and using the fact that convexity of f means that

$$\langle \nabla f(x_t), x_t - x^*\rangle \ge f(x_t) - f(x^*),$$

we get

$$L'(t) \le (3-\alpha)t(f(x_t) - f(x^*)) \tag{7}$$

Since $L(t) \geq 0$ for $t \geq 0$, we get that

$$\int_0^t t(f(x_t) - f(x^*))dt \leq \frac{L(0)}{\alpha - 3} < \infty. \tag{8}$$

□

The same proof immediately shows that

$$t^2\big(f(x_t) - f(x^*)\big) \leq L(t) \leq L(0) \quad \Rightarrow \quad f(x_t) - f(x^*) \leq \frac{(\alpha-1)^2\,\|x_0 - x^*\|^2}{2t},$$

where choosing the minimal value $\alpha = 3$ yields the optimal bound. We further note an improvement of "physical" nature. Namely, if $\alpha \geq 3$, then the total energy (potential and kinetic) of a particle x_t with mass $m = 1/2$ satisfies

$$\begin{aligned}
t^2\left(f(x_t) - f(x^*) + \frac{m}{2}\|\dot{x}_t\|^2\right) &= t^2\left(f(x_t) - f(x^*)\right) + \frac{1}{4}\,\|t\dot{x}_t\|^2 \\
&\leq t^2\left(f(x_t) - f(x^*)\right) + \frac{1}{2}\left\|t\dot{x}_t + (\alpha-1)(x_t - x^*)\right\|^2 + \frac{(\alpha-1)^2}{2}\|x_t - x^*\|^2 \\
&\leq L(t) + \frac{(\alpha-1)^2}{2}\|x_t - x^*\|^2 \\
&\leq L(0) + \frac{(\alpha-1)^2}{2}\|x_t - x^*\|^2.
\end{aligned}$$

In particular, in any situation in which the sublevel sets of f are compact, we find that not only the potential, but also the kinetic energy of x_t decays at least as fast as t^{-2}.

4.2 *The Nesterov Oscillator*

As a special example, we consider the "Nesterov oscillator" equation:

$$\ddot{x} = -\frac{\alpha}{t}\,\dot{x} - \mu x, \qquad x_0 = x_0, \qquad \dot{x}_0 = 0, \tag{9}$$

based on a dampened harmonic oscillator with Nesterov-type friction. This equation arises as the heavy-ball ODE for the convex function $f(x) = \frac{\mu}{2}\,x^2$. The coefficient of friction is not chosen optimally for this function, which is indeed strongly convex, but the example provides us with valuable intuition and tools for the proof below.

The Nesterov oscillator has time-variable friction which transitions from the overdampened to the underdampened regime when

$$\left(\frac{\alpha}{2t}\right)^2 = \mu \qquad \Leftrightarrow \qquad t^2 = \frac{\alpha^2}{4\mu}.$$

Recall that solutions to the classical harmonic oscillator equation

$$\ddot{x} = -\beta\,\dot{x} - \mu x, \qquad x_0 = x_0, \qquad \dot{x}_0 = 0$$

are given by

$$x(t) = \exp\left(-\frac{\beta}{2}t\right)\left(c_1\,\cos(\omega t) + c_2\,\sin(\omega t)\right), \qquad \omega := \sqrt{\mu - \left(\frac{\beta}{2}\right)^2}$$

in the underdampened regime where $(\beta/2)^2 < \mu$ and

$$x(t) = c_1\,\exp\left(\lambda_+ t\right) + c_2\,\exp(\lambda_- t), \qquad \lambda_\pm = -\frac{\beta}{2} \pm \sqrt{\left(\frac{\beta}{2}\right)^2 - \mu}$$

in the overdampened regime where $(\beta/2)^2 > \mu$ (see, e.g., [29, Section 10.4] for a derivation). Overdampened solutions can overshoot the minimizer if the initial velocity is high enough, but at most once, and approach exponentially fast in a monotone fashion afterwards, while an underdamped oscillator changes sign an infinite number of times.

With Nesterov friction, the coefficient of friction is initially infinitely strong. In the initial phase, we expect that solutions to the ODE gradually approach the minimizer $x = 0$ in a monotone fashion. At time $t = \alpha/2\sqrt{\mu}$, the dampening changes type, and we expect the oscillator to change sign an infinite number of times with frequency approaching $\sqrt{\mu}$ as $t \to \infty$. In the long term, $f(x) = \frac{\mu}{2}x^2$ decays on average as $t^{-\alpha}$ according to [5]. More generally, the energy decays as $t^{-2\alpha/3}$ for general strongly convex functions due to [3]. In the one-dimensional case, it crosses the global minimizer many times along this trajectory (at least in continuous time), but it does not come to rest. In the higher-dimensional case, the minimizer is not usually attained exactly at any finite time. More general results under more complicated curvature conditions are given in [5].

A numerical investigation is given in Fig. 1, where we approximated $x(n\sqrt{h})$ by the Nesterov scheme:

$$y_0 = x_0, \qquad x_{n+1} = y_n - h\,\nabla f(y_n), \qquad y_{n+1} = x_{n+1} + \frac{n}{n+\alpha}(x_{n+1} - x_n), \tag{10}$$

which discretizes the heavy-ball equation due to [43]. Analytically, we focus on the initial overdamped phase. Since x is not expected to change sign during this

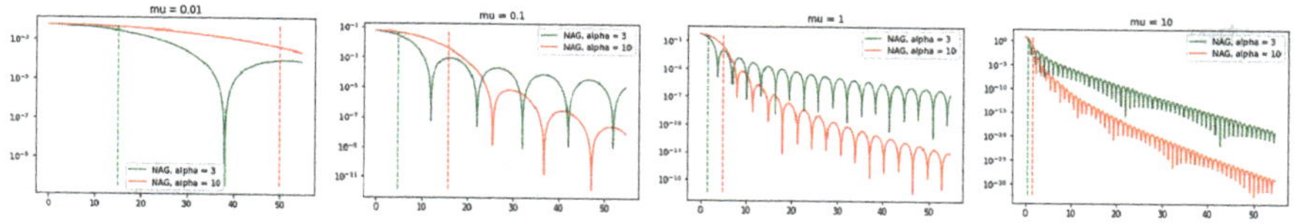

Fig. 1 We consider numerical solutions of the Nesterov oscillator Eq. (9) by the Nesterov algorithm (10) with step size $h = 0.003$ and compare $f(x_t)$ for friction parameters $\alpha = 3$ (green) and $\alpha = 10$ (red) given the function $f(x) = \frac{\mu}{2} x^2$ where $\mu \in \{0.001, 0.1, 1, 10\}$ (left to right). The vertical dashed lines signal the critical time at which the oscillator transitions from the overdamped to the underdamped regime. We see that indeed, an initial slow overdamped period is followed by a period of oscillations whose length is essentially constant and only depends on μ, but not α. At the spikes, $f(x_t)$ behaves like $c^* \sqrt{|t - t^*|}$, since $\dot{x}_t \neq 0$, i.e., x_t crosses the minimizer with positive velocity. If $\dot{x}_t$ were zero, the kinetic energy would be zero at the minimum, making the total energy zero and the system would be at equilibrium, meaning that there would be no crossing. This observation does not transfer to the higher-dimensional case

relatively brief period, we can make the ansatz

$$x(t) = \exp\left(-\int_0^t \lambda(s)\,\mathrm{d}s\right)$$

and compute

$$\dot{x} = -\lambda\, x, \qquad \ddot{x} = (-\lambda' + \lambda^2)x$$

such that

$$0 \stackrel{!}{=} \ddot{x} + \frac{\alpha}{t}\dot{x} + \mu x = \left(-\lambda' + \lambda^2 - \frac{\alpha}{t}\lambda + \mu\right) x.$$

The initial condition $0 = \dot{x}_0 = -\lambda(0)\, x_0$ induces the corresponding condition that $\lambda(0) = 0$. We note that the functions

$$\lambda_0(t) = 0, \qquad \lambda_+(t) = \frac{\alpha}{2t} - \sqrt{\left(\frac{\alpha}{2t}\right)^2 - \mu}$$

satisfy

$$\lambda_0^2 - \frac{\alpha}{t}\lambda_0 + \mu = \mu > 0, \qquad \lambda_+^2 - \frac{\alpha}{t}\lambda_+ + \mu \equiv 0.$$

In particular, by the comparison principle, we have $\lambda_0 \leq \lambda \leq \lambda_+$ for all $t \leq \frac{\alpha^2}{2\mu}$, at which point λ_+ is no longer defined. We conclude that

$$x_+(t) := x_0 \exp\left(\int_0^t \sqrt{\left(\frac{\alpha}{2s}\right)^2 - \mu} - \frac{\alpha}{2s}\,\mathrm{d}s\right) \qquad \text{for } 0 < t < \frac{\alpha}{2\sqrt{\mu}}$$

satisfies $|x_+(t)| \leq |x(t)|$. Since $\sqrt{1-z} \geq 1-z$ for $z \in (0,1)$, we have

$$0 \leq \int_0^t \frac{\alpha}{2s} - \sqrt{\left(\frac{\alpha}{2s}\right)^2 - \mu}\,\mathrm{d}s = \int_0^t \frac{\alpha}{2s}\left(1 - \sqrt{1 - \frac{4\mu s^2}{\alpha^2}}\right)\mathrm{d}s$$

$$\leq \int_0^t \frac{\alpha}{2s} \cdot \frac{4\mu}{\alpha^2} s^2\,\mathrm{d}s = \frac{\mu}{\alpha} t^2$$

for $t \in (0, \alpha/2\sqrt{\mu})$. In particular, $\lim_{t\to 0} x_+(t) = x_0$. In particular, we see that

$$|x(t)| \geq |x_+(t)| \geq |x_0|\,\exp\left(-\frac{\mu}{\alpha}t^2\right) \geq |x_0|\,\exp\left(-\frac{\alpha}{4}\right) \qquad \forall\, t \in \left(0, \frac{\alpha^2}{4\mu}\right).$$

This settles the existence of an initial "slow" phase.

Notably, the heavy-ball ODE with a momentum parameter scaling as α/t leads to much slower risk decay $t^{-2\alpha/3}$ in the strongly convex case than the gradient flow, which achieves the exponential rate $e^{-\mu t}$. With constant friction

$$\ddot{x} = -2\sqrt{\mu}\,\dot{x} - \nabla f(x)$$

where μ is the strong convexity constant of f, the decay is faster than the gradient flow with a decay as $e^{-\sqrt{\mu}t}$; see, e.g., [41, Theorem 1]. This dampening is in fact optimal as seen in study of the harmonic oscillator.

4.3 Heavy-Ball Dynamics: Quadratic Forms in Hilbert Spaces

We show that Theorem 1 is essentially optimal by mimicking the argument of Lemma 6.

Lemma 13 *Let H be a separable Hilbert space. Then there exists a strictly convex quadratic function $F : H \to [0,\infty)$ such that $F(u) = 0$ if and only if $u = 0$ and such that the following is true: For any monotone decreasing function $g : (0,\infty) \to (0,\infty)$ such that*

$$\int_0^\infty t\, g(t)\,\mathrm{d}t < \infty,$$

there exists a solution u to the heavy-ball ODE

$$\ddot{u} = -\frac{\alpha}{t}\,\dot{u} - \nabla F(u)$$

such that $F(u(t)) \geq g(t) \quad \forall\, t \geq 1$.

Proof **Step 1** Again, we identify H isometrically with $L^2(1, \infty)$ and consider

$$F : H \to \mathbb{R}, \qquad F(u) = \frac{1}{2}\int_1^\infty \frac{u^2(s)}{s}\,\mathrm{d}s.$$

The heavy-ball ODE acts pointwise in s, so by the analysis for the Nesterov oscillator with $\mu = 1/s$, we see that

$$u(t)(s) \geq \exp(-\alpha/4)\,u_0(s) \qquad \forall\, s > 0,\ t \in \left(0, \frac{\alpha\sqrt{s}}{2}\right).$$

In particular, we have

$$F(u(t)) \geq \exp(-\alpha/4)\int_{4t^2/\alpha^2}^\infty \frac{u_0^2(s)}{s}\,\mathrm{d}s$$

Step 2 From the proof of Lemma 6, we recall that for any monotone decreasing integrable function $\tilde{g} : \mathbb{R} \to \mathbb{R}$, there exists $u_0 \in H$ such that

$$\int_t^\infty \frac{u_0^s(s)}{s}\,\mathrm{d}s \geq \tilde{g}(t).$$

Combining these arguments, we note that for any monotone decreasing integrable function $\tilde{g} : (0, \infty) \to (0, \infty)$, there exists $u_0 \in H$ such that

$$F(u(t)) \geq \exp(-\alpha/4)\int_{4t^2/\alpha^2}^\infty \frac{u_0^2(s)}{s}\,\mathrm{d}s \geq \tilde{g}(t^2).$$

Now note that

$$\int_0^\infty g(\sqrt{t})\,\mathrm{d}t = 2\int_0^\infty g(\sqrt{t})\,\frac{\sqrt{t}}{2\sqrt{t}}\,\mathrm{d}t = 2\int_0^\infty g(s)\,s\,\mathrm{d}s,$$

i.e., $\tilde{g}(t) := g(\sqrt{t})$ is integrable if and only if $t \cdot g(t)$ is integrable. The decrease condition concerns $\tilde{g}$ and thus g, not $t \cdot g$. □

A slight mismatch remains between Theorem 1 and Lemma 13: We have shown that $t \cdot (f(x_t) - \inf f)$ is integrable (but possibly non-monotone) and on the other hand that $f(x_t) - \inf f$ can decay as slowly as any *monotone* function for which $\int_0^\infty t(f(x_t) - \inf f)\,\mathrm{d}t < +\infty$.

4.4 Heavy-Ball Dynamics: No Minimizers

In Sect. 2.6, we have shown that the risk decay along a gradient flow can be arbitrarily slow for convex functions without minimizers. Here, we prove the same for heavy-ball dynamics.

Lemma 14 *Let $g : [0, \infty) \to \mathbb{R}$ be a convex function such that* $\lim_{x\to\infty} g(x) = 0$. *Then for any $\alpha > 0$, the solution to the Nesterov ODE*

$$\ddot{x} = -\frac{\alpha}{t} - f'(x), \quad x_0 = 0, \quad \dot{x}_0 = 0, \qquad f(x) = g\left(\frac{x}{2\sqrt{g(0)}}\right)$$

satisfies $f(x(t)) \geq g(t)$.

The strategy of proof is as follows: There is only so much total (kinetic and potential) energy in the system, and it is dissipated by the dynamics. Even if energy were conserved and totally translated into kinetic energy, this would bound the speed at which we move towards $\pm\infty$. Thus, if the function $x \mapsto f(x)$ decays slowly, so does $t \mapsto f(x_t)$.

Proof Let $f : \mathbb{R}^d \to \mathbb{R}$ be any convex function and x a solution to the Nesterov ODE:

$$\ddot{x} = -\frac{\alpha}{t}\dot{x} - \nabla f(x), \qquad x_0 = x_0, \qquad \dot{x}_0 = 0.$$

Note that

$$\frac{d}{dt}\left(\frac{|\dot{x}|^2}{2} + f(x)\right) = \dot{x}\,\ddot{x} + \nabla(x)\,\dot{x} = (\ddot{x} + \nabla f(x)) \cdot \dot{x} = -\frac{\alpha}{t}\,\|\dot{x}\|^2 \leq 0.$$

In particular

$$\|\dot{x}\|^2 \leq 2\left(f(x) + \frac{\|\dot{x}\|^2}{2}\right) \leq 2\,f(x_0)$$

so

$$\|x(t) - x_0\| \leq \int_0^t \|\dot{x}(s)\|\,\mathrm{d}s \leq \sqrt{2\,f(x_0)}t$$

and hence

$$f(x(t)) \geq \inf\left\{f(z) : z \in B\big(x_0, \sqrt{2\,f(x_0)}t\big)\right\}.$$

If $f : [0, \infty) \to \mathbb{R}$ is such that $\lim_{x\to\infty} f(x) = 0$, then $0 = \inf f$ and f is strictly monotone decreasing. Hence, if $x_0 = 0$, then

$$f(x(t)) \geq f\left(\sqrt{2\, f(0)}\, t\right).$$

□

We have seen in the proof of Lemma 10 that there is no difference between the decay rate at infinity achievable by monotone decreasing or convex functions. In particular, also solutions to the heavy-ball ODE with Nesterov momentum can decay arbitrarily slowly at infinity.

Appendix: A Comparison Principle for Convex Functions with Derivatives in $L^{1/2}$

Lemma 8 *Let $g, G : [0, \infty) \to [0, \infty)$ be decreasing, differentiable convex functions such that*

$$\lim_{t\to\infty} G(t) = \lim_{t\to\infty} g(t) = 0, \qquad \liminf_{t\to\infty} \frac{G(t)}{g(t)} > 0.$$

Then

$$\int_1^\infty \sqrt{-g'(t)}\,\mathrm{d}t = +\infty \qquad \Rightarrow \qquad \int_1^\infty \sqrt{-G'(t)}\,\mathrm{d}t = +\infty.$$

We remark that it is evident from the proof that this lemma holds more broadly for any increasing concave function of the negative derivative, not just the square root function. For simplicity, we do not state or prove the result in this generality because the square root function is all we need for the analysis of gradient flows.

As the proof is somewhat technical, we present two proofs of Lemma 8 of very different flavor—one combinatorial, one variational—in order to make the result approachable to readers from different backgrounds. The key to the combinatorial proof is the following well-known proposition from the theory of majorization (see, for instance, [33]).

Proposition 1 *Let $a_1 \geq a_2 \geq \cdots a_n \geq 0$ and $b_1 \geq b_2 \geq \cdots b_n \geq 0$ be decreasing sequences which satisfy*

$$\sum_{j=i}^n b_j \geq \sum_{j=i}^n a_j \tag{11}$$

for all $i = 1, \ldots, n$. Then the sequence b_i is pointwise greater than an average of permutations of the sequence a_i. Specifically, letting S_n denote the symmetric group *on n elements, there exists a map* $\alpha : S_n \to \mathbb{R}_{\geq 0}$ *satisfying*

$$\sum_{\pi \in S_n} \alpha(\pi) = 1, \tag{12}$$

such that

$$b_i \geq \sum_{\pi \in S_n} \alpha(\pi) a_{\pi(i)}. \tag{13}$$

for $i = 1, \ldots, n$.

Although this proposition is well known, we give the simple proof for the reader's convenience.

Proof When $n = 1$ the statement is trivial. We proceed by induction on n.

Suppose first that $b_i \geq a_i$ for $i = 2, \ldots, n$. Let $\tau_i \in S_n$ denote the transposition which swaps the ith element and the first element for $i = 2, \ldots, n$ and let $e \in S_n$ denote the identity permutation. If $b_1 \geq a_1$, we can simply take $\alpha(e) = 1$ and $\alpha = 0$ for all other permutations.

Otherwise, we must have $a_1 > b_1 \geq a_i$ for all $i \geq 2$. In this case, Condition (11) implies that

$$\sum_{i=2}^{n} (b_i - a_i) \geq a_1 - b_1. \tag{14}$$

Choose any numbers $0 \leq q_i \leq b_i - a_i$ such that $\sum_{i=2}^{n} q_i = a_1 - b_1$ and set

$$\alpha(\tau_i) = \frac{q_i}{a_1 - a_i} \tag{15}$$

for $i = 2, \ldots, n$, $\alpha(e) = 1 - \sum_{i=2}^{n} \alpha(\tau_i)$, and $\alpha = 0$ for all other permutations. We claim that α defined in this way is nonnegative, for which we must check that

$$\sum_{i=2}^{n} \alpha(\tau_i) = \sum_{i=2}^{n} \frac{q_i}{a_1 - a_i} \leq \sum_{i=2}^{n} \frac{q_i}{a_1 - b_1} = 1, \tag{16}$$

which follows from the choice of q_i. For each $i = 2, \ldots, n$ we calculate that

$$\sum_{\pi \in S_n} \alpha(\pi) a_{\pi(i)} = a_i + \alpha(\tau_i)(a_1 - a_i) = a_i + q_i \leq b_i, \tag{17}$$

since $q_i \leq b_i - a_i$. For $i = 1$ we calculate

$$\sum_{\pi \in S_n} \alpha(\pi) a_{\pi(1)} = a_1 - \sum_{i=2}^{n} \alpha(\tau_i)(a_1 - a_i) = a_1 - \sum_{i=2}^{n} q_i = b_1. \tag{18}$$

This completes the proof in the case where $b_i \geq a_i$ for $i = 2, \ldots, n$.

For the general case, the inductive assumption applied to the tail sequences $a_2, \ldots, a_n$ and $b_2, \ldots, b_n$ implies that there exists a map $\gamma : S_n \to \mathbb{R}_{\geq 0}$ such that $\sum_{\pi \in S_n} \gamma(\pi) = 1$, $\gamma(\pi) = 0$ if $\pi(1) \neq 1$ (i.e., the map is supported on the set of permutations of $\{2, \ldots n\}$), and

$$b_i \geq \sum_{\pi \in S_n} \gamma(\pi) a_{\pi(i)} \tag{19}$$

for $i = 2, \ldots, n$. Set

$$\tilde{a}_i := \sum_{\pi \in S_n} \gamma(\pi) a_{\pi(i)}$$

for $i = 1, \ldots, n$ (note that $\tilde{a}_1 = a_1$ since $\gamma(\pi) = 0$ if $\pi(1) \neq 1$). We complete the proof by applying the previous part to the sequences $b_1, \ldots, b_n$ and $\tilde{a}_1, \tilde{a}_2, \ldots, \tilde{a}_n$ (denoting the resulting averaging map by β) and noting that for each $i = 1, \ldots, n$

$$\begin{aligned} b_i \geq \sum_{\sigma \in S_n} \beta(\sigma) \tilde{a}_{\sigma(i)} &= \sum_{\sigma \in S_n} \beta(\sigma) \sum_{\pi \in S_n} \gamma(\pi) a_{\pi(\sigma(i))} \\ &= \sum_{\pi \in S_n} \left(\sum_{\sigma \in S_n} \beta(\sigma) \gamma(\pi \sigma^{-1}) \right) a_{\pi(i)}. \end{aligned} \tag{20}$$

Setting

$$\alpha(\pi) = \sum_{\sigma \in S_n} \beta(\sigma) \gamma(\pi \sigma^{-1}), \tag{21}$$

noting that $\alpha(\pi) \geq 0$ since β and γ are nonnegative and that

$$\sum_{\pi \in S_n} \alpha(\pi) = \sum_{\sigma \in S_n} \beta(\sigma) \sum_{\pi \in S_n} \gamma(\pi \sigma^{-1}) = \sum_{\sigma \in S_n} \beta(\sigma) = 1 \tag{22}$$

completes the proof. □

Proof (Combinatorial Proof of Lemma 8) Note first that since G and g are convex, they are differentiable almost everywhere [37]. By considering the interval $[\epsilon, \infty)$ with $\epsilon \to 0$, it suffices to consider the case where $-g'(0), -G'(0) < \infty$.

The result follows from the stronger statement that if g, G are nonnegative convex functions with $\lim_{t \to \infty} g(t) = \lim_{t \to \infty} G(t) = 0$ and $G(t) \geq g(t)$, then

$$\int_0^\infty \sqrt{-g'(t)}\, \mathrm{d}t \leq \int_0^\infty \sqrt{-G'(t)}\, \mathrm{d}t.$$

Assume on the contrary that there exist g, G satisfying these assumptions but for which

$$\int_0^\infty \sqrt{-g'(t)}\,\mathrm{d}t > \int_0^\infty \sqrt{-G'(t)}\,\mathrm{d}t.$$

Since $\lim_{t\to\infty} G(t) = 0$ and $g \geq 0$, this implies that there exists a $T < \infty$ such that

$$\int_0^T \sqrt{-g'(t)}\,\mathrm{d}t + \sqrt{g(T)} > \int_0^T \sqrt{-G'(t)}\,\mathrm{d}t + \sqrt{G(T)}. \tag{23}$$

Since $G(T) \geq g(T)$ by assumption, we also have

$$\int_0^T \sqrt{-g'(t)}\,\mathrm{d}t + c\sqrt{g(T)} > \int_0^T \sqrt{-G'(t)}\,\mathrm{d}t + c\sqrt{G(T)} \tag{24}$$

for any $c \leq 1$.

The functions $-g'$, $-G'$, $\sqrt{-g'}$, and $\sqrt{-G'}$ are bounded and decreasing (due to convexity) and so are Riemann integrable on $[0, T]$. This means that for any $\epsilon > 0$ we can choose an N sufficiently large so that

$$\frac{1}{N}\sum_{i=1}^{N} f(Ti/N) - \epsilon \leq \int_0^T f\,\mathrm{d}t \leq \frac{1}{N}\sum_{i=0}^{N-1} f(Ti/N) + \epsilon \tag{25}$$

for each of the functions $f = -g', -G', \sqrt{-g'}, \sqrt{-G'}$.

Since the function $-g'(t)$ is decreasing, we have

$$\frac{-1}{N} g'(T(i+1)/N) \leq \int_{Ti/N}^{T(i+1)/N} -g'(t)dt \leq \frac{-1}{N} g'(Ti/N).$$

Combined with the estimate (25), this implies that

$$\left| \int_0^T \sqrt{-g'(t)}\,\mathrm{d}t - \sum_{i=0}^{N-1} \frac{1}{\sqrt{N}} \sqrt{\int_{Ti/N}^{T(i+1)/N} -g'(t)dt} \right| < \epsilon + \frac{1}{N} \sup_{t>0} \sqrt{-g'(t)}. \tag{26}$$

We obtain an analogous bound for G.

Consider two sequences defined by

$$\begin{aligned} a_i &= \int_{T(i-1)/N}^{Ti/N} -g'(t)dt,\ i = 1, \ldots, N,\ a_{N+1} = \int_T^\infty -g'(t)dt = g(T) \\ b_i &= \int_{T(i-1)/N}^{Ti/N} -G'(t)dt,\ i = 1, \ldots, N,\ b_{N+1} = \int_T^\infty -G'(t)dt = G(T). \end{aligned} \tag{27}$$

The sequences a_i and b_i are decreasing by the convexity of g and G, and since $G(t) \geq g(t)$, we have

$$\sum_{j=i}^{N+1} b_j \geq \sum_{j=i}^{N+1} a_j \tag{28}$$

for all $i = 1, \ldots, N+1$. However, using (24) and (26), choosing $c = 1/\sqrt{N}$ and N large enough (i.e., ϵ small enough), we get

$$\sum_{i=1}^{N+1} \sqrt{b_i} < \sum_{i=1}^{N+1} \sqrt{a_i}. \tag{29}$$

However, this contradicts Proposition 1 and Jensen's inequality, which completes the proof. □

We now present an alternative proof in the spirit of the calculus of variations.

Proof (Variational Proof of Lemma 8) **Preliminaries** Since g, G are convex, they are (twice) differentiable almost everywhere on $(0, \infty)$ by the Alexandrov theorem [18, Theorem 6.4.1]. Since $\lim_{x\to\infty} g(x) = \lim_{x\to\infty} G(x) = 0$, they are monotone decreasing. Taking both together, we see that the integrands $\sqrt{-g'}$, $\sqrt{-G'}$ are well-defined. Since $\sqrt{-g'}$, $\sqrt{-G'}$ are monotone decreasing functions, they are (Riemann and Lebesgue) integrable on finite intervals. Their integrals over $(0, \infty)$ are well-defined (albeit potentially infinite).
Step 1: Reduction. We implicitly assume that $g(t) > 0$ for all $t > 0$ since otherwise $g \equiv 0$ on an interval $[T, \infty)$ and thus $\int_0^\infty \sqrt{-g'(t)}\,\mathrm{d}t < \infty$. Up to a rescaling and translation, we may assume that $G \geq g > 0$ on $[0, \infty)$ and that $g(0)$ and $g'(0)$ are finite. Under these stronger assumptions, we prove the stronger statement that

$$\int_0^\infty \sqrt{-g'(t)}\,\mathrm{d}t \leq \int_0^\infty \sqrt{-G'(t)}\,\mathrm{d}t. \tag{30}$$

Step 2: Representation by derivatives. We note that

$$g(t) = -\int_t^\infty g'(s)\,\mathrm{d}s, \qquad G(t) = -\int_t^\infty G'(s)\,\mathrm{d}s,$$

and rewrite the problem in terms of the nonnegative monotone decreasing functions $\phi := -g'$ and $\psi := -G'$. For this, we consider the convex set

$$K_\phi := \left\{ \psi \in L^1(0, \infty) : \psi \text{ nonincreasing}, \int_t^\infty \psi\,\mathrm{d}s \geq \int_t^\infty \phi\,\mathrm{d}s \quad \forall\, t > 0 \right\}$$

and the "energy" functional

$$E : K_\phi \to (0, \infty), \qquad E(\psi) = \int_0^\infty \sqrt{\psi}\,\mathrm{d}t.$$

For the bijection between G and $\psi = -G'$, denote $G_\psi(t) = \int_t^\infty \psi(s)\,\mathrm{d}s$.
Intermezzo: Proof strategy We will show that the functional E has a minimizer ψ^* in K_ϕ (Step 3). By constructing energy competitors, we will argue that if $G_\psi > g$ somewhere, then there exists $\tilde{\psi}$ such that $E(\tilde{\psi}) < E(\psi)$, i.e., $\psi \neq \psi^*$. The construction of energy competitors is the content of Step 4. This immediately implies that ψ^* is such that $G_{\psi^*} = g = G_\phi$ everywhere on $(0, \infty)$, which concludes the proof.
Step 3: Existence of minimizers. If there exists no $\psi \in K_\phi$ such that $E(\psi) < +\infty$, then any function is a minimizer. In particular, the main statement of the Lemma, which can be phrased as

$$E(\phi) = +\infty \qquad \Rightarrow \qquad E(\psi) = +\infty \quad \forall\, \psi \in K_\phi,$$

holds. We now consider the case where there exists some $\psi \in K_\phi$ such that $E(\psi) < +\infty$. In this step, we prove that then there also exists $\psi^* \in K_\phi$ such that $E(\psi^*) = \inf_{\psi \in K_\phi} E(\psi) < +\infty$.

The existence of a minimizer is established by the direct method of the calculus of variations. Let $\psi_n \in K_\phi$ be a sequence such that $\lim_{n\to\infty} E(\psi_n) = \inf_{\psi\in K_\phi} E(\psi)$. Passing to a subsequence, we may even assume that

$$E(\psi_n) \leq 1 + \inf_{\psi \in K_\phi} E(\psi) =: C \qquad \forall\, n \in \mathbb{N}.$$

Then $|\{t : \psi_n(t) > R\}| \leq C/\sqrt{R}$ by Chebyshev's inequality, and since ψ is monotone decreasing, we find that $0 \leq \psi_n(t) \leq R$ for $t > C/\sqrt{R}$ and for all $n \in \mathbb{N}$. Since ψ_n is bounded on $[\varepsilon, \infty)$ for all $\varepsilon > 0$ and monotone decreasing, we find that the sequence ψ_n is bounded in $BV(a, b)$ for any $0 < a < b < \infty$. In particular, there exists ψ^* such that $\psi_n \to \psi^*$ in $L^p(a, b)$ for any $p < \infty$ due to the compact embedding of BV into all Lebesgue spaces L^p for finite p in one dimension. We deduce that

$$\int_a^b \sqrt{\psi^*(t)}\,\mathrm{d}t = \lim_{n\to\infty} \int_a^b \sqrt{\psi_n(t)}\,\mathrm{d}t \leq \liminf_{n\to\infty} \int_0^\infty \sqrt{\psi_n(t)}\,\mathrm{d}t.$$

Since the inequality holds independently of a, b, we can send $a \to 0$ and $b \to \infty$ to obtain

$$E(\psi^*) = \int_0^\infty \sqrt{\psi^*(t)}\,\mathrm{d}t \leq \liminf_{n\to\infty} \int_0^\infty \sqrt{\psi_n(t)}\,\mathrm{d}t = \inf_{\psi\in K_\phi} E(\psi).$$

It remains to show that $\psi^* \in K_\phi$. Observe first that for any $\psi \in K_\phi$ we have

$$G_\psi(T) = \int_T^\infty \psi(t)\,\mathrm{d}t \le \sqrt{\psi(T)} \int_T^\infty \sqrt{\psi(t)}\,\mathrm{d}t$$
$$\le \frac{\int_0^T \sqrt{\psi(t)}\,\mathrm{d}t}{T} \cdot \int_T^\infty \sqrt{\psi(t)}\,\mathrm{d}t \le \frac{E(\psi)^2}{T}.$$

In particular, for every $\varepsilon > 0$ there exists $T > 0$ such that $G_\psi(T) < \varepsilon$ for all $\psi \in K_\phi$ satisfying the energy bound $E(\psi) \le C$. We conclude that

$$G_{\psi^*}(t) \ge \int_t^T \psi^*(t)\,\mathrm{d}t = \lim_{n\to\infty} \int_t^T \psi_n(t)\,\mathrm{d}t \ge \liminf_{n\to\infty} \int_t^\infty \psi_n(t)\,\mathrm{d}t - \varepsilon$$
$$\ge \liminf_{n\to\infty} g_{\psi_n}(t) - \varepsilon \ge g(t) - \varepsilon$$

for all $t < T$. Since this holds for any $\varepsilon > 0$ and we can choose T larger if we desire, we have $G_{\psi^*}(t) \ge g(t)$ for all $t \in (0, \infty)$.

Step 4: Identifying the minimizer. In the following, we will show that $\psi^* = \phi$ to conclude the proof. This step will be partitioned into several arguments:

1. First, we show that ψ^* must be piecewise constant in the set $\{G_{\psi^*} > g\}$.
2. Then, we show that ψ^* is piecewise constant with at most one jump in connected components of $G_{\psi^*} > g$.
3. Finally, we see that the piecewise constant function is not energy-optimal unless $\psi^* = \phi$ (i.e., unless ϕ itself is piecewise constant with one jump).

In every step, we require a similar but slightly different "energy competitor" argument.

Step 4.1: Step function structure. The two functions

$$g(t) = \int_t^\infty \phi(s)\,\mathrm{d}s, \qquad G^*(t) = \int_t^\infty \psi^*(s)\,\mathrm{d}s$$

are continuous, so the coincidence set $I^* = \{t > 0 : g(t) = G^*(t)\}$ is closed. Assume that $t^* \in (0, \infty) \setminus I^*$, i.e., $G^*(t) > g(t)$ in an interval $(t^* - \varepsilon, t^* + \varepsilon)$. Since both functions are continuous and decreasing, we may assume that

$$\inf_{t\in(t^*-\varepsilon,t^*+\varepsilon)} G^*(t) = G^*(t^* + \varepsilon) > g(t^* - \varepsilon) = \sup_{t\in(t^*-\varepsilon,t^*+\varepsilon)} g(t)$$

for sufficiently small $\varepsilon > 0$. This gives us great leeway to modify ψ^* inside the interval $(t^* - \varepsilon, t^* + \varepsilon)$. We will show in this step that ψ^* must be a step function with a single jump in $(t^* - \varepsilon, t^* + \varepsilon)$.

Namely, consider

$$\tilde{\psi}(t) = \begin{cases} \psi^*(t^* - \varepsilon) & t^* - \varepsilon \leq t \leq t^\sharp \\ \psi^*(t^* + \varepsilon) & t^\sharp \leq t \leq t^* + \varepsilon \\ \psi^*(t) & \text{else} \end{cases}$$

where $t^\sharp \in (t^* - \varepsilon, t^* + \varepsilon)$ is chosen such that

$$\int_{t^*-\varepsilon}^{t^*+\varepsilon} \tilde{\psi}(t)\,\mathrm{d}t = \int_{t^*-\varepsilon}^{t^*+\varepsilon} \psi^*(t)\,\mathrm{d}t.$$

In particular, we note that

$$\tilde{G}(t) := \int_t^\infty \tilde{\psi}(s)\,\mathrm{d}s$$

satisfies

$$\tilde{G}(t) \geq \tilde{G}(t^* + \varepsilon) = G^*(t^* + \varepsilon) > g(t^* - \varepsilon) \geq g(t) \qquad \forall\, t \in (t^* - \varepsilon, t^* + \varepsilon)$$

and $G(t) = G^*(t)$ for all other t. In other words: $\tilde{\psi} \in K_\phi$. We claim that $E(\tilde{\psi}) \leq E(\psi^*)$ and that the inequality is strict unless $\tilde{\psi} \not\equiv \psi^*$. To see this, we write $\alpha = \psi^*(t^*)$, $\beta := \psi^*(t^* - \varepsilon)$ and

$$\psi^*(t) = \lambda^*(t)\alpha + \big(1 - \lambda^*(t)\big)\beta, \qquad \tilde{\psi}(t) = \tilde{\lambda}(t)\alpha + \big(1 - \tilde{\lambda}(t)\big)\beta$$

for functions $\tilde{\lambda}, \lambda^* : (t^* - \varepsilon, t^*) \to \mathbb{R}$. The fact that $\int \tilde{\psi} = \int \psi^*$ is equivalent to observing that $\int \tilde{\lambda} = \int \lambda^*$. Since the square root function is concave, we have

$$\begin{aligned} \int_{t^*-\varepsilon}^{t^*+\varepsilon} \sqrt{\psi^*}\,\mathrm{d}s &= \int_{t^*-\varepsilon}^{t^*+\varepsilon} \sqrt{\lambda^*\alpha + (1-\lambda^*)\beta}\,\mathrm{d}s \geq \int_{t^*-\varepsilon}^{t^*+\varepsilon} \lambda^*\sqrt{\alpha} + (1-\lambda^*)\sqrt{\beta}\,\mathrm{d}s \\ &= \int_{t^*-\varepsilon}^{t^*+\varepsilon} \tilde{\lambda}\sqrt{\alpha} + (1-\tilde{\lambda})\sqrt{\beta}\,\mathrm{d}s = \int_{t^*-\varepsilon}^{t^*+\varepsilon} \sqrt{\tilde{\psi}}\,\mathrm{d}s. \end{aligned}$$

The inequality is strict unless $\alpha = \beta$ (which implies that $\psi \equiv \psi^*$) or $\lambda^* \in \{0, 1\}$ almost everywhere. Since ψ^* and thus also λ^* is monotone, together with the integral constraint this means that $\psi^* \equiv \tilde{\psi}$. A strict inequality is excluded since $E(\psi^*) \leq E(\tilde{\psi})$ since ψ^* minimizes E in K_ϕ. Thus, ψ^* must be a step function with only one step in $(t^* - \varepsilon, t^* + \varepsilon)$.

Step 4.2: Only one step. Assume that (a, b) is a connected component of the open set $\mathbb{R} \setminus I^*$. Step 4.1 shows that ψ^* is a step function on (a, b) with at most a finite number of jumps in any subinterval (a', b') with $a < a' < b' < b$. If there were an *infinite* number of jumps, we could choose an accumulation point for t^* in Step 4.1 and obtain a contradiction.

Assume that there are at least *two* jumps in (a, b), i.e., there exist $a \leq t_1 < t_2 < t_3 < t_4 \leq b$ and $z_1 > z_2 > z_3$ such that

$$\psi^*(t) = \begin{cases} z_1 & t_1 < t < t_2 \\ z_2 & t_2 < t < t_3 \\ z_3 & t_3 < t < t_4 \end{cases}.$$

We can see by the same strategy as in Step 4.1 that shifting t_2 right and t_3 left reduces the energy since z_2 is a convex combination of z_1 and z_3. The details are left to the reader. Since $G_{\psi^*} \geq g + \varepsilon$ for some $\varepsilon > 0$ on the compact subset $[t_2, t_3]$, small perturbations of this type are admissible. By contradiction, we find that on every connected component (a, b), ψ^* can jump only once.

Step 4.3: No unbounded component. Since $G \geq g > 0$ and $\lim_{t\to\infty} G(t) = 0$, we note $G'(t) < 0$ for all t. In particular, by the step function structure we find that $(0, \infty) \setminus I^*$ cannot have an unbounded connected component.

Step 4.4: Coincidence at the origin. We argue that without loss of generality, we may assume that $G(0) = g(0)$. If this is not the case, we can modify g in the spirit of G_r (see the passage after the statement of Lemma 8) to increase $g(0)$ with an arbitrarily small increase in $\int \sqrt{-g'}$.

Step 4.5: Conclusion. In this step, we will show that $(0, \infty) \setminus I^* = \emptyset$, i.e., that $G_{\psi^*} \equiv g$. Otherwise, this open set has at least one connected component.

Assume that the interval (a, b) is a connected component of $\mathbb{R} \setminus I^*$. Since $0 \in I^*$ and there are no unbounded connected components, we find that $a, b \in I^*$. Thus,

$$\int_t^b \psi^*(s)\,ds > \int_t^b \phi(s)\,ds \quad \forall\, t \in (a, b) \qquad \text{and} \qquad \int_a^b \psi^*(s)\,ds = \int_a^b \phi(s)\,ds$$

and consequently

$$\int_a^t \psi^*(s)\,ds < \int_a^t \phi(s)\,ds \quad \forall\, t \in (a, b).$$

The minimizer ψ^* satisfies $\psi^*(t) = \alpha\, 1_{(a,t^\sharp)}(t) + \beta\, 1_{(t^\sharp,b)}$ for $t^\sharp \in (a, b)$ and $\alpha \geq \beta$. We immediately conclude that

$$\phi(a^+) := \lim_{t\searrow a} \phi(t) \geq \alpha, \qquad \phi(b^-) := \lim_{t\nearrow b} \phi(t) \leq \beta.$$

We distinguish three cases.

1. $\alpha = \phi(a^+)$. Since $\phi = -g'$ is monotone decreasing, we have

$$G(t) = G(a) - \int_a^t \psi^*(s)\,ds = g(a) - \int_a^t \alpha\,ds \leq g(a) - \int_a^t \phi(s)\,ds = g(t).$$

for all $t \in (a, t^\sharp)$. Since $G \geq g$, this implies that $G \equiv g$ in $(a, t^\sharp)$ and hence $[a, t^\sharp] \subseteq I$. This contradicts the choice of $(a, b) \subseteq (0, \infty) \setminus I^*$.

2. $\beta = \phi(b^-)$. A contradiction follows by the same argument.
3. So far, we have only used the monotonicity of ϕ, the step function properties of ψ^* and the fact that $g \leq G_{\psi^*}$. If we have strict inequalities

$$\alpha < \phi(a^+), \qquad \phi(b^-) < \beta,$$

we are using an energy argument, i.e., we show that the ψ^* cannot be energy optimal on (a, b). This is not surprising—we can increase α and decrease β slightly without violating the constraints of our problem. As in previous arguments, this reduces the energy. The remainder of this proof is dedicated to the finer details of this argument.

Let $\xi > 0$ be such that $a + \xi < t^\sharp < b - \xi$ and consider an energy competitor

$$\psi^\varepsilon(t) = \begin{cases} \alpha + \varepsilon & a < t < a + \xi \\ \beta - \varepsilon & b - \xi < t < b \\ \psi^*(t) & \text{else} \end{cases}.$$

By construction, we have $G_{\psi^\varepsilon} = G_{\psi^*}$ outside (a, b) since $\int_a^b \psi^\varepsilon = \int_a^b \psi^*$ by construction. If $\varepsilon > 0$ is sufficiently small, then $\psi^\varepsilon(a^+) < \phi(a^+)$ and $\psi^\varepsilon(b^-) > \phi(b^-)$. By definition of $\phi(b^-)$, we conclude that $\psi^\varepsilon(t) > \phi(t)$ for t close to b and thus

$$\begin{aligned} G_{\psi^\varepsilon}(t) &= \int_t^b \psi^\varepsilon(s)\, \mathrm{d}s + G_{\psi^\varepsilon}(b) \\ &= \int_t^b \psi^\varepsilon(s)\, \mathrm{d}s + g(b) \geq \int_t^b \phi(s)\, \mathrm{d}s + g(b) = g(t). \end{aligned}$$

By a similar argument, we observe that $G_{\psi^\varepsilon}(t) \geq g(t)$ in a neighborhood of a. Without loss of generality, we choose $\varepsilon_0, \delta > 0$ such that $G_{\psi^\varepsilon} \geq g$ on $(0, a + \delta] \cup [b - \delta, \infty)$. On the compact interval $[a + \delta, b - \delta]$, we have

$$\min_{t \in [a+\delta, b-\delta]} (G_{\psi^*} - g)(t) > 0 \qquad \Rightarrow \qquad G_{\psi^*} \geq g + \rho$$

for some small $\rho > 0$. By continuity, we conclude that $G_{\psi^\varepsilon} \geq g$ also on $[a+\delta, b-\delta]$ for sufficiently small $\varepsilon > 0$. In total, we have shown that $\psi^\varepsilon \in K_\phi$ for all sufficiently small ε, i.e., that ψ^ε is a valid energy competitor. Since

$$\frac{d}{d\varepsilon}\bigg|_{\varepsilon=0} E(\psi^\varepsilon) = \frac{d}{d\varepsilon}\bigg|_{\varepsilon=0} \left(\int_a^{a+\xi} \sqrt{\alpha + \varepsilon}\, \mathrm{d}t + \int_{b-\xi}^b \sqrt{\beta - \varepsilon}\, \mathrm{d}t \right)$$

$$= \xi \, \frac{d}{d\varepsilon} \left(\sqrt{\alpha + \varepsilon} + \sqrt{\beta - \varepsilon} \right) = \frac{\xi}{2} \left(\frac{1}{\sqrt{\alpha}} - \frac{1}{\sqrt{\beta}} \right) < 0$$

unless $\alpha = \beta$ since $\alpha \geq \beta$ due to the monotonicity of ψ^*. If $\alpha = \beta$, the first derivative vanishes: Constant ψ^* is a *maximum* in the space of monotone decreasing functions:

$$\frac{d^2}{d\varepsilon^2}\bigg|_{\varepsilon=0} E(\psi^{\varepsilon}) = \frac{d}{d\varepsilon}\bigg|_{\varepsilon=0} \frac{\xi}{2} \left(\frac{1}{\sqrt{\alpha + \varepsilon}} - \frac{1}{\sqrt{\alpha - \varepsilon}} \right) = -\frac{\xi}{2} \alpha^{-3/2} < 0.$$

□

References

1. Guillaume Alain, Nicolas Le Roux, and Pierre-Antoine Manzagol. Negative eigenvalues of the Hessian in deep neural networks. *arXiv preprint arXiv:1902.02366*, 2019.
2. Sanjeev Arora, Simon S Du, Wei Hu, Zhiyuan Li, Russ R Salakhutdinov, and Ruosong Wang. On exact computation with an infinitely wide neural net. *Advances in neural information processing systems*, 32, 2019.
3. Hedy Attouch, Zaki Chbani, Juan Peypouquet, and Patrick Redont. Fast convergence of inertial dynamics and algorithms with asymptotic vanishing viscosity. *Mathematical Programming*, 168:123–175, 2018.
4. Hedy Attouch and Juan Peypouquet. The rate of convergence of Nesterov's accelerated forward-backward method is actually faster than $1/k^2$. *SIAM Journal on Optimization*, 26(3):1824–1834, 2016.
5. Jean-Francois Aujol, Charles Dossal, and Aude Rondepierre. Optimal convergence rates for Nesterov acceleration. *SIAM Journal on Optimization*, 29(4):3131–3153, 2019.
6. JB Baillon. Un exemple concernant le comportement asymptotique de la solution du problème $du/dt + \partial\phi(u) \ni 0$. *Journal of Functional Analysis*, 28(3):369–376, 1978.
7. Raef Bassily, Mikhail Belkin, and Siyuan Ma. On exponential convergence of SGD in non-convex over-parametrized learning. *CoRR*, abs/1811.02564, 2018.
8. Sébastien Bubeck. Convex optimization: Algorithms and complexity. *Foundations and Trends® in Machine Learning*, 8(3-4):231–357, 2015.
9. Yaim Cooper. Global minima of overparameterized neural networks. *SIAM Journal on Mathematics of Data Science*, 3(2):676–691, 2021.
10. Aris Daniilidis, Mounir Haddou, and Olivier Ley. A convex function satisfying the łojasiewicz inequality but failing the gradient conjecture both at zero and infinity. *Bulletin of the London Mathematical Society*, 54(2):590–608, 2022.
11. Aris Daniilidis, Olivier Ley, and Stéphane Sabourau. Asymptotic behaviour of self-contracted planar curves and gradient orbits of convex functions. *Journal de mathématiques pures et appliquées*, 94(2):183–199, 2010.
12. Klaus Deimling. *Ordinary differential equations in Banach spaces*, volume 596. Springer, 2006.
13. Steffen Dereich, Arnulf Jentzen, and Sebastian Kassing. On the existence of minimizers in shallow residual relu neural network optimization landscapes. *arXiv preprint arXiv:2302.14690*, 2023.
14. Simon Du, Jason Lee, Haochuan Li, Liwei Wang, and Xiyu Zhai. Gradient descent finds global minima of deep neural networks. In *International conference on machine learning*, pages 1675–1685. PMLR, 2019.

15. Simon S Du, Xiyu Zhai, Barnabas Poczos, and Aarti Singh. Gradient descent provably optimizes over-parameterized neural networks. *arXiv preprint arXiv:1810.02054*, 2018.
16. Estibalitz Durand-Cartagena and Antoine Lemenant. Self-contracted curves are gradient flows of convex functions. *Proceedings of the American Mathematical Society*, 147(6):2517–2531, 2019.
17. Weinan E, Chao Ma, and Lei Wu. A comparative analysis of optimization and generalization properties of two-layer neural network and random feature models under gradient descent dynamics. *Sci. China Math*, 2019.
18. Lawrence C Evans. *Measure theory and fine properties of functions*. Routledge, 2018.
19. Lawrence C Evans. *Partial differential equations*, volume 19. American Mathematical Society, 2022.
20. Alexandr Nikolaevich Godunov. Peano's theorem in an infinite-dimensional hilbert space is false even in a weakened formulation. *Mathematical notes of the Academy of Sciences of the USSR*, 15(3):273–279, 1974.
21. Chirag Gupta, Sivaraman Balakrishnan, and Aaditya Ramdas. Path length bounds for gradient descent and flow. *The Journal of Machine Learning Research*, 22(1):3154–3216, 2021.
22. Kanan Gupta, Jonathan Siegel, and Stephan Wojtowytsch. Achieving acceleration despite very noisy gradients. *arXiv:2302.05515 [stat.ML]*, 2023.
23. Petr Hájek and Michal Johanis. On Peano's theorem in Banach spaces. *Journal of Differential Equations*, 249(12):3342–3351, 2010.
24. Like Hui and Mikhail Belkin. Evaluation of neural architectures trained with square loss vs cross-entropy in classification tasks. *arXiv preprint arXiv:2006.07322*, 2020.
25. Arthur Jacot, Franck Gabriel, and Clément Hongler. Neural tangent kernel: Convergence and generalization in neural networks. *Advances in neural information processing systems*, 31, 2018.
26. Arnulf Jentzen and Adrian Riekert. On the existence of global minima and convergence analyses for gradient descent methods in the training of deep neural networks. *arXiv preprint arXiv:2112.09684*, 2021.
27. Hamed Karimi, Julie Nutini, and Mark Schmidt. Linear convergence of gradient and proximal-gradient methods under the polyak-łojasiewicz condition. In *Machine Learning and Knowledge Discovery in Databases: European Conference, ECML PKDD 2016, Riva del Garda, Italy, September 19–23, 2016, Proceedings, Part I 16*, pages 795–811. Springer, 2016.
28. A. Klenke. *Probability Theory: A Comprehensive Course*. Universitext. Springer, 2013.
29. Konrad Königsberger. *Analysis 1*. Springer, 1999.
30. Konrad Königsberger. *Analysis 2*. Springer, 2013.
31. Xin Liu, Zhisong Pan, and Wei Tao. Provable convergence of Nesterov's accelerated gradient method for over-parameterized neural networks. *Knowledge-Based Systems*, 251:109277, 2022.
32. Paolo Manselli and Carlo Pucci. Maximum length of steepest descent curves for quasi-convex functions. *Geometriae Dedicata*, 38(2):211–227, 1991.
33. Albert W Marshall, Ingram Olkin, and Barry C Arnold. *Inequalities: theory of majorization and its applications*. Springer, 1979.
34. Yurii Nesterov. *Introductory lectures on convex optimization: A basic course*, volume 87. Springer Science & Business Media, 2003.
35. Yurii Evgen'evich Nesterov. A method of solving a convex programming problem with convergence rate $o(1/k^2)$. *Doklady Akademii Nauk*, 269(3):543–547, 1983.
36. Constantin P Niculescu and Florin Popovici. A note on the behavior of integrable functions at infinity. *Journal of mathematical analysis and applications*, 381(2):742–747, 2011.
37. R Tyrrell Rockafellar. *Convex analysis*, volume 11. Princeton university press, 1997.
38. Itay Safran and Ohad Shamir. Spurious local minima are common in two-layer relu neural networks. In *Proceedings of the 35th International Conference on Machine Learning, ICML 2018, Stockholmsmässan, Stockholm, Sweden, July 10–15, 2018*, volume 80 of *Proceedings of Machine Learning Research*, pages 4430–4438. PMLR, 2018.

39. Levent Sagun, Leon Bottou, and Yann LeCun. Eigenvalues of the Hessian in deep learning: Singularity and beyond. *arXiv preprint arXiv:1611.07476*, 2016.
40. Levent Sagun, Utku Evci, V Ugur Guney, Yann Dauphin, and Leon Bottou. Empirical analysis of the Hessian of over-parametrized neural networks. *arXiv preprint arXiv:1706.04454*, 2017.
41. Jonathan W Siegel. Accelerated first-order methods: Differential equations and Lyapunov functions. *arXiv preprint arXiv:1903.05671*, 2019.
42. Eugene Stepanov and Yana Teplitskaya. Self-contracted curves have finite length. *Journal of the London Mathematical Society*, 96(2):455–481, 2017.
43. Weijie Su, Stephen Boyd, and Emmanuel Candès. A differential equation for modeling Nesterov's accelerated gradient method: theory and insights. *Advances in neural information processing systems*, 27, 2014.
44. Gerald Teschl. *Ordinary differential equations and dynamical systems*, volume 140. American Mathematical Soc., 2012.
45. Luca Venturi, Afonso S Bandeira, and Joan Bruna. Spurious valleys in two-layer neural network optimization landscapes. *arXiv preprint arXiv:1802.06384*, 2018.
46. Yifei Wang, Jonathan Lacotte, and Mert Pilanci. The hidden convex optimization landscape of regularized two-layer relu networks: an exact characterization of optimal solutions. In *International Conference on Learning Representations*, 2021.
47. Stephan Wojtowytsch. Stochastic gradient descent with noise of machine learning type. Part I: Discrete time analysis. *arXiv:2105.01650 [stat.ML]*, 2021.
48. Pengyun Yue, Cong Fang, and Zhouchen Lin. On the lower bound of minimizing Polyak-łojasiewicz functions. In *The Thirty Sixth Annual Conference on Learning Theory*, pages 2948–2968. PMLR, 2023.

Applied and Numerical Harmonic Analysis (110 volumes)

1. A. I. Saichev and W. A. Woyczyński: *Distributions in the Physical and Engineering Sciences* (ISBN: 978-0-8176-3924-2)
2. C. E. D'Attellis and E. M. Fernandez-Berdaguer: *Wavelet Theory and Harmonic Analysis in Applied Sciences* (ISBN: 978-0-8176-3953-2)
3. H. G. Feichtinger and T. Strohmer: *Gabor Analysis and Algorithms* (ISBN: 978-0-8176-3959-4)
4. R. Tolimieri and M. An: *Time-Frequency Representations* (ISBN: 978-0-8176-3918-1)
5. T. M. Peters and J. C. Williams: *The Fourier Transform in Biomedical Engineering* (ISBN: 978-0-8176-3941-9)
6. G. T. Herman: *Geometry of Digital Spaces* (ISBN: 978-0-8176-3897-9)
7. A. Teolis: *Computational Signal Processing with Wavelets* (ISBN: 978-0-8176-3909-9)
8. J. Ramanathan: *Methods of Applied Fourier Analysis* (ISBN: 978-0-8176-3963-1)
9. J. M. Cooper: *Introduction to Partial Differential Equations with MATLAB* (ISBN: 978-0-8176-3967-9)
10. Procházka, N. G. Kingsbury, P. J. Payner, and J. Uhlir: *Signal Analysis and Prediction* (ISBN: 978-0-8176-4042-2)
11. W. Bray and C. Stanojevic: *Analysis of Divergence* (ISBN: 978-1-4612-7467-4)
12. G. T. Herman and A. Kuba: *Discrete Tomography* (ISBN: 978-0-8176-4101-6)
13. K. Gröchenig: *Foundations of Time-Frequency Analysis* (ISBN: 978-0-8176-4022-4)
14. L. Debnath: *Wavelet Transforms and Time-Frequency Signal Analysis* (ISBN: 978-0-8176-4104-7)
15. J. J. Benedetto and P. J. S. G. Ferreira: *Modern Sampling Theory* (ISBN: 978-0-8176-4023-1)

S. Foucart, S. Wojtowytsch (eds.), *Explorations in the Mathematics of Data Science*, Applied and Numerical Harmonic Analysis,
https://doi.org/10.1007/978-3-031-66497-7

16. D. F. Walnut: *An Introduction to Wavelet Analysis* (ISBN: 978-0-8176-3962-4)
17. A. Abbate, C. DeCusatis, and P. K. Das: *Wavelets and Subbands* (ISBN: 978-0-8176-4136-8)
18. O. Bratteli, P. Jorgensen, and B. Treadway: *Wavelets Through a Looking Glass* (ISBN: 978-0-8176-4280-80)
19. H. G. Feichtinger and T. Strohmer: *Advances in Gabor Analysis* (ISBN: 978-0-8176-4239-6)
20. O. Christensen: *An Introduction to Frames and Riesz Bases* (ISBN: 978-0-8176-4295-2)
21. L. Debnath: *Wavelets and Signal Processing* (ISBN: 978-0-8176-4235-8)
22. G. Bi and Y. Zeng: *Transforms and Fast Algorithms for Signal Analysis and Representations* (ISBN: 978-0-8176-4279-2)
23. J. H. Davis: *Methods of Applied Mathematics with a MATLAB Overview* (ISBN: 978-0-8176-4331-7)
24. J. J. Benedetto and A. I. Zayed: *Sampling, Wavelets, and Tomography* (ISBN: 978-0-8176-4304-1)
25. E. Prestini: *The Evolution of Applied Harmonic Analysis* (ISBN: 978-0-8176-4125-2)
26. L. Brandolini, L. Colzani, A. Iosevich, and G. Travaglini: *Fourier Analysis and Convexity* (ISBN: 978-0-8176-3263-2)
27. W. Freeden and V. Michel: *Multiscale Potential Theory* (ISBN: 978-0-8176-4105-4)
28. O. Christensen and K. L. Christensen: *Approximation Theory* (ISBN: 978-0-8176-3600-5)
29. O. Calin and D.-C. Chang: *Geometric Mechanics on Riemannian Manifolds* (ISBN: 978-0-8176-4354-6)
30. J. A. Hogan: *Time-Frequency and Time-Scale Methods* (ISBN: 978-0-8176-4276-1)
31. C. Heil: *Harmonic Analysis and Applications* (ISBN: 978-0-8176-3778-1)
32. K. Borre, D. M. Akos, N. Bertelsen, P. Rinder, and S. H. Jensen: *A Software-Defined GPS and Galileo Receiver* (ISBN: 978-0-8176-4390-4)
33. T. Qian, M. I. Vai, and Y. Xu: *Wavelet Analysis and Applications* (ISBN: 978-3-7643-7777-9)
34. G. T. Herman and A. Kuba: *Advances in Discrete Tomography and Its Applications* (ISBN: 978-0-8176-3614-2)
35. M. C. Fu, R. A. Jarrow, J.-Y. Yen, and R. J. Elliott: *Advances in Mathematical Finance* (ISBN: 978-0-8176-4544-1)
36. O. Christensen: *Frames and Bases* (ISBN: 978-0-8176-4677-6)
37. P. E. T. Jorgensen, J. D. Merrill, and J. A. Packer: *Representations, Wavelets, and Frames* (ISBN: 978-0-8176-4682-0)
38. M. An, A. K. Brodzik, and R. Tolimieri: *Ideal Sequence Design in Time-Frequency Space* (ISBN: 978-0-8176-4737-7)
39. S. G. Krantz: *Explorations in Harmonic Analysis* (ISBN: 978-0-8176-4668-4)

40. B. Luong: *Fourier Analysis on Finite Abelian Groups* (ISBN: 978-0-8176-4915-9)
41. G. S. Chirikjian: *Stochastic Models, Information Theory, and Lie Groups, Volume 1* (ISBN: 978-0-8176-4802-2)
42. C. Cabrelli and J. L. Torrea: *Recent Developments in Real and Harmonic Analysis* (ISBN: 978-0-8176-4531-1)
43. M. V. Wickerhauser: *Mathematics for Multimedia* (ISBN: 978-0-8176-4879-4)
44. B. Forster, P. Massopust, O. Christensen, K. Gröchenig, D. Labate, P. Vandergheynst, G. Weiss, and Y. Wiaux: *Four Short Courses on Harmonic Analysis* (ISBN: 978-0-8176-4890-9)
45. O. Christensen: *Functions, Spaces, and Expansions* (ISBN: 978-0-8176-4979-1)
46. J. Barral and S. Seuret: *Recent Developments in Fractals and Related Fields* (ISBN: 978-0-8176-4887-9)
47. O. Calin, D.-C. Chang, and K. Furutani, and C. Iwasaki: *Heat Kernels for Elliptic and Sub-elliptic Operators* (ISBN: 978-0-8176-4994-4)
48. C. Heil: *A Basis Theory Primer* (ISBN: 978-0-8176-4686-8)
49. J. R. Klauder: *A Modern Approach to Functional Integration* (ISBN: 978-0-8176-4790-2)
50. J. Cohen and A. I. Zayed: *Wavelets and Multiscale Analysis* (ISBN: 978-0-8176-8094-7)
51. D. Joyner and J.-L. Kim: *Selected Unsolved Problems in Coding Theory* (ISBN: 978-0-8176-8255-2)
52. G. S. Chirikjian: *Stochastic Models, Information Theory, and Lie Groups, Volume 2* (ISBN: 978-0-8176-4943-2)
53. J. A. Hogan and J. D. Lakey: *Duration and Bandwidth Limiting* (ISBN: 978-0-8176-8306-1)
54. G. Kutyniok and D. Labate: *Shearlets* (ISBN: 978-0-8176-8315-3)
55. P. G. Casazza and P. Kutyniok: *Finite Frames* (ISBN: 978-0-8176-8372-6)
56. V. Michel: *Lectures on Constructive Approximation* (ISBN : 978-0-8176-8402-0)
57. D. Mitrea, I. Mitrea, M. Mitrea, and S. Monniaux: *Groupoid Metrization Theory* (ISBN: 978-0-8176-8396-2)
58. T. D. Andrews, R. Balan, J. J. Benedetto, W. Czaja, and K. A. Okoudjou: *Excursions in Harmonic Analysis, Volume 1* (ISBN: 978-0-8176-8375-7)
59. T. D. Andrews, R. Balan, J. J. Benedetto, W. Czaja, and K. A. Okoudjou: *Excursions in Harmonic Analysis, Volume 2* (ISBN: 978-0-8176-8378-8)
60. D. V. Cruz-Uribe and A. Fiorenza: *Variable Lebesgue Spaces* (ISBN: 978-3-0348-0547-6)
61. W. Freeden and M. Gutting: *Special Functions of Mathematical (Geo-)Physics* (ISBN: 978-3-0348-0562-9)
62. A. I. Saichev and W. A. Woyczyński: *Distributions in the Physical and Engineering Sciences, Volume 2: Linear and Nonlinear Dynamics of Continuous Media* (ISBN: 978-0-8176-3942-6)

63. S. Foucart and H. Rauhut: *A Mathematical Introduction to Compressive Sensing* (ISBN: 978-0-8176-4947-0)
64. G. T. Herman and J. Frank: *Computational Methods for Three-Dimensional Microscopy Reconstruction* (ISBN: 978-1-4614-9520-8)
65. A. Paprotny and M. Thess: *Realtime Data Mining: Self-Learning Techniques for Recommendation Engines* (ISBN: 978-3-319-01320-6)
66. A. I. Zayed and G. Schmeisser: *New Perspectives on Approximation and Sampling Theory: Festschrift in Honor of Paul Butzer's 85th Birthday* (ISBN: 978-3-319-08800-6)
67. R. Balan, M. Begue, J. Benedetto, W. Czaja, and K. A. Okoudjou: *Excursions in Harmonic Analysis, Volume 3* (ISBN: 978-3-319-13229-7)
68. H. Boche, R. Calderbank, G. Kutyniok, and J. Vybiral: *Compressed Sensing and its Applications* (ISBN: 978-3-319-16041-2)
69. S. Dahlke, F. De Mari, P. Grohs, and D. Labate: *Harmonic and Applied Analysis: From Groups to Signals* (ISBN: 978-3-319-18862-1)
70. A. Aldroubi: *New Trends in Applied Harmonic Analysis* (ISBN: 978-3-319-27871-1)
71. M. Ruzhansky: *Methods of Fourier Analysis and Approximation Theory* (ISBN: 978-3-319-27465-2)
72. G. Pfander: *Sampling Theory, a Renaissance* (ISBN: 978-3-319-19748-7)
73. R. Balan, M. Begue, J. Benedetto, W. Czaja, and K. A. Okoudjou: *Excursions in Harmonic Analysis, Volume 4* (ISBN: 978-3-319-20187-0)
74. O. Christensen: *An Introduction to Frames and Riesz Bases, Second Edition* (ISBN: 978-3-319-25611-5)
75. E. Prestini: *The Evolution of Applied Harmonic Analysis: Models of the Real World, Second Edition* (ISBN: 978-1-4899-7987-2)
76. J. H. Davis: *Methods of Applied Mathematics with a Software Overview, Second Edition* (ISBN: 978-3-319-43369-1)
77. M. Gilman, E. M. Smith, and S. M. Tsynkov: *Transionospheric Synthetic Aperture Imaging* (ISBN: 978-3-319-52125-1)
78. S. Chanillo, B. Franchi, G. Lu, C. Perez, and E. T. Sawyer: *Harmonic Analysis, Partial Differential Equations and Applications* (ISBN: 978-3-319-52741-3)
79. R. Balan, J. Benedetto, W. Czaja, M. Dellatorre, and K. A. Okoudjou: *Excursions in Harmonic Analysis, Volume 5* (ISBN: 978-3-319-54710-7)
80. I. Pesenson, Q. T. Le Gia, A. Mayeli, H. Mhaskar, and D. X. Zhou: *Frames and Other Bases in Abstract and Function Spaces: Novel Methods in Harmonic Analysis, Volume 1* (ISBN: 978-3-319-55549-2)
81. I. Pesenson, Q. T. Le Gia, A. Mayeli, H. Mhaskar, and D. X. Zhou: *Recent Applications of Harmonic Analysis to Function Spaces, Differential Equations, and Data Science: Novel Methods in Harmonic Analysis, Volume 2* (ISBN: 978-3-319-55555-3)
82. F. Weisz: *Convergence and Summability of Fourier Transforms and Hardy Spaces* (ISBN: 978-3-319-56813-3)
83. C. Heil: *Metrics, Norms, Inner Products, and Operator Theory* (ISBN: 978-3-319-65321-1)

84. S. Waldron: *An Introduction to Finite Tight Frames: Theory and Applications.* (ISBN: 978-0-8176-4814-5)
85. D. Joyner and C. G. Melles: *Adventures in Graph Theory: A Bridge to Advanced Mathematics.* (ISBN: 978-3-319-68381-2)
86. B. Han: *Framelets and Wavelets: Algorithms, Analysis, and Applications* (ISBN: 978-3-319-68529-8)
87. H. Boche, G. Caire, R. Calderbank, M. März, G. Kutyniok, and R. Mathar: *Compressed Sensing and Its Applications* (ISBN: 978-3-319-69801-4)
88. A. I. Saichev and W. A. Woyczyński: *Distributions in the Physical and Engineering Sciences, Volume 3: Random and Fractal Signals and Fields* (ISBN: 978-3-319-92584-4)
89. G. Plonka, D. Potts, G. Steidl, and M. Tasche: *Numerical Fourier Analysis* (978-3-030-04305-6)
90. K. Bredies and D. Lorenz: *Mathematical Image Processing* (ISBN: 978-3-030-01457-5)
91. H. G. Feichtinger, P. Boggiatto, E. Cordero, M. de Gosson, F. Nicola, A. Oliaro, and A. Tabacco: *Landscapes of Time-Frequency Analysis* (ISBN: 978-3-030-05209-6)
92. E. Liflyand: *Functions of Bounded Variation and Their Fourier Transforms* (ISBN: 978-3-030-04428-2)
93. R. Campos: *The XFT Quadrature in Discrete Fourier Analysis* (ISBN: 978-3-030-13422-8)
94. M. Abell, E. Iacob, A. Stokolos, S. Taylor, S. Tikhonov, J. Zhu: *Topics in Classical and Modern Analysis: In Memory of Yingkang Hu* (ISBN: 978-3-030-12276-8)
95. H. Boche, G. Caire, R. Calderbank, G. Kutyniok, R. Mathar, P. Petersen: *Compressed Sensing and its Applications: Third International MATHEON Conference 2017* (ISBN: 978-3-319-73073-8)
96. A. Aldroubi, C. Cabrelli, S. Jaffard, U. Molter: *New Trends in Applied Harmonic Analysis, Volume II: Harmonic Analysis, Geometric Measure Theory, and Applications* (ISBN: 978-3-030-32352-3)
97. S. Dos Santos, M. Maslouhi, K. Okoudjou: *Recent Advances in Mathematics and Technology: Proceedings of the First International Conference on Technology, Engineering, and Mathematics, Kenitra, Morocco, March 26-27, 2018* (ISBN: 978-3-030-35201-1)
98. Á. Bényi, K. Okoudjou: *Modulation Spaces: With Applications to Pseudodifferential Operators and Nonlinear Schrödinger Equations* (ISBN: 978-1-0716-0330-7)
99. P. Boggiato, M. Cappiello, E. Cordero, S. Coriasco, G. Garello, A. Oliaro, J. Seiler: *Advances in Microlocal and Time-Frequency Analysis* (ISBN: 978-3-030-36137-2)
100. S. Casey, K. Okoudjou, M. Robinson, B. Sadler: *Sampling: Theory and Applications* (ISBN: 978-3-030-36290-4)

101. P. Boggiatto, T. Bruno, E. Cordero, H. G. Feichtinger, F. Nicola, A. Oliaro, A. Tabacco, M. Vallarino: *Landscapes of Time-Frequency Analysis: ATFA 2019* (ISBN: 978-3-030-56004-1)
102. M. Hirn, S. Li, K. Okoudjou, S. Saliana, Ö. Yilmaz: *Excursions in Harmonic Analysis, Volume 6: In Honor of John Benedetto's 80th Birthday* (ISBN: 978-3-030-69636-8)
103. F. De Mari, E. De Vito: *Harmonic and Applied Analysis: From Radon Transforms to Machine Learning* (ISBN: 978-3-030-86663-1)
104. G. Kutyniok, H. Rauhut, R. J. Kunsch, *Compressed Sensing in Information Processing* (ISBN: 978-3-031-09744-7)
105. P. Flandrin, S. Jaffard, T. Paul, B. Torresani, *Theoretic Physics, Wavelets, Analysis, Genomics: An Indisciplinary Tribute to Alex Grossmann* (ISBN: 978-3-030-45846-1)
106. G. Plonka-Hoch, D. Potts, G. Steidl, M. Tasche, *Numerical Fourier Analysis, Second Edition* (ISBN: 978-3-031-35004-7)
107. P. Alonso Ruiz, M. Hinz, K. Okoudjou, L. Rogers, A. Teplyaev, *From Classical Analysis to Analysis on Fractals: A Tribute to Robert Strichartz, Volume 1* (ISBN: 978-3-031-37799-0)
108. S. Casey, M. Dodson, P. Ferreira, A. Zayed, *Sampling, Approximation, and Signal Analysis: Harmonic Analysis in the Spirit of J. Rowland Higgins* (ISBN: 978-3-031-41129-8)
109. J. Feuto, B. A. Kpata, *Harmonic Analysis and Partial Differential Equations: Proceedings of the Workshop in Abidjan, Côte de'Ivoire, May 22-26, 2023* (ISBN: 978-3-031-66374-1)
110. S. Foucart, S. Wojtowytsch, *Explorations in the Mathematics of Data Science: The Inaugural Volume of the Center for Approximation and Mathematical Data Analytics* (ISBN: 978-3-031-66496-0)

For an up-to-date list of ANHA titles, please visit http://www.springer.com/series/4968

Printed in the USA
CPSIA information can be obtained
at www.ICGtesting.com
CBHW051057141024
15701CB00024B/222

9 783031 664960